Introduction to
GENERAL AND APPLIED
ENTOMOLOGY

Introduction to
GENERAL AND APPLIED
ENTOMOLOGY

Dr. Abhishek Shukla

Assistant Professor (Entomology)
Fruit Research Station
Navsari Agricultural University
Gandevi – 396 360, Gujarat, India

Dr. Sushil Kumar Saxena

Associate Professor (Entomology)
ASPEE College of Horticulture & Forestry,
Navsari Agricultural University
Navsari – 396 450, Gujarat, India

2012
DAYA PUBLISHING HOUSE®
New Delhi - 110 002

Published by   : **Daya Publishing House®**
                            A Division of
                            Astral International Pvt. Ltd.
                            – ISO 9001:2008 Certified Company –
                            4760-61/23, Ansari Road, Darya Ganj
                            New Delhi-110 002
                            Ph. 011-43549197, 23278134
                            E-mail: info@astralint.com
                            Website: www.astralint.com

Laser Typesetting   : **Classic Computer Services**, Delhi - 110 035

Printed at   : **Chawla Offset Printers**, Delhi - 110 052

PRINTED IN INDIA

Preface

Although, there are good number of books on this subject but they are either too exhaustive or too brief or mainly deal with only one or two aspects of the Entomology hence do not fulfill the need of the under graduate or post graduate studies. Thus, there is a great need of a balanced account of general and applied aspects of Entomology to fulfill the expectations of the students of Entomology.

This book provides balanced and latest information on insect morphology, physiology, taxonomy, ecology and different aspects of insect pest management. This book also covers information on biotechnological approaches of pest management. Vectors of plant diseases play an important role in spreading disease in crops, this aspect along with the management is covered in detail in this book. Due to indiscriminate and injudicious uses of insecticides, insect pests develop resistance against many of them. The chapter on insect resistance to insecticides covers detailed mechanism of resistance and its management strategies.

The major pests of agricultural and horticultural crops and their management have also been covered along with latest developments. Stored grain pests and their management have been also discussed in a separate chapter.

The insects are not always harmful, they are some time beneficial for mankind. Beneficial insects *viz.*, honey bees, silk worm and lac insect have also been discussed in this book in detail for the benefit of students.

A large number of illustrations have been given to make this fascinating subject easy and interesting. The basic aim of this publication is to provide the basic information on insect- structure, function, taxonomy, ecology, management as well as benefits drawn out of them.

Since this book contains information collected during my post graduation and doctoral study period, so information provided by several pioneer workers has been incorporated and it is not possible to thank them all individually. I (first author) must express my most humble and profound gratitude to all of them. It was their hard work which enabled me to collect and compile useful information in such a manner that would be useful to students of Entomology.

We hope that the student of Entomology will find this book useful for whom it is meant and take pain to know more and more on the subject and about insects.

We are extremely grateful to Mr. Anil Mittal and the team of Daya Publishing House, New Delhi, for showing their keen interest in our work and whole hearted support in the publication of this book.

Dr. Shukla would like to thank his wife Dr. Ruchira Shukla, for going through the manuscript and making necessary improvements. He also wish to thank his son Shourya Shukla for his cooperation.

Suggestions for the improvement of the book from students as well as teachers will be highly appreciated.

Abhishek Shukla
Sushil Kumar Saxena

Contents

Appendices

1

Importance of Entomology

Entomology is a branch of zoological sciences dealing with a specific group of organisms, the insects. Man originated about a million years ago, and insects at least 250-500 million years ago. Insects constitute the largest class of the whole living organisms and about 72 per cent of all living animals are insects with 9-15 lakh known species. Insects are omnipresent and each crop we cultivate is being attacked by at least a dozen of insect species called as pests. Apart from the insect pests, there are several productive and useful insects. Insects are considered as one of the major constraints in increasing agricultural productivity. In general way, the insects are important to man due to the following reasons:

1. The damage they cause to field crops, orchards, forests, products and household materials.

2. Their interference with man's health and comfort.

3. Their value as: (*a*) pollinators and agents of fruit setting in orchards and crops, (*b*) producer of honey, silk and lac, (*c*) scavengers (*d*) useful in biological control *i.e.* parasites and predators.

4. Their remarkable suitability and utility as material for purely scientific studies *e.g.* genetics and inheritance, of which *Drosophila* is well known; sociology and population growth, for which ants, termites, wasps and bees have been very useful, adaptability, mimicry, etc., for which the leaf butterfly *Kallima* and the stick insect are the classical examples.

Hence, it is important to understand the insects, their biology, classification and management. Our earliest knowledge about insects dates back to 6000 years due to the fact that our Indian ancestors were well versed in the art of rearing silk worms and weaving silk cloth. Even during 3870 BC, an Indian king sent various silken materials as presents to a Persian king. There are lot of mentions about insects in our mythological epics *viz.*, the Ramayana (2550-2150 BC) and the Mahabharata (1424-1366 BC). Mentions about silk, honey and lac were found in these epics. The first detailed classification of insects was done by *Umaswati* (0-100 AD). Classification of bees by the Indian physician *Charaka* (1200-1000 BC) and classification of ants, flies and mosquitoes by the surgeon *Sushruta* (100-200 AD) are the evidences of our earliest knowledge of insects. Entomology in modern India must have taken its place in Natural History some time after the 16th century. It is possible to define the modern history of entomology in India under the following two headings.

1. From the advent of the early Christian Missions and the establishment of East India Company at Calcutta. The contributions made by several eminent amateur entomologists, otherwise serving in the army, Indian Forest Service, Indian Medical Service and Indian Civil Service.

2. Entomology in India during 1900-1950.

The beginning of the modern era of Indian Entomology came with the publication of the tenth edition of Carl Linnaeus's *"Systema Naturae"* in 1758. This contained the earliest record of 28 species of Indian insects. The first entomologist who made any extensive study of Indian Insects was J. C. Fabricius. He was a Danish Professor of political economy turned in to a systematist and classified the insects in to 13 orders based on type of mouth parts. J. G. Koenig (1767–1779), a medical Officer from Denmark, a student of Carl Linnaeus and a friend of Fabricius for the first time during the 18th century collected number of insects from Coromandal as well as other Peninsular areas of India. His collections were studied and named by Professor Linnaeus himself. He also published a special account of the termites of Thanjavur District. Fabricius made Koenig's name remembered for ever by naming the well known and destructive red cotton bug of this country as *Dysdercus koenigi*. In 1782 Dr. Kerr published on account of lac insect. Asiatic Society of Bengal started in Calcutta in 1785. In 1791 Dr. J. Anderson issued a monograph on Cochineal scale insects. During 1799 Dr. Horsfield, an American doctor and first Keeper of the East India Museum published his famous book "A catalogue of the Lepidopterous Insects in the Museum of the Honourable East India Company", 2 vols. (1857, 1858-59). In 1800 Buchanan (Traveler studying the natural wealth of India) wrote on the cultivation of lac in India and on sericulture in some parts of South India. Edward Denovan published an illustrated book entitled "An epitome of the natural history of insects of India and the Islands in the Indian seas" which was the first pictorial documentation on the insects of Asia and was revised in 1842 by West Wood. During 1875 foundation of the "Indian Museum" at Calcutta. "Bombay Natural History Society" was started in 1883 . After the foundation of these two organisations scientific studies received greater attention in India. Numerous contributions of Indian insects were published in the "Journal of the Bombay Natural History". In year 1883 commencement of "Fauna of British India" series under the editorship of W. T. Blandford. Entomological part of the "Fauna of British India" (now Fauna of India) series started in 1892 with Sir George Hampson contributed first of the four volumes on the moths of India. Rothney (1893) publication on Indian Ants (earliest record of biological pest control in India) *i.e.* White ants attack on stationary items was kept free by red ants. From 1889-1903 Indian Museum, Calcutta published the "Indian Museum Notes" in five volumes, which contributed much on economic entomology and applied entomology in India. Nineteenth century marks the major progress and expansions in the field of Applied Entomology. In 1901 Lionel de Nicevelle was posted as the first entomologist to the Government of India. While Professor Maxwell Lefroy succeeded Nicevelle as Government Entomologist (1903). In 1905 Imperial Agricultural Research Institute at Pusa, Bihar was established and Professor Lefroy became the first Imperial Entomologist. He convened a series of entomological meetings on all India basis to bring together all the entomologists of the country. From 1915, five such meetings were held at the Imperial Agricultural Research Institute at Pusa. While the Proceedings of the first meeting was not published, proceedings of subsequent four meetings became a treasure of entomological knowledge, which can never be overlooked by any student of Entomology in India. Another contribution of Lefroy is publication of the series under the name "Catalogue of Indian Insects" which marked another era in history of Indian Insect Taxonomy. Professor Lefroy published 'Indian Insect Pests' in 1906. Subsequently, State Governments also initiated entomological work. Madras, Punjab and Uttar Pradesh appointed their first State/ Provincial Government Entomologists in 1912, 1919 and 1922, respectively. In 1912 Plant Quarantine Act was enforced. In 1914 Destructive Insects and Pests Act was enforced. Dr. T.B. Fletcher (1914), the first Government Entomologist to Madras State published his book "Some South Indian Insects". In 1914, 'Indian Forest Insects of Economic Importance: Coleoptera' was published by the first Imperial Forest Entomologist E. P. Stebbing. In 1916, Imperial Forest Research Institute was established at Dehra Dun, and E.P. Stebbing was appointed as Forest Zoologist. The Natural History Section of the Indian Museum was formed as the Zoological Survey of India in 1916. In 1921, Indian Central Cotton Committee was formed to investigate on pests of cotton. Indian Lac Research Institute was established at Ranchi during 1925. Dr Hem Singh Pruthi who succeeded Fletcher as Imperial Entomologist in 1934, made efforts to bring all the entomological workers still closer resulting into the foundation of 'Entomological Society of India' in 1938. Mian Afzal Hussain was the first President of the Entomological Society of India and the Vice-Presidents were Hem Singh Pruthi and Ramakrishna Ayyar. The first official publication of the Society was named the 'Indian Journal of Entomology'. Locust Warning Organisation was established at Jodhpur in the year 1939. Dr. T.V. Ramakrishna Ayyar published the book "Handbook of Economic Entomology for South India" and the revised edition was published in 1963. In 1941, C.F.C. Beeson published "The Ecology and Control of the Forest Insects of India and Neighboring Countries". H.S. Pruthi published "Text book of Agricultural Entomology" (1963). Likewise in 1968 Dr. M.S. Mani published "General Entomology". In 1969, Dr. S. Pradhan's book entitled "Insect Pests of Crops" was published. Government of India established 'Directorate of Plant Protection, Quarantine and Storage' at Faridabad, Haryana in the year 1946. Dr. Y.R. Rao (1960) published the monograph "The Desert Locust in India". In 1968 the Govt. of India enacted 'Central Insecticide Act' which came into force from 1st January, 1971. "The monograph on Indian Thysanoptera" was published by Dr. T. N. Ananthakrishnan in the year 1969.

2

Position of Insect in Animal Kingdom and its Relationship with Other Arthropods

Insects are invertebrates grouped in the phylum Arthropoda (Arthro-joint, poda-foot) and subphylum Uniramia.

Characters of the Phylum Arthropoda:

1. Organ-system level of body organization.
2. Body bilaterally symmetrical, triploblastic and metamerically segmented.
3. Appendages jointed, usually one pair to a somite and with varied functions as jaws, gills, legs, etc.
4. Exoskeleton of dead chitinous cuticle shed at intervals is called ecdysis or moulting *i.e.* for growth and development.
5. Body divided into head, thorax and abdomen. Head and thorax often fused to form a cephalothorax.
6. True coelom reduced and largely replaced by a blood-filled haemocoel.
7. Muscles mostly striated usually capable of rapid contraction.
8. Digestive system complete with mouth parts and anus. Mouth parts adopted for various mode of feeding.
9. Circulatory system open with a dorsal often many chambered heart, arteries and blood sinuses or haemocoel.

10. Respiration by general body surface, gills, or book lungs.

11. Excretory organs are green glands or malpighian tubules.

12. Nervous system typical with a dorsal brain connected to a nerve ring subsequently to a double ventral nerve cord.

13. Sensory organs comprise of eyes (simple and compound), chemo- and tactile receptors, and balancing and auditory organs.

14. Sexes are usually separate (dioecious). Reproductive organs and ducts paired. Fertilization usually internal. Oviparous or ovoviparous.

15. Development direct or indirect with one to many larval stages. Parthenogenesis in some individuals is a common feature.

16. Parental care often well-marked.

There are five principal classes of this phylum:

1. Myriaprda

(a) Chilopoda (Chilo–lip; poda–appendage)

This class includes centipedes (hundred legged worms). They are flattened worm like creatures, with a distinct head bearing two antennae. They have one pair of legs on every segment of the body. The first pair just behind the head, has poison glands whose secretion has the property of paralyzing the prey on which it feeds, *e.g.* Centipedes, *Scolopendra moristiana*

(b) Diplopoda (Diplo–two; poda- appendage)

This class includes the millipedes (thousand-legged worms). These are also elongated worm like creatures, but unlike the centipedes, their body is rounded and the segments are internally fused. The antennae are short and there are no poison claws. The reproductive organs open on the anterior end of the body, close to head. They generally feed on decaying vegetable matter, *e.g.* Millipede, *Thyroglutus malayus*

2. Crustacea (Crusta–shell)

The head often jointed with thorax to form cephalothorax. Exoskeleton is chitinous, hard and calcareous (limy). Head is 5 segmented bearing 2 pairs of antennae, 1 pair of mandibles and 2 pairs of maxillae. They breath by means of gills or body surface. Excretion by antennal gills. Sexes are usually separate, *e.g.* Prawn, crab, wood louse

3. Arachnida (Arachne–spider)

Body has two distinct regions, cephalothoraxe and abdomen. They do not have antennae and compound eyes but have simple eyes. Cephalothorax (prosoma) with 2 chelicerae, 2 pedipalps and 4 pairs of walking legs. Abdomen generally without appendages. Respiration by tracheae, book lungs or both. Excretion by coxal glands and malpighian tubes. Dioecious, mostly oviparous, courtship before mating. *e.g.* Scorpion, spider, tick, mite

Table 1: Relationship of Insects with Other Arthropods

Characters	Chilopoda	Diplopoda	Crustacea	Arachnida	Insecta
Habit	Terrestrial	Terrestrial	Aquatic and some terrestrial	Terrestrial	Many terrestrial and few aquatic
Body regions	Two-Head and multi-segmented trunks	Two-Head and multi-segmented trunks	Two-Cephalo-thorax and abdomen	Two in spiders and three in scorpion	Three-head, thorax and abdomen
Antennae	One pair	One pair	2 pair-Antennule and Antenna	No antenna	One pair
Visual organ	One pair-simple eyes	One pair-simple eyes	One pair-Stalked compound eyes	One pair-simple eyes	Both simple eyes and compound eyes (one pair)
Locomotor organs	One pair per segment (first pair of legs modified as poison claw)	Two pair per segment (No poison claw)	Minimum five pairs of biramous legs	Four pairs	Three pairs of legs on three thoracic segments and two pairs of wings on meso and metathorax
Mouth parts	Mandibulate (1 pair)	Mandibulate (1 pair)	Mandibulate (1 pair)	No mandibulate, but possess chelioerae	Mandibulate (1 pair)
Respiration	Tracheal	Tracheal	Gill breathing	Book lungs and tracheals	Tracheal
Body fluid	Haemolymph	Haemolymph	Haemolymph	Haemolymph	Haemolymph
Circulatory system	Heart with ostia	Heart with ostia	Heart with ostia	Heart with ostia	Heart with ostia
Development	Metamor-phosis	Metamor-phosis	Anamor-phosis	Metamor-phosis Absent (scorpion) and metamorphosis present in	Metamor-phosis
Habit	Carnivorous	Herbivorous	Herbivorous and Carnivorous	Phytopha-gous and predatory mites	Phytopha-gous, predatory and parasites

4. Insecta (In- internal; sect–cut) or Hexapoda (Hexa- six; poda-legs)

Body made of head (6 fused segments), thorax (3 segments) and abdomen (up to 11 segments). Head with compound eyes (1 pair), antennae (1 pair), mandibles (1 pair) and maxillae (2 pairs). Mouth parts modified for different feeding habits. Thorax with 3 pairs of jointed legs and 1 or 2 pair of wings which may absent. Respiration by tracheae. Spiracles lateral. Excretion by malpighian tubules. Uni–sexual, fertilization internal. Development usually with metamorphosis, *e.g.* Insects.

3

Insect Dominance

It is certainly no accident that insects are the most abundant and most diverse group of organisms on earth. They have maintained a position of ecological pre-eminence for over 400 million years. They have witnessed the rise and fall of dinosaurs, they have survived at least four major cataclysms that resulted in planet-wide extinctions, and they continue to thrive despite mankind's best efforts of eradication. While no single ecological or physiological attribute can account for this unparalleled success, the insects do have a unique combination of characteristics which, as a whole, have given them an unusual survival advantage.

The following are the measures of dominance of insects.

1. Number of Species

There are several ways to measure or assess the dominance of insects. Perhaps, the simplest approach is to compare the total number of insect species with similar totals for other taxonomic groups. By such an accounting, entomologists estimate that over 800,000 insect species have been named and described since Carolus Linnaeus (1707-1778) founded the modern system of nomenclature (genus and species) for all plants and animals. The entire plant kingdom, by comparison, contains between 400,000 and 500,000 species. In the animal kingdom, the "lower" invertebrates account for around 200,000 species, while vertebrates total 50,000 species only. Thus, insects represent more than half (about 53 per cent) of the 1.5 million species of living organisms known to science. But the catalogue of life is not yet complete! Discovery of a new species of vertebrate is a rare event–only a handful have been found in the past 20 years. But millions of undiscovered insect species are still hiding in the soil, in the canopies of tropical rain forests, and even right in our own backyards. Entomologists describe hundreds of these new species each year, and still estimate that only one-half to one-third of the earth's total insect fauna has even yet been discovered. In the final analysis, two of every three living species may be insects.

2. Number of Individuals

Another way to look at the dominance of insects is to consider the distribution and density of individual species or populations. Because of their small size and high reproductive potential, insect populations often grow to epic proportions. A swarm of migratory locusts may contain up to 10 billion individuals, covering an area of several thousand hectares, and have a total biomass of over 30,000 metric tons. A single colony of termites can swell to several million individuals within an earthen mound 20-25 feet tall. Above the arctic circle, hordes of mosquitoes and black flies can make life miserable for any warm-blooded animal. In some cases, these biting flies are so abundant and persistent that large animals, such as reindeer, have been known to die from the massive loss of blood (desanguination). In the tropics, ants are the most pervasive and diverse of all animal species. Colonies of driver ants (*Dorylus* sp.) on the African savannah may contain 20 million individuals. Even temperate forests and grasslands support unbelievably large populations of insects. As many as 40,000 cicadas (*Magicicada septendecim*) may feed on the roots of a large fruit tree. A "supercolony" of ants (*Formica* sp.) that covered 250 hectares (nearly a square mile) of coastal real estate near Sapporo, Japan, reportedly contained over a million queens and 300 million workers in a network of 45,000 interconnected nests. Admittedly, these are extreme examples, but even one stink bug per meter of row in a soybean field equals a population of more than 1,25,000 per hectare. In fact, the insects are so numerous that if they were divided equally among each one of the earth's 6 billion human inhabitants, each of us would be allotted 1×10^{18} insects– that's a billion billion-1,000,000,000,000,000,000.

3. Distribution

A third way to assess the dominance of insects is to examine their abundance and diversity in a wide range of ecological habitats. Indeed, insects are found in virtually every terrestrial and fresh-water environment on the face of the earth. They live everywhere from mountain tops in the Himalayas to tide pools at the seashore. Dozens of eyeless species spend their entire lives in the total darkness of underground caverns. At least 40 species live on the polar ice caps of Antarctica, others exist on the surface of glaciers, or near the edge of melting snow fields. A few species can survive in hot springs where the water temperature is 50°C; others inhabit deserts where water is scarce and daytime temperatures exceed 60°C. Insects live both above and below the surface of rivers, streams, lakes, and ponds. The petroleum fly (*Helaeomyia petrolei*), develops in pools of crude oil while the brine fly (*Ephydra cinera*), thrives in the high salinity of Utah's Great Salt Lake. In fact, the only place on earth where insects are not plentiful is in the ocean depths. Most marine environments are filled by another successful group of arthropods, the Crustaceans.

Reasons for Dominance in Animal Kingdom

There are several structural, morphological and physiological factors responsible for insect dominance. They are:

1. Structural Perfection

1.1. Exoskeleton

Unlike vertebrates, an insect's supporting skeleton is located on the outside of its body. This exoskeleton is a marvelous structure that not only gives shape and support to the body's soft tissues, but also provides protection from attack or injury, minimizes the loss of body fluids in both arid and freshwater environments, and assures mechanical advantage to muscles for strength and ability in movement. As a "suit of armor", the exoskeleton can resist both physical and chemical attack. It is covered by an impervious layer of wax that prevents desiccation. Much of the exoskeleton is fabricated from chitin, a polysaccharide that binds with various protein molecules to form a body wall that may be as flexible and elastic as rubber or as hard and rigid as some metals. Freedom of movement is ensured by membranes and joints in the exoskeleton. Muscles that attach directly to the body wall combine maximum strength with near-optimum mechanical advantage (leverage). The result is an ant, for example, that can lift up to 50 times its own body weight.

1.2. Small Size

In general, the insects are marvels of miniaturization. Most species are between 2 and 20 mm (0.1–1.0 inch) in length, although they range in size from giant moths. *Dichomorpha echmepterygis* is the smallest of the small. The wasp (family Mymaridae) is a parasite of other insects' eggs. Adult males may be only 0.139 mm (0.00055 inch) in length–nearly 1/3 smaller than some single-celled protozoa (*e.g., Paramecium caudatum*). For an animal with an exoskeleton, small size is a distinct advantage. If insects were as large as cows or elephants, their exoskeleton would have to be proportionately thicker to support the additional mass of body tissue. But a thicker exoskeleton would also be heavier and more cumbersome. Even the simplest movements would require a larger muscle volume and consume more energy. Since a two-fold increase in body length typically results in a four-fold increase in surface area and an eight-fold increase in volume and mass, there is an upper limit to how large insects can become (somewhere around 125-150 grams). Beyond this size, the insect's surface area is just too small for attachment of all the additional muscle tissue. Another advantage of small size is the minimal resources needed for survival and reproduction. A crumb is a feast; a dewdrop quenches thirst; a pebble provides shade. In some cases, food requirements are so modest that an insect may live on a single plant or animal for its entire life and never exhaust its food supply. A leaf miner, for example, spends its entire larval stage tunnelling within the paper-thin layer of cells (parenchyma) between the upper and lower epidermis of a single leaf. In some ant species, an entire colony may live inside a single acorn or plant gall. Parasitic wasps in at least seven families of Hymenoptera complete their entire development within the eggs of other insects. Finally, small size is a big advantage to insects that must avoid predation. They can hide in the cracks of a rock, beneath the bark of a tree, behind the petal of a flower, or under a blade of grass. The exoskeleton is hard enough for them to burrow between individual grains of sand, yet flexible enough to let them squeeze through the tiniest of cracks. Small size, together with adaptations in body shape and coloration, gives many species the ability to blend so well with their environment that they become virtually undetectable.

1.3. Quicker Speciation

As a result of the exoskeleton and small size, there has been quick speciation. It has been concluded that the peculiar combination of the 2 main characters *viz.*, the chitinous exoskeleton and the aerial mode of life has probably been the largest contributor to the extraordinarily high degree of speciation in insects. Owing to the mechanical rigidity of the chitinous exoskeleton, comparatively minute variations in the external genitalia have led to reproductive isolation and consequent fixation of the species.

1.4. Capacity of Flight

Insects are the only invertebrates that can fly. Judging from the fossil record, they acquired this ability about 300 million years ago–nearly 100 million years before the advent of the first flying reptiles. Flight gave these insects a highly effective mode of escape from predators that roamed the prehistoric landscape. It was also an efficient means of transportation, allowing populations to expand more quickly into new habitats and exploit new resources. The forces of natural selection have shaped insect flight into a remarkably efficient process employing principles of aerodynamics that human engineers do not yet fully understand. Although the metabolic cost of flight (calories per unit of lift) is similar to that of birds and bats, an insect's flight musculature produces at least 2X more power per unit of muscle mass. This high efficiency is largely due to elasticity of the thorax–90-95 per cent of the potential energy absorbed by flexion of the exoskeleton is released as kinetic energy during the wing's downstroke. Efficient use of energy allows some insects to travel great distances or remain airborne for long periods of time. More than 200 species, including moths, dragonflies, locusts, flies, and beetles are known to migrate over long distances by air. The migratory locust, *Schistocerca gregaria*, can fly for up to 9 hours without stopping. Speed and manuverability are the traits that are usually associated with highly advanced orders of insects, but some primitive insects (notably the dragonflies) are remarkably agile fliers. They can hover, turn while hovering, and even fly backwards or sideways. For their size, some species are remarkably swift. Large hawker dragonflies (family Ashnidae) have been clocked at a top speed of 58 km/hr (36 mph) over level ground. By comparison, the fastest human sprinters run only about 36 km/hr (22.5 mph).

1.5. Hexapod Locomotion

Jointed nature of the legs is a characteristic which insects share with other arthropods, but in insects the number of legs has reached an ideal stage of evolution. The 6 legs of insects is the optimum number which no other group of animal has. The larger number of legs as in millipedes is bound to make locomotion rather cumbersome, but when the number of walking legs becomes less than 6 it may create a problem of balancing during locomotion. During insect locomotion, the body always rests on 3 legs (a tripod) while the other 3 move forward. Thus, 6 is the smallest number for a stable equilibrium during all phases of terrestrial locomotion.

1.6. Compound Eyes

Insect usually have a pair of compound eyes in nymphal and adult stages and some times simple eyes too. Insects share these characteristics only with crustaceans. The compound eyes are of very complicated structure, not only with in other groups of animal. They are called compound eyes because each instead of having a single large cornea, consist of a number of hexagonal areas each representing the cornea of a discrete visual organ called *ommatidium*, all of which are compacted together. Thus, ants have 50 to 400 facets or corneas in each eye, house-fly has 4,000, a swallow-tail butterfly 17,000, and certain sphinx moths and dragonflies more than 50,000. It is obvious that an insect will not lose the power of vision completely if a few ommatidia are injured.

1.7. Scattered Sense Organs

Except the eyes, none of the sense organs are invariably concentrated on the head. The organs of hearing are sometimes associated with antennae. The organs of taste and smell may be on the antennae, on mouthparts, or even on tarsi or cersi. Such diffused and scattered nature of sense organs is bound to be advantageous to the insects. By this arrangement there is very less chances of all being injured at the same time.

1.8. Decentralized Nervous System

The central nervous system is a ladder like chain of ganglia along with ventral side of the body, except one complex nerve centre in the head, *i.e.*, the brain. This system despite the highly evolved stage of insect as a class is so decentralized that brainless insects can be artificially stimulated to walk, fly or even feed, and according to some, the abdomen of the living female silkworm moth when separated from the thorax can be fertilized by the male and stimulated to lay fertile eggs.

1.9. Direct Respiration

This ensures direct transfer of adequate oxygen to actively breathing tissues. Spiracles through their closing mechanism admit air and restrict water loss.

2. Developmental Characters (Metamorphosis)

Most insects undergo significant developmental changes as they grow from immatures to adults. These changes, collectively known as metamorphosis, may involve physical, biochemical, and/or behavioral alterations that promote survival, dispersal, and reproduction of the species. In the more primitive insects, most of these changes occur gradually as the animal matures *viz.*, organs of reproduction and flight develop incrementally during the immature stages and become functional only in adults. Since this transformation process is slow and does not include all body tissues (incomplete metamorphosis) the immatures and adults share many characteristics–they often live in similar habitats and feed on similar types of food. More advanced insects, however, undergo complete metamorphosis that is a dramatic transformation in form and function between the immature (larval) and adult stages of development. In these insects, a larva is primarily adapted for feeding and growth.

It assimilates energy reserves which in some cases will sustain the insect for the rest of its life. When fully grown, a larva molts into a transitional stage, called the pupa, and begins a period of massive internal and external reorganization. Body organs and tissues encoded by larval DNA are disassembled and rebuilt according to a second DNA blueprint that had been repressed during larval life. An adult insect (imago) eventually emerges from within the pupal exoskeleton bearing little or no resemblance to its larval form. Its primary function is dispersal and reproduction. In class Insecta, only 9 out of 28 orders undergo complete metamorphosis, yet these 9 orders represent about 86 per cent of all insect species alive today. The obvious advantage to this type of development lies in the compartmentalization of the life cycle. Through natural selection, larval form and function can be optimized for growth and feeding without compromising adaptations of the adult for dispersal and reproduction. Each stage of the life cycle is entirely free to adapt to its own ecological role. In some cases, this means that immatures and adults may consume different types of food, exploit different environmental resources, and even occupy different habitats.

3. High Fecundity and Reproduction Potential

Reproductive success is one of the most significant measures of an organism's fitness. In insect populations, females often produce large numbers of eggs (high fecundity), most of the eggs hatch (high fertility), and life cycle is relatively short (often as little as 2-4 weeks). Together, these three characteristics enable insects to produce remarkably large numbers of offspring. A typical female lays 100-500 eggs in her lifetime, but numbers in the thousands are not uncommon. The queen of a termite colony may be the mother of more than ten million workers during her 20-25 years lifespan. Antoni van Leeuwenhoek, the Dutch scientist who first discovered single-celled organisms, was probably the first person to comprehend the reproductive potential of insects. In 1687, he reared blow flies (probably *Calliphora erythrocephala*) on a diet of owl meat and found that females produced an average of 144 progeny (half male, half female). Using this data (and assuming no mortality), he calculated that a single pair of flies could give rise to 10,368 offspring in the third generation, 746,496 in the fourth generation, 53,747,712 in the fifth generation, and 3,869,835,264 in the sixth generation. At this rate, the fly population would grow to more than 1×10^{17} individuals in one year (10 generations). In two years of continuous reproduction, the flies would fill a ball nearly 8 million kilometers in diameter. Calculations like this are completely unrealistic because they ignore all the limiting factors (like food supply, predation, climate, and disease) that keep natural populations in check. But the numbers do illustrate how quickly insect populations can grow, and they help explain why pest outbreaks often seem to develop so suddenly. Since most insects die before they ever have an opportunity to reproduce, a high reproductive potential is the species' best chance for survival. Many adaptations help maximize this potential. Most females, for example, can store sperm for months or years within the spermatheca, a special region of the reproductive system. A single mating can supply a female with enough sperm to fertilize all the eggs that will produce in her lifetime. An unbalanced sex ratio, where females outnumber males, is another way to maximize reproductive potential. Since most insects are not monogamous, a few males can supply sperm for a large number of females, and finally, there are many species (*e.g.* aphids, scale insects, thrips, and midges) where males are entirely absent–all members of the population are female and contribute offspring through a process of asexual reproduction.

4. Feeding Specificity

While there is a vast diversity of food habits among insects, there being often some specific preference for one kind of food or the other, there are insects which have overcome interspecific competition by becoming polyphagous by living on almost all kind of vegetative materials, *e.g.*, locust. Some insects taking such food for which there is no competition, *e.g.*, cellulose-eating termites. There is sometimes extreme degree of specificity not only for different plant species or varieties but also for different parts of the same plant, as in case of root-borer, stem–borer, and top-borer of sugarcane.

5. Adaptability

A combination of large and diverse populations, high reproductive potential, and relatively short life cycles, has equipped most insects with the genetic resources to adapt quickly in the face of a changing environment. Their record of achievement is impressive *viz.*, they were among the first creatures to invade the arid expanses of dry land and exploit green plants as a source of food, they were the first animals to use flight as an escape from predators, and they were the first organisms to develop a complex social hierarchy with division of labor and cooperative care of the young. As a group, they have endured 400 million years of climatological and geophysical upheaval, including the evaporation of inland seas, formation of mountain ranges, shifts in continental plates, onset of ice ages, and the fallout from cosmic impacts. Adaptation is an ongoing process. Populations must continually change as new resources appear and old ones disappear. Just within the few thousand years since humans began roaming the earth, insects have acquired a taste for new products that would never be a part of their "natural" environment: *e.g.*, glue and wallpaper paste, book bindings, cardboard, paintbrushes, tanned leather products, the corks of wine bottles, mummies, stuffed museum specimens, chocolate, ginger, yeast cakes, tobacco, pepper, and even potent drugs like marijuana and opium. In the early 1900's, a powder post beetle (*Scobicia declivis*) made itself a nuisance by gnawing holes through the lead sheath around telephone and telegraph cables. Once inside the cable, these "short-circuit beetles" ate the fiber insulator around the copper conductor. Once enough insulation was gone, the wire would touch the lead sheath and create a short-circuit in the line. The high cost of replacing damaged cable eventually forced the telephone company to develop non-fibrous insulators for their long-distance cable. Perhaps, the most remarkable example of insect adaptation in this century has been the speed with which pest populations have developed resistance to a broad range of chemical and biological insecticides. Significant levels of pesticide resistance have now been reported in over 500 insect species, and many of these animals are resistant to compounds from more than one chemical family.

6. Zenith of Evolution

The zenith of evolution depicted by such social insects as bees, ants and termites shows the degree of specialization which the insects have attained. They have evolved their own agriculture, dairying, division of labour and social order, and even their own language. This all round efficiency is very effective in the struggle for existence.

4

Insect Structure and Function

External Structure and Function

Insect Integument

Insect body wall is called as Integument or Exoskeleton. It is the external covering of the body which is ectodermal in origin. It is rigid, flexible, lighter, stronger and modified into different body parts to suit different modes of life.

Structure

The integument consists of the following layers (1) Cuticle (2) the hypodermis (3) the basement membrane.

(1) Cuticle

It is a complex, non-cellular layer secreted largely by the hypodermis and it is considered as outer non-living material. The insect cuticle comprising of three sub layers:

(i) Endocuticle

It is the innermost and thickest layer. This layer is made up of chitin and arthropodin. This layer is colourless, soft and flexible.

(ii) Exocuticle

Outer layer, much thicker with the composition of chitin and sclerotin. This layer is dark in colour and rigid. Endocuticle and exocuticle put together are called as procuticle.

(iii) Epicuticle

Outer most layer is very thin. Pore canals present in the exocuticle helps in the deposition of epiculticle. This layer is differentiated into the following layers:

(a) Inner epicuticle: It contains wax filaments.

(b) Outer epicuticle: It makes the contact with cuticulin.

(c) Cuticulin: Non chitinous polymerised lipoprotein layer.

(d) Wax layer: It contains closely packed wax molecules which prevents desiccation.

(e) Cement layer: Outer most layer formed by lipid and tanned protein. It protects wax layer.

(2) The Hypodermis

It is an inner unicellular layer formed from polygonal cells. The polygonal cells may modify into columnar cells, having indistinct cell boundaries at the time of moulting. The hypodermal cells differentiate in some regions of the integument and constitute various types of mechano- and chemoreceptor organs. The hypodermis has the following functions:

(i) Cuticle secretion

(ii) Digestion and absorption of old cuticle

(iii) Wound repairing

(iv) Provides surface look

(3) The Basement Membrane

It is formed from the degenerated hypodermal cells and appears as a non- living, amorphous, granular inner lining of the integument. It is about 0.5 micron thick.

Composition of Cuticle

(i) Chitin

It is the main constituent of cuticle, which is nitrogenous polysaccharide and polymer of N-acetylglucosamine. It is water insoluble but soluble in dilute acids, alkalies and organic solvents.

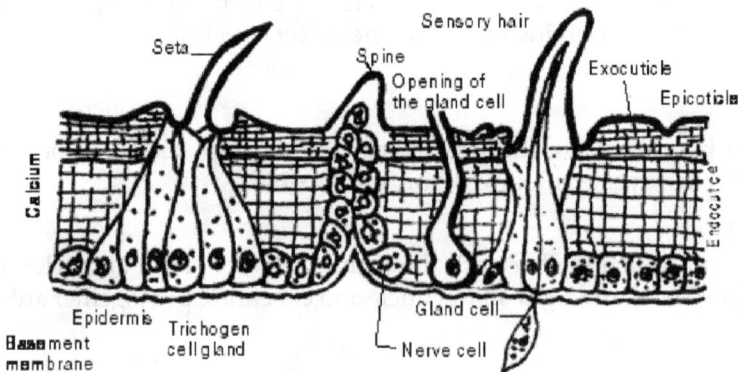

Figure 4.1: Structure of Insect Cuticle

(*ii*) Arthropodin

An untanned protein, which is water soluble.

(*iii*) Sclerotin

Tanned protein, which is water insoluble.

(*iv*) Resilin

An elastic protein responsible for the flexibility of wing sclerites.

Cuticular Appendages

Non-cellular

Non cellular appendages have no epidermal association, but rigidly attached, *e.g.*, minute hairs and thorns.

Cellular

Cellular appendages have epidermal association and it may be unicellular or multicellular.

(*i*) Unicellular Structures

    (*a*) Clothing hairs, plumose hairs, *e.g.* Honey bee.

    (*b*) Bristles, *e.g.*flies.

    (*c*) Scales–flattened out growth of body wall, *e.g.* Moths and butterflies

    (*d*) Glandular seta, *e.g.* caterpillar

    (*e*) Sensory setae–associated with sensory neuron or neurons.

    (*f*) Seta–hair like out growth from epidermis. Epidermal cell generating seta is known as Trichogen, while the socket forming cell housing trichogen is known as Tormogen. Study of arrangement of seta is known as Chaetotaxy.

(*ii*) Multicellular Structures

e.g. Spur–movable; Spine- immovable.

Endoskeleton

Cuticular in growth of body wall providing space for muscle attachment is known as endoskeleton. There are two types:

(*i*) Apodeme

Hollow invagination of body wall (ridge like).

(*ii*) Apophysis

Solid invagination of body wall (spine like).

Coloration

The colour of adult and immature insects may be grouped into three classes:

(*i*) Pigmentary Colour (Chemical Colour)

These substances have defined chemical composition which have the property of absorbing some light waves and of reflecting others. These substances are by-products of metabolism and in some cases may be of an excretory nature. They may be classified into cuticular, hypodermal and subhypodermal colours according to their location.

Cuticular Colours

Cuticular colours are mostly contained in the exocuticle, they consist of brown, black and yellows, which are permanent. Some common cuticular colours are:

1. Melanins
2. Carotenoids
3. Pterins
4. Insectorubin (Ommochromes)
5. Anthoxanthins (Flavones)
6. Anthocyanins
7. Miscellaneous pigments

Hypodermal Colours

Hypodermal colours are present in the form of granules or drops of fat in the cells of the hypodermis. They may be red, orange, yellow or green and are very evanescent after death.

Subhypodermal Colours

Subhypodermal colours are contained in the fat body and blood.

(*ii*) Structural Colours

Structural colours differ from those due to pigments in that thay are changed or destroyed by physical changes in the cuticle such as results from shrinkage, swelling, distortion or permeation with liquids of the same refractive index as the cuticle. They may also be duplicated by physical methods. Four types of structural colours may distinguished:

1. Structural white
2. Tyndall blue
3. Interference colours
4. Diffraction colours

(*iii*) Combination Colours

These are produced by a structural modification in conjuction with a layer of pigment and are much commoner than purely structural colours.

Moulting (Ecdysis)

Periodical process of shedding the old cuticle accompanied by the formation of new cuticle is known as moulting or ecdysis. The cuticular part discarded during moulting is known as exuvia. Moulting occurs many times in an insect during the immatured stages before attaining the adult-hood. The time interval between the two subsequent moultings is called as stadium and the form assumed by the insect in any stadium is called as instar.

Steps in Moulting

1. *Behaviroual changes*: Larva stops feeding and become inactive.
2. *Changes in epidermis*: In the epidermis cell size, its activity, protein content and enzyme level increases. Cells divide miotically and increases the tension, which results in loosening of cells of cuticle.
3. *Apolysis*: Detachment of cuticle from the epidermis
4. *Formation of subcuticular space.*
5. Secretion of moulting gel in the sub cuticular space which is rich with chitinase and protease.
6. *New epicuticle formation*: Cuticulin layer is laid over the epidermis.
7. *Procuticle formation*: Procuticle is formed below the epicuticle.
8. *Activation of moulting gel*: Moulting gel is converted into moulting fluid which is rich in enzymes. This activates endocuticle digestion and absorption.
9. *Wax layer formation*: Wax threads of pore canals secrete wax layer.
10 *Cement layer formation*: Dermal glands secretes cement layer (Tectocuticle).
11. *Moulting*: This involves two steps
    (i) *Rupturing of old cuticle*: Insect increases its body volume through intake of air or water which enhances the blood flow to head and thorax. Thereby the old cuticle ruptures along the predetermined line of weakness, known as ecdysial line.
    (ii) *Removal of old cuticle*: Peristaltic movement of body and lubricant action of moulting fluid helps in the removal of old cuticle. During each moulting, the cuticular coverings of body, legs, internal linings of foregut, hindgut and trachea are discarded.
12. *Formation of exocuticle*: The upper layer of procuticle develops as exocuticle through addition of protein and tanning by phenolic substance.
13. *Formation of endocuticle*: The lower layer of procuticle develops as endocuticle through addition of chitin and protein. This layer increases in thickness.

Control of Moulting

It is controlled by endocrine glands like prothoracic gland(PTG) which secrete moulting hormone. Endocrine glands are activated by prothoracico-tropic hormones produced by neurosecretory cells (NCC) of brain.

Insect Head

Based on the inclination of long axis of the head and orientation of mouth parts, there are three types of insect's heads.

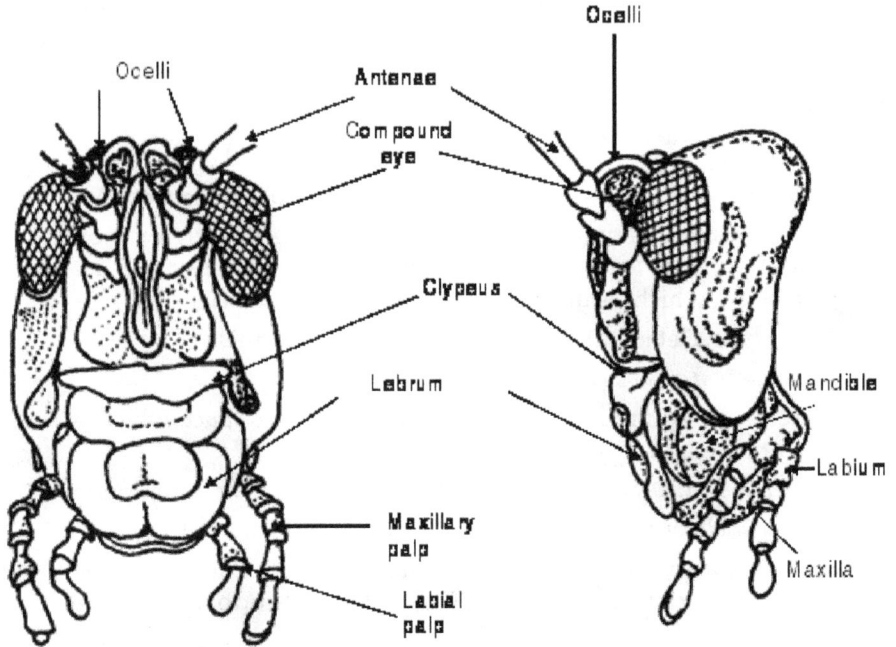

Figure 4.2(a): Insect Head

1. Hypognathous: (Hypo-below; gnathous-jaw)

This type is called orthopteroid type. The long axis of the head is vertical, it is at right angles to the long axis of the body. Mouth parts are ventrally placed and project downwards, *e.g.* grasshopper, cockroach.

2. Prognathous: (Pro-infront; gnathous-jaw)

This type is also called coleopteroid type. The long axis of the head is horizontal. It is in line with the long axis of the body. Mouth parts are directed forward. *e.g.* ground beetle.

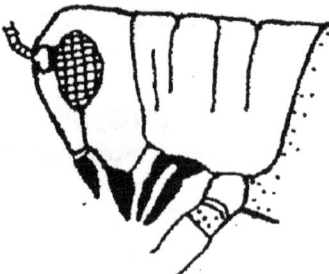

Figure 4.2(b): Hypognathous Head

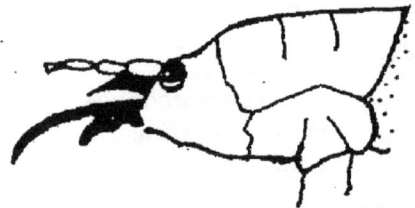

Figure 4.3: Prognathous Head

3. Opisthognathuos: (Opistho-behind; gnathuos-jaw)

This is also called hemipteroid type or opisthorhynchous type. Head is deflexed. Mouthparts are directed backwards and held in between the forelegs. *e.g.* stink bug.

Head Appendages

Structure of Insect Antenna

Antennae are also called feelers. They are paired, highly mobile and segmented. Antennae are located between or behind the compound eyes. All insects except protura have a pair of antennae. Antennae are well developed in adults and poorly developed in immature stages. The antenna is set in a socket of the cranium called antennal socket. The base of the antenna is connected to the edge of

Figure 4.4: Opisthognathuos Head

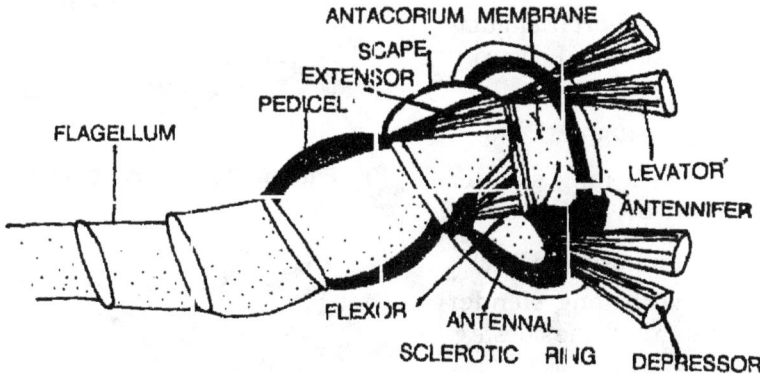

the socket by an articulatory membrane. This permits free movement of antennae. The basal segment is called *scape*. It is conspicuously larger than succeeding segments. The second antennal segment is called *pedicel* which immediately follow the scape.

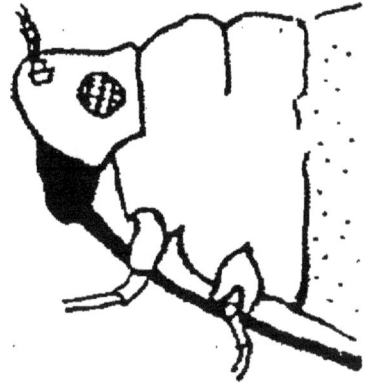

Figure 4.5: The Basic Structure of Antennae

A mass of sense cells called *Johnston's organ* is present in the pedicel, which is used as a chordatonal organ in some of the insects like mosquitoes. Both scape and pedicel are provided with intrinsic muscles. The remaining annuli or flagellomeres are known as *flagellum* or *clavola* which lack individual muscle. Surface of the flagellum is supplied with many sensory receptors that are innervated by the duetocerebrum of brain. Flagellum may vary in size and form.

Function

Antenna is used to detect chemicals including food and pheromones (chemicals secreted into air by opposite sex). It perceives smell, humidity changes, variation in temperature, vibration, wind velocity and direction. Antenna is useful to perceive the forward environment and detect danger. It is useful for hearing in mosquitoes and communication in ants. Rarely, it is also useful to clasp the mate (*e.g.* Flea) and grasp the prey.

Types of Antennae

1. Setaceous (Bristle like)

Size of the segments decreases from base to apex, *e.g.* Leafhopper, Dragonfly, Damselfly.

Figure 4.6: Setaceous Antennae

2. Filiform (Thread like)

Segments are usually cylindrical. Thickness of segments remains same, *e.g.* Grasshopper.

Figure 4.7: Filiform Antennae

3. Moniliform (Beaded)

Segments are either globular or spherical with prominent constriction in between, *e.g.* Termite.

Figure 4.8: Moniliform Antennae

4. Serrate (Saw like)

Segments have short triangular projections on one side, *e.g.* Longicorn bettle.

Figure 4.9: Serrate Antennae

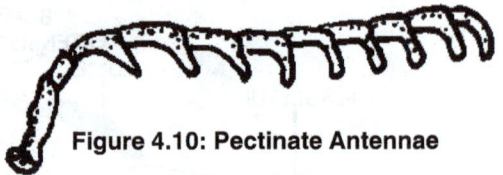

5. Pectinate (Comb like)

Segments with long slender processes on one side, *e.g.* Sawfly.

Figure 4.10: Pectinate Antennae

6. Bipectinate (Double comb like)

Segments with long slender lateral processes on both the sides, *e.g.* Silkworm moth

Figure 4.11: Bipectinate Antennae

7. Clavate (Clubbed)

Antenna enlarges gradually towards the tip, *e.g.* Blister beetle.

Figure 4.12: Clavate Antennae

8. Capitate (Knobbed)

Terminal segments become enlarged suddenly, *e.g.* butterfly.

9. Lamellate (Plate like)

Antennal tip is expanded laterally on one side to form flat plates, *e.g.* lamellicorn beetle.

Figure 4.13: Capitate Antennae

Figure 4.14: Lamellate Antennae

Figure 4.15: Aristate Antennae

10. Aristate

The terminal segment is enlarged. It bears a conspicuous dorsal bristle called arista, *e.g.* House fly.

11. Stylate

Figure 4.16: Stylate Antennae

Terminal segment bears a style like process, *e.g.* Horse fly, Robber fly.

12. Plumose (Feathery)

Segments with long whorls of hairs, *e.g.* male mosquito.

13. Pilose (Hairy)

Figure 4.17: Plumose Antennae

Antenna is less feathery with few hairs at the junction of flagellomeres, *e.g.* Female mosquito.

14. Geniculate (Elbowed)

Scape is a long remaining segment which is small and arranged at an angle to the first resembling an elbow joint, *e.g.* Ant, weevil and honey bee.

Insect Mouthparts

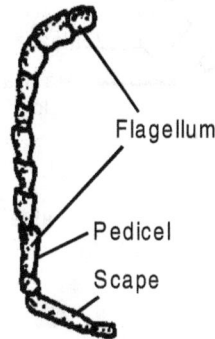

Figure 4.18: Geniculate Antennae

Mouthparts of insects vary to a great extent among insects of different groups depending upon their feeding habits. They are mainly of two types *viz.*, Mandibulate (feeding mainly on solid food) and haustellate (feeding mainly on liquid food).

1. Biting and Chewing Type

E.g. Cockroach and grasshopper. It is the primitive type of mouth part and consists of the following parts.

(*i*) Labrum (Upper lip)

It is flap like, bilobed and attached to the clypeus by an articular membrane. It is movable. It covers the mouth cavity from above. It helps to pull the food into the mouth. It holds the food in position so that mandibles can act on it. It forms the roof of the pre oral food cavity.

(*ii*) Labrum-epipharynx

Inner surface of the labrum is referred to as epipharynx. It is frequently membranous and continuous with the dorsal wall of pharnyx. It is an organ of taste.

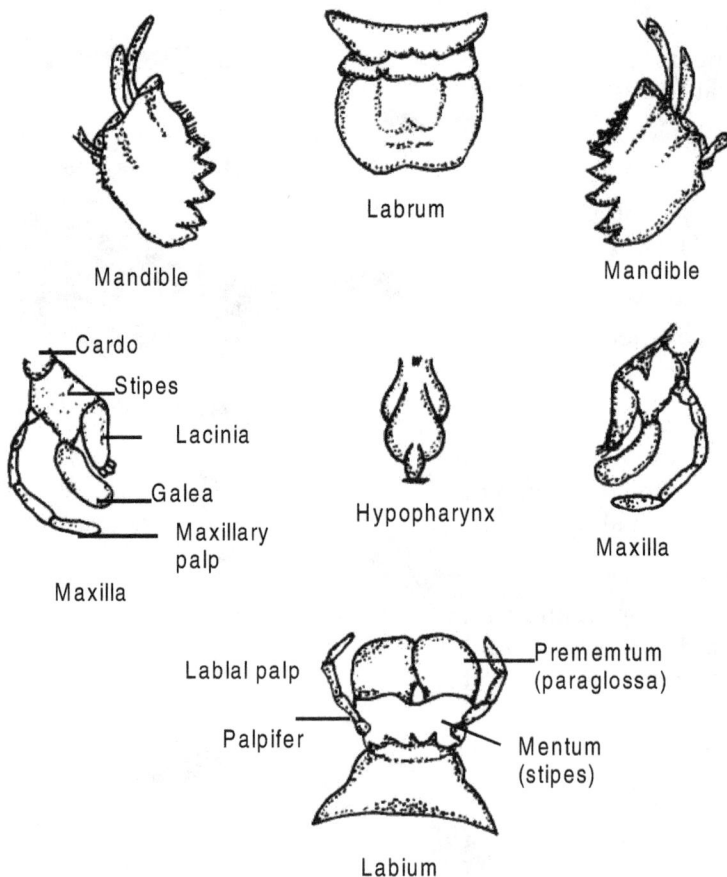

Figure 4.19: Biting and Chewing Type

(*iii*) Mandibles

There is a pair of mandibles. They are the first pair of jaws. They are also called as primary jaws or true jaws. Mandibles articulate with the cranium at two points. They are heavily sclerotised. They are toothed on their inner border. There are two types of teeth. Distal are sharply pointed and are called incisor or cutting teeth and proximal teeth are called molar or grinding teeth. They act transversely to bite and grind the food into small fragments.

(*iv*) Maxillae

They are paired and more complicated than mandibles. They are called secondary jaws or accessory jaws. At proximal end, the first sclerite cardo, joins the maxilla to head. The second sclerite is called stipes which articulates with cardo. Stipes carries a lateral sclerite called palpifer which bears a five segmented antenna like maxillary palp. On the distal end of the stipes, there are two lobes. The outer lobe is called galea and inner lobe is lacinia which is toothed. Maxille direct the food into the mouth. They hold the food in place when the mandibles are in action. They act as auxillary jaws and assist in mastication of food. Sense organs connected with the perception of touch, smell and taste are abundantly found in palpi.

(*v*) Hypopharynx

It is a tongue like organ. It is located centrally in the preoral cavity. Salivary gland duct opens through it.

(*vi*) Labium/Lower Lip

It is a composite structure formed by the fusion of two primitive segmented appendages. It bounds the mouth cavity from below or behind. It forms the base of the preoral cavity. It consists of three median sclerites *viz.*, submentum (large basalsclerite), mentum (middle sclerite) and prementum (apical sclerite). On the lateral side of the prementum, there are two small lateral sclerites called palpifer bearing three segmented labial palpi. Distally prementum bears two pairs of lobes. The other pair of lobes is called paraglossa and inner pair of lobes, glossa. Both pairs when fused are called ligula.

2. Piercing and Sucking/Hemipterous/Bug Type *e.g.* Plant Bugs

Labium projects downwards from the anterior part of the head like a beak. Beak is four segmented and grooved throughout its entire length. At the base of the labium there is a triangular flap like structure called labrum. Labium is neither involved in piercing nor sucking. It functions as a protective covering for the four stylets (fascicle) found within the groove. Both mandibles and maxillae are modified into long slender sclerotized hair like structure called stylets. They are lying close together and suited for piercing and sucking. The tips of the stylets may have minute teeth for piercing the plant tissue. The inner maxillary stylets are doubly grooved on their inner faces. When these are closely opposed, they form two canals *viz.*, food canal and salivary canal through sap and saliva, respectively. Saliva contains enzymes or toxins that can distort plant cell wall to permit the stylets to penetrate down and reach phloem for suking the sap. Both palps are absent.

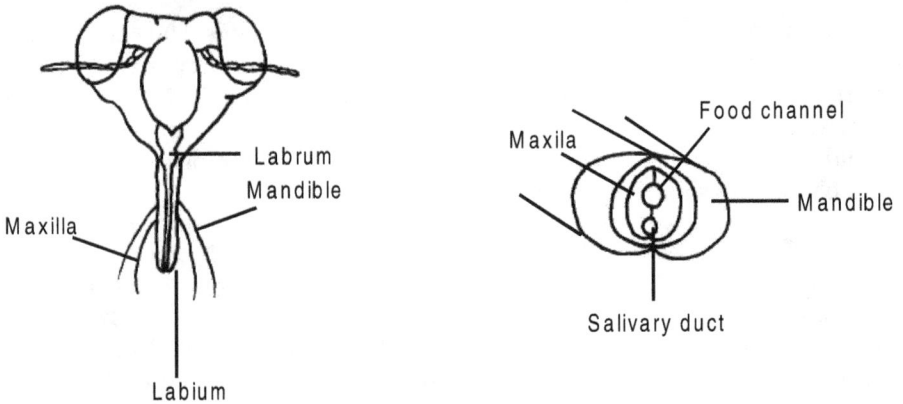

Figure 4.20: Piercing and Sucking Type (Bug Type) Mouth Parts

3. Piercing and Sucking/Dipterous/Mosquito Type: *e.g.* Female Mosquito

Mouth parts of female mosquito consists of an elongate labium which is grooved forming a gutter which encloses six stylets. The stylets are composed of labrum–epipharynx (enclosing the food canal), the hyphophrynx (containing the salivary canal), two maxillae and two mandibles. Both the ends of maxillary stylets and mandibular stylets are saw like and suited to piercing flesh. The stylets are inserted into hosts skin by a strong downward and forward thrust of body. Both mandibles and maxillae are reduced in male as they feed on plant nectar and juices of decaying

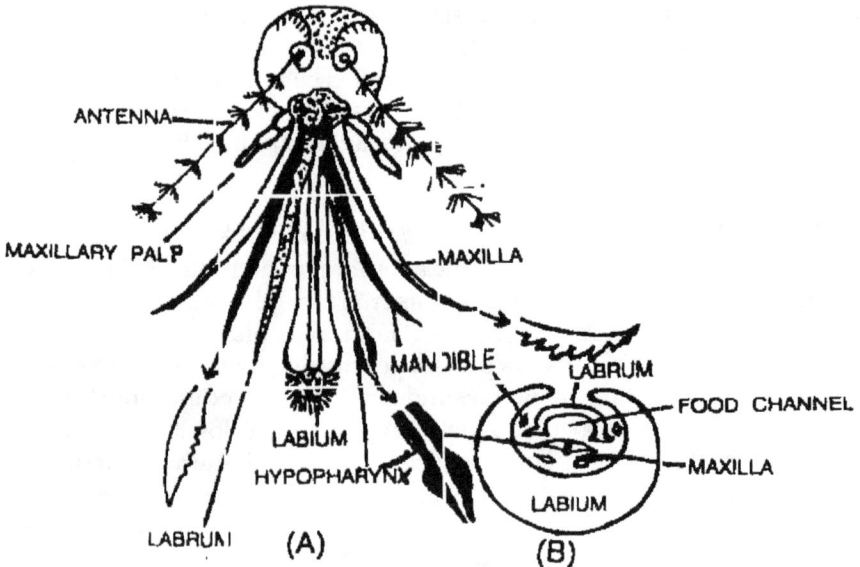

Figure 4.21: Piercing and Sucking Mouth Parts
(A) Frontal View; (B) Cross Section

fruits. Female pierces the skin of human beings into which it injects saliva containing an anticoagulant (to keep the blood flowing without clotting) and an anesthetic (to keep the victim unaware of the bite) and sucks up the blood. Labium does not pierce but folds up or back as stylets pierce. Maxillary palpi are present.

4. Chewing and Lapping Type: *e.g.* honey Bee

Labrum and mandibles are as in biting and chewing type of mouth parts. But mandibles are blunt and not toothed. They are useful to crush and shape wax for comb building; ingest pollen grains and other manipulative functions. Maxillolabial structures are modified to form the lapping tongue. The tongue unit consists of two galea of maxillae, two labial palpi and elongated flexible hairy glossa of labium. The glossa terminates into a small circular spoon shaped lobe called spoon or bouton or flabellum which is useful to lick the nectar.

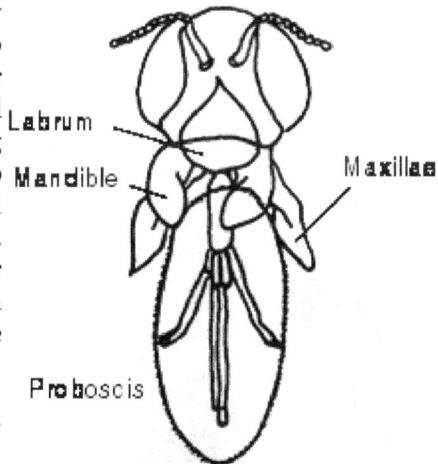

Figure 4.22: Chewing and Lapping Mouth Parts

5. Rasping & Sucking Type: *e.g.* Thrips

Mouth cone consists of labrum, labium and maxillae. There are three stylets derived from two maxillae and a left mandible. Right mandible is absent. Stylets are used to lacerate the plant tissue and the oozing sap is sucked up by the mouth cone. Both maxillary palpi and labial palpi are present.

6. Mandibulosuctorial Type: *e.g.* Grub of Antlion

Mandibles are elongate sickle shaped and grooved on the inner surface. Each maxilla is elongated and fits against the mandibular groove to from a closed food canal. The body of the insect victim is pierced by the opposing mandibles and fluids are extracted.

7. Sponging Type: *e.g.* House Fly

The proboscis is fleshy, elbowed, retractile and projects downwards from head. The proboscis can be differentiated into basal rostrum and distal haustellum. The proboscis consists of labium which is grooved on its anterior surface. Within this groove, lie the labrum-epiphraynx (enclosing the food canal) and slender hypopharynx (containing the salivary canal). Mandibles are absent. Maxillae are represented by single segmented maxillary

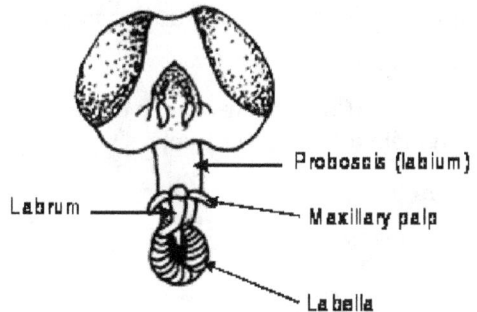

Figure 4.23: Sponging Type Mouth Parts

palpi. The end of the proboscis is enlarged, sponge like and two lobed which acts as suction pads. They are called oral discs or labella. The surfaces of labella are transvered by capillary canals called pseudotracheae which collect the liquid food and convey it to the canal. Labella function as sponging organ and is capable of taking exposed fluids. These insects often spit enzyme containing saliva onto solid foods to liquify them.

8. Siphoning Type: *e.g.* Moths and Butterflies

Mouth parts consists of elongate sucking tube or proboscis. It is formed by two greatly elongated galeae of maxillae which are zippered together by interlocking spines and hooks. Galeae are grooved on their inner surface and when they are fitting together closely, they form a suctorial food canal through which the nectar is sucked up. The proboscis is coiled up like watch spring and kept beneath the head when it is not in use. By pumping blood into galeae, the proboscis is extended. The other mouth parts are reduced or absent except the labial palpi and smaller maxillary palpi.

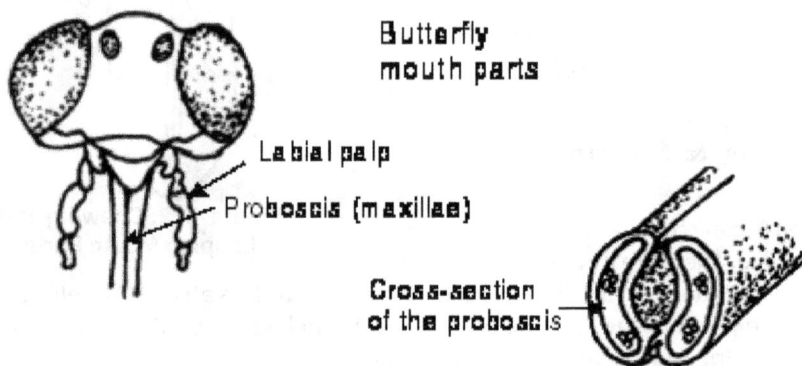

Figure 4.24: Siphoning Type Mouth Parts

Thorax

Second and middle tagma which is three segmented, namely prothorax, mesothorax and metathorax. Meso and metathorax which bear wings are called as Pterothorax. Thoracic segments are made up of three sclerites namely, dorsal body plate tergum or nota, ventral body plate sternum and lateral plate pleuron.

Thoracic Nota

Dorsal body plate of each thoracic segments are called as pronotum, mesonotum and metanotum respectively.

Pronotum

This sclerite is undivided and saddle shaped in grasshopper and shield like in cockroach.

Pterothoracic Notum

Have 3 transverse sutures (antecostal, prescutal and scuto-scutellar) and 5 tergites (acrotergite, prescutum, scutum, scutellum and post-scutellum).

Thoracic Sterna

Vental body plate of each thoracic segments are called as prosternum, mesosternum and metasternum. Thoracic sterna is made up of a segmental plate called eusternun and an intersternite called spinasternum. Eusternum is made up of three sternites *viz.,* presternum, basisternum and sternellum.

Thoracic Pleura

It is the lateral body wall of thoracic segment between notum and sternum. Sclerites of pleuron is called as pleurite and they fuse to form pleural plate. Pleural plate is divided into anterior episternum and posterior epimeron by pleural suture. Pterothoracic pleuron provides space for articulation of wings and legs. Thoracic appendages are three pairs of legs and two pairs of wings. Two pairs of spiracles are also present in the mesopleuron and metapleuron.

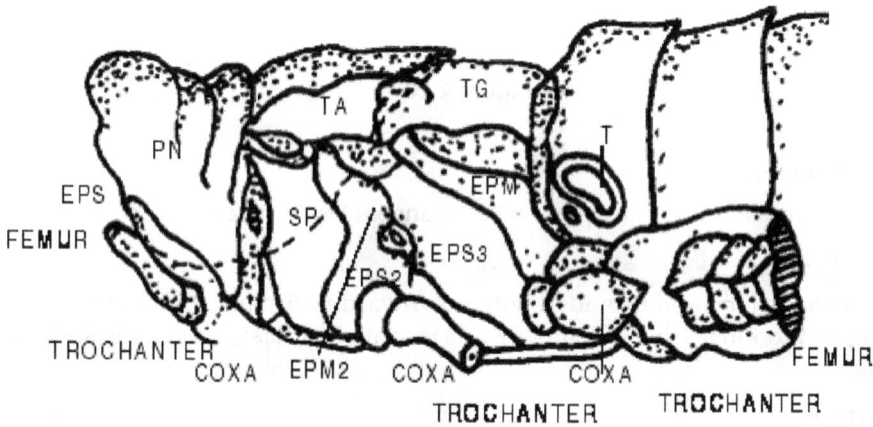

Figure 4.25: Structure of Insect Thorax

Functions of Thorax

Mainly concerned with locomotion.

Legs

In almost all insects, all the three thoracic segments *viz.,* pro-, meso- and metathorax bear a pair of segmented legs. Each leg consists of five segments *viz.,* coxa, trochanter, femur, tibia and tarsus.

Coxa (Pl. coxae)

It is the first or proximal leg segment. It articulates with the cup like depression on the thoracic pleuron. It is generally freely movable.

Trochanter

It is the second leg segment. It is usually small and single segmented. Trochanter seems to be two segmented in dragonfly, dameselfly and Ichneumonid wasp. The apparent second trochanter is in fact a part of femur, which is called trochantellus.

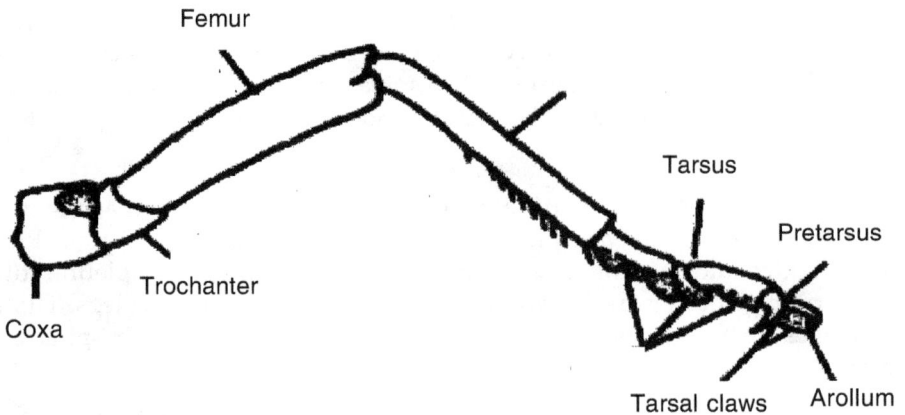

Figure 4.26

Femur (Pl. femora)

It is the largest and stoutest part of the leg and is closely attached to the trochanter.

Tibia (Pl. tibiae)

It is usually long and provided with downward projecting spines which aid in climbing and footing. Tibia of many insects is armed with large movable spur near the apex.

Tarsus (Pl. tarsi)

It is further sub-divided. The sub segment of the tarsus is called tarsomere. The number of tarsomeres vary from one to five. The basal tarsal segment is often larger than others and is named as basitarsus.

Pretarsus

Beyond the tarsus there are several structures, collectively known as pretarsus. Tarsus terminates in a pair of strongly curved claws with one or two pads of cushions at their base between them. A median pad between the claws is usually known as arolium and a pair of pads at their base is called pulvilli (Pulvillus-singular). Leg pads are useful while walking on smooth surface and claws give needed grip while walking on rough surface. When one structure is used, the other is bent upwards.

Types or Modifications

Legs are modified into several types based on habitat and food habit of insect and used for a wide variety of functions.

1. *Ambulatorial* (Ambulate–to walk; Walking leg) *e.g.* Fore leg and middle leg of grasshopper. Femur and tibia are long. Legs are suited for walking.

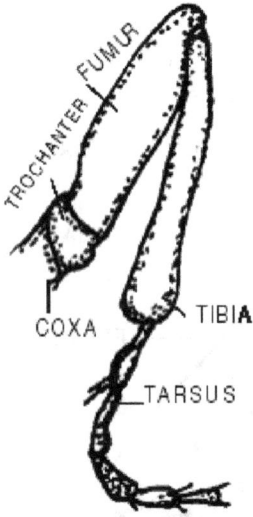

Figure 4.27: Ambulatorial Leg

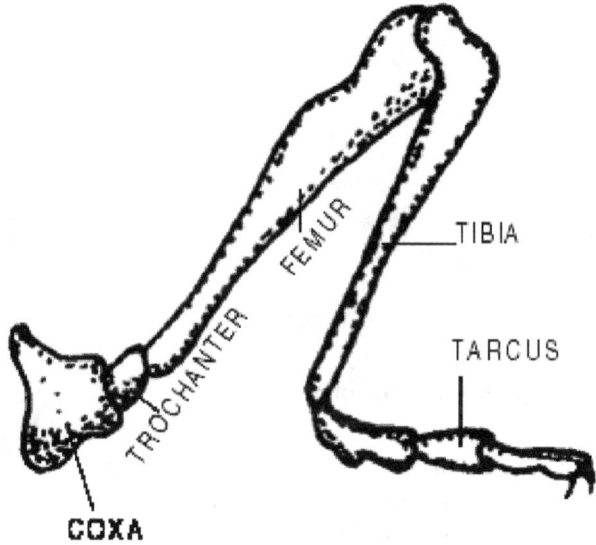

Figure 4.28: Cursorial Leg

2. *Cursorial*: (Cursorial = adapted for running: Running leg) *e.g.*All the three pairs of legs of cockroach. Legs are suited for running. Femur is not swollen.
3. *Saltatorial*: (Salatorial = Leaping: Jumping Leg) *e.g.* hind leg of grasshopper.
4. *Scansorial*: (Scansorial = Climbing; climbing or clinging leg) *e.g.* all the three pairs of legs of head louse.
5. *Fussorial*: (Fussorial= Digging; Burrowing leg) *e.g.* Fore legs of mole cricket.

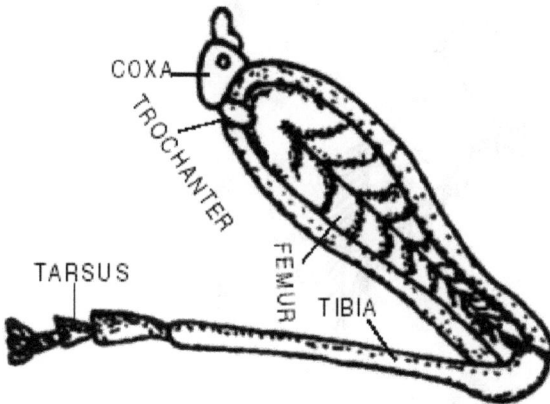

Figure 4.29: Saltatorial Leg

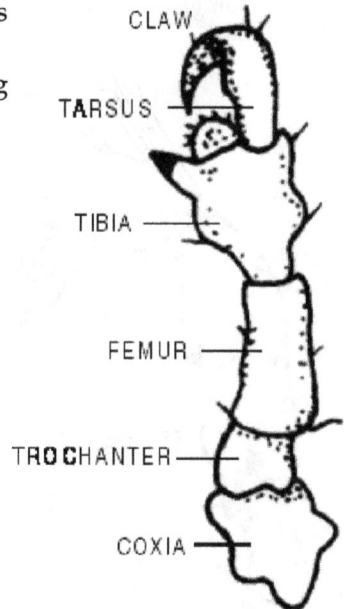

Figure 4.30: Scansorial Leg

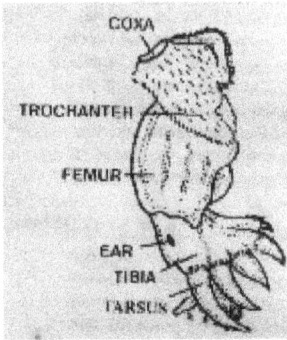

Figure 4.31: Fussorial Leg

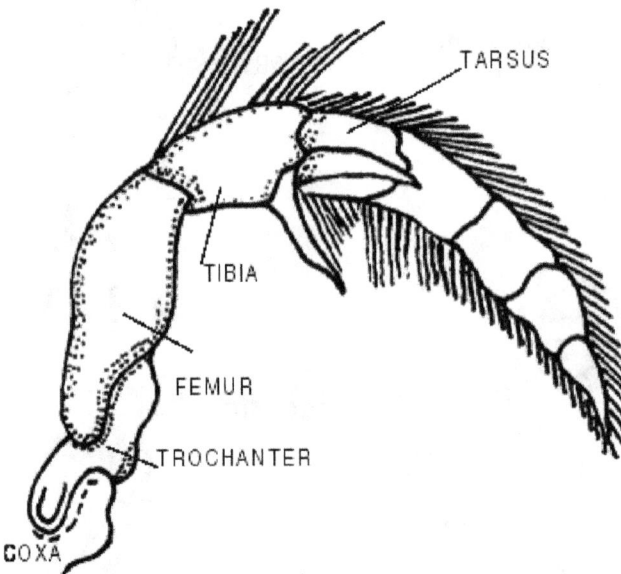

6. *Raptorial*: (Raptorial = predatory; Grasping leg) *e.g.* Forelegs of preying mantis.

7. *Natatorial*: (Natatorial = pertaining to swimming; Swimming leg) *e.g.* hing legs of water bug and water beetle.

Figure 4.32: Raptorial Leg

8. *Sticking leg*: *e.g.* all the three pairs of legs of house fly.

9. *Basket like leg*: *e.g.* Legs of dragonfly and damselfly.

10. *Clasping leg*: *e.g.* Forelegs of male water beetle.

11. *Foragial leg*: (Forage = to collect food material) *e.g.* Legs of honey bee.

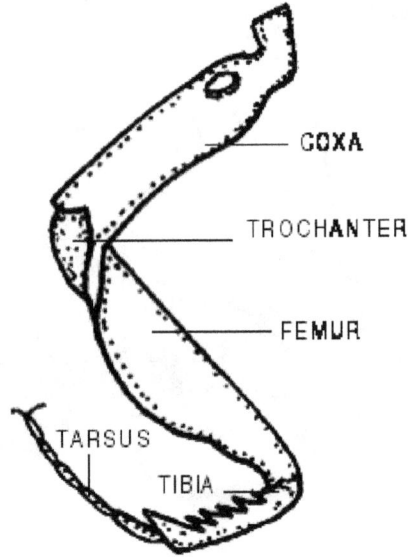

Figure 4.33: Natatorial Leg

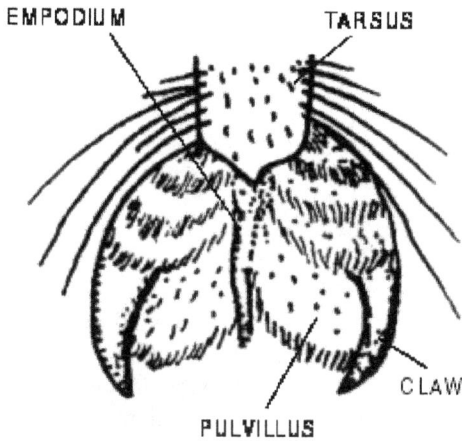

EMPODIUM TARSUS

CLAW

PULVILLUS

Figure 4.34: Sticky Leg

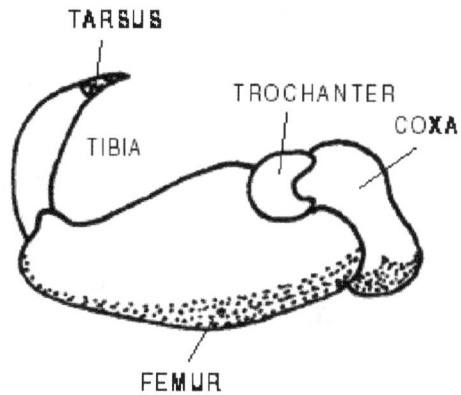

TARSUS

TROCHANTER

COXA

TIBIA

FEMUR

Figure 4.35: Claspy Leg

(i) *Forelegs*: The foreleg has three important structures (Eye brush

(a) Antenna cleaner or strigillis:

(b) Pollen brush:

(ii) *Middle legs*: It has two important structures.

(a) *Pollen brush*: Stiff hairs on basitarsus form pollen brush which is used to collect pollen from middle part of their body.

(b) *Tibial spar*: At the distal end of the tibia, a movable spur is present which is used to loosen the pellets of pollen from the pollen basket of hind legs and to clean wings and spiracles.

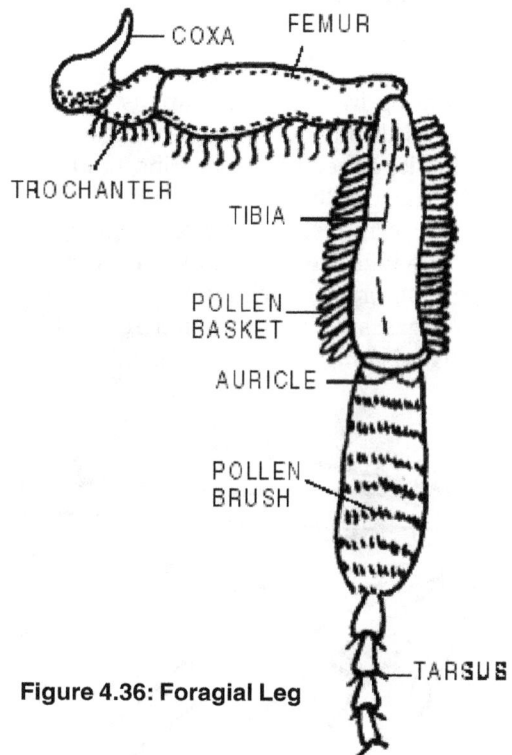

COXA FEMUR

TROCHANTER

TIBIA

POLLEN BASKET

AURICLE

POLLEN BRUSH

TARSUS

Figure 4.36: Foragial Leg

(iii) *Hind legs*: It has three important structures *viz.*, pollen basket, pollen packer and pollen comb.

(a) *Pollen basket*: It is also called corbicula. The outer surface of the hind tibia contains a shallow cavity. The edges of the cavity are fringed with long hairs. The pollen basket enables the bee to carry a larger load of pollen and propolis from field to the hive.

(*b*) *Pollen packer*: It is also called pollen press. It consists of pecten and auricle. Pecten is a row of stout bristles at the distal end of tibia. Auricle is a small plate fringed with hairs at the basal end of basitarsus. Pollen packer is used to load pollen in corbicula.

(*c*) *Pollen comb*: About ten rows of stiff spines are present on the inner side of hind basitarsus. The pollen comb is used to collect pollen from middle legs and from posterior part of the body.

12. *Prolegs or False legs or Pseudolegs*: *e.g.* abdominal legs of caterpillar. There are two to five pairs of abdominal legs termed prolegs in caterpillar. Prolegs are thick, fleshy and not segmented. They are shed with last larval moult. One pair of prolegs on the last abdominal segment are called anal prolegs or claspers. The tip of proleg is called planta upon which are borne hooks or claws known as crochets which are useful in crawling or clinging to surface.

Wings

Among invertebrate animals, only insects possess wings. Wings are present only in adult stage. Number of wings vary from two pairs to none. Certain primitive insects like silverfish and spring tail have no wings (apterous). Ectoparasites like head louse, poultry louse and flea are secondarily wingless. Wings are deciduous in ants and termites. There is only one pair of wings in the true flies. Normally, two pairs of wings are present in insects and they are borne on pterothoracic segments *viz.*, mesothorax and metathorax. Wings are moved by thoacic flight muscles attached to their bases. Wing is a flattened double–layered expansion of body wall with a dorsal and ventral lamina having the same structure as the integument. Both dorsal

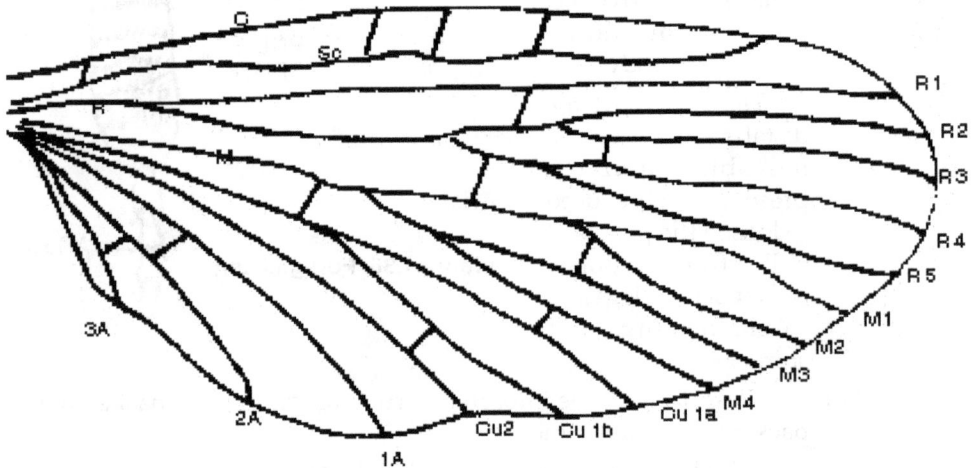

Figure 4.37: Typical Wing Venation of Insect

and ventral laminane grow, meet and fuse except along certain lines. Thus, a series of channels is formed. These channels serve for the passage of tracheae, nerves and blood. Wing is nourished by blood circulating veins. Later the walls of these channels become thickened to form veins or nervures. The arrangement of veins on the wings is called venation which is extensively used in insect classification. The principal longitudinal veins arranged in order from the anterior margin are costa (C), sub costa (Sc), radius (R), median (M), cubitus (Cu) and anal veins (A). Small veins often found inter connecting the longitudinal veins are called cross veins. Due to the presence of longitudinal veins and cross veins, the wing surface gets divided into a number of enclosed spaces termed cells. In insects like dragonfly and damselfly, there is an opaque spot near the coastal margin of the wing called pterotigma.

Margins and Angles

The wing is triangular in shape and has therefore three sides and three angles. The anterior margin strengthened by the costa is called coastal margin and the lateral margin is called apical margin and the posterior margin is called anal margin. The angle by which the wing is attached to the thorax is called humeral angle. The angle between the coastal and apical margins is called apical angle. The angle between apical and anal margins is anal angle.

Wing Regions

The anterior area of the wing supported by veins is usually called remigium. The flexible posterior area is termed vannus. The two regions are separated by vannal fold. The proximal part of vannus is called jugum, when well developed it is separated by a jugal fold. The area containing wing articulation sclerites, pteralia is called axilla.

Wing Types

1. Tegmina: (Singular: Tegmen)

Wings are leathery or parchment like. They are protective in function. They are not used for flight, *e.g.* Forewings of cockroach and grasshopper.

2. Elytra: (Sigular: Elytron)

The wing is heavily sclerotised. Wing venation is lost. Wing is tough and it is protective in function. It protects hind wings and abdomen. It is not used during flight. But during flight, they are kept at an angle allowing free movement of hind wings, *e.g.* Forewings of beetles and weevils.

3. Hemelytra: (Singular: Hemelytron)

The basal half of the wing is thick and leathery and distal half is membranous. They are not involved in flight and are protective in function, *e.g.* Forewing of heteropteran bugs.

4. Halteres: (Singular: Haltere)

In true flies, the hind wings are modified into small knobbed vibrating organs called haltere. Each haltere is a slender rod clubbed at the free end (capitellum) and enlarged at the base (scabellum). On the basal part, two large group of sensory bodies forming the smaller hick's papillae and the large set of scapel plate. They act as balancing organs and provide the needed stability during flight, *e.g.* true flies, mosquito, male scale insect.

5. Fringed Wings

Wings are usually reduced in size. Wing margins are fringed with long setae. These insects literally swim through the air, *e.g.* Thrips.

6. Scaly Wings

Wings of butterfly and moths are covered with small coloured scales. Scales are unicellular flattened outgrowth of body wall. Scales are inclined to the wing surface and overlap each other to form a complete covering. Scales are responsible for colour. They are important in smoothing the air flow over wings and body.

7. Membranous Wings

They are thin, transparent wings and supported by a system of tubular veins. In many insects, either forewings (true flies) or hind wings (grass hopper, cockroach, beetles and earwig) or both fore wings and hind wings (wasp, bees, dragonfly and damselfly) are membranous. They are useful in flight.

Wing Coupling

Among the insects with two pairs of wings, the wings may work separately as in the dragonflies and damselflies. But in higher pterygote insects, fore and hind wings are coupled together as a unit, so that both pairs move synchronously. By coupling the wings, the insects become functionally two winged.

Types of Wing Coupling

1. Hamulate

A row of small hooks is present on the coastal margin of the hind wing which is known as hamuli. These engage the folded posterior edge of fore wing, *e.g.* bees.

HEMULI

Figure 4.38: Hamulate Wing Coupling

2. Amplexiform

It is the simplest form of wing coupling. A linking structure is absent. Coupling is achieved by broad overlapping of adjacent margins, *e.g.* butterflies.

3. Frenate

There are two sub types, *e.g.* Fruit sucking moth.

(i) Male Frenate

Hindwing bears near the base of the coastal margin a stout bristle called frenulum which is normally held by a curved process, retinaculum arising from the subcostal vein found on the surface of the forewing.

ANTERIOR WING

VEINS

POSTERIOR WING

VEINS

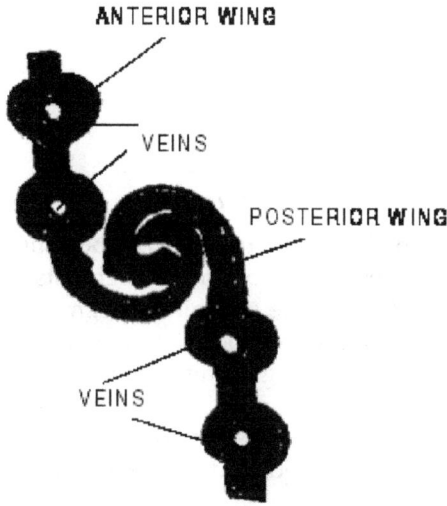

Figure 4.39: Amplexiform Wing Coupling

FRENULUM

Figure 4.40: Frenulum and Retinaculum: Male Butterfly

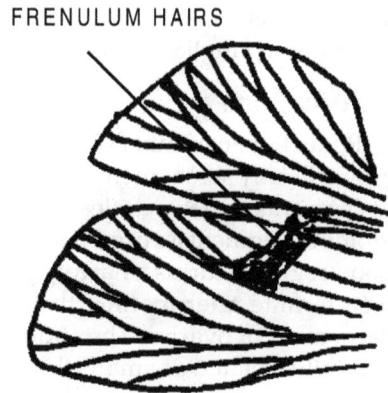

FRENULUM HAIRS

Figure 4.41: Frenulum and Retinaculum: Female Butterfly

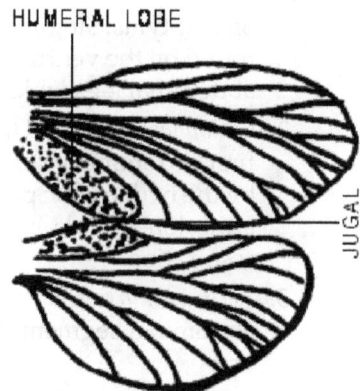

(ii) Female Frenate

Hindwing bears near the base of the costal margin a group of stout bristle (frenulum) which lies beneath extended forewing and engages there in a retinaculum formed by a patch of hairs near cubitus.

4. Jugate

Jugam of the forewings are lobe like and it is locked to the coastal margin of the hindwings, *e.g.* Hepialid moths.

Abdominal Structures in Insects

Third and posterior tagma of insect body. This tagma is made up of 9-11 uromeres (segments) and is highly flexible. Abdominal segments are telescopic in nature and are interconnected by a membrane called conjunctiva. Each abdominal segment is made up of only two sclerites namely dorsal body plate (tergum) and ventral body plate (sternum). In grass hopper, eight pairs of spiracles are present in the first eight segments, in addition to a pair of tympanum in the first segment. Eighth and ninth abdominal segments bears

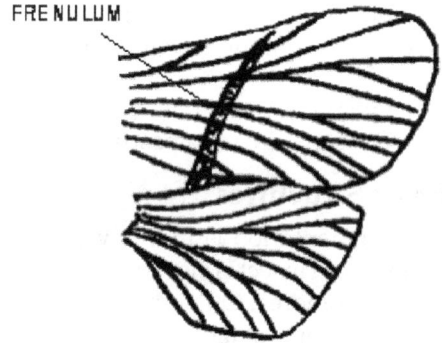

HUMERAL LOBE

JUGAL

Figure 4.42: Jugal and Humeral Lobs

the female genital structure and ninth segment bears a male genital structure. Abdominal appendages in adult insects are genital organs and cerci.

Function

Concerned with reproduction and metabolism.

Basic Structures

Segmentation is more evident in abdomen. The basic number of abdominal segments in insect is eleven plus a telson which bears anus. Abdominal segments are called uromeres. On eighth and ninth segment of female and ninth segment of male, the appendages are modified as external organs of reproduction or genitalia. These segments are known as genital segments. Usually eight pairs of small lateral openings (spiracles) are present on the first eight abdominal segments. In grasshoppers, a pair of tympanum is found one on either side of the first abdominal segment. It is an auditory organ. It is obliquely placed and connected to the metathoracic ganglia through auditory nerve.

Modifications

Reduction in number of abdominal segments has taken place in many insects. In spring tail, only six segments are present. In house fly, only segments 2 to 5 are visible and segments 6 to 9 are telescoped within others. In ants, bees and wasps, the first abdominal segment is fused with the metathorax and is called propodeum. Often the second segment forms a narrow petiole. The rest of the abdomen is called gaster. In queen termite after mating, the abdomen becomes gradually swollen due to the enlargement of ovaries. The abdomen becomes bloated and as a result sclerites are eventually isolated as small islands. Obesity of abdomen of queen termite is called physogastry.

Abdominal Appendages

(*i*) Pregenital Abdominal Appendages in Wingless Insects

1. *Styli*: (Stylus: Singular) Varying number of paired tube like outgrowths are found on the ventral side of the abdomen of silverfish. These are reduced abdominal legs which help in locomotion.

2. *Collophore or ventral tube or glue peg*: It is located on the ventral side of the first abdominal segment of spring tail. It is cylindrical. It is protruded out by the hydrostatic pressure of haemolymph. It might serve as an organ of adhesion. It aids in water absorption from the substratum and also in respiration.

3. *Retinaculum or tenaculum or catch*: It is present on the ventral side of the third abdominal segment. It is useful to hold the springing organ when not in use.

4. *Furcula or Furca*: This is a 'Y' shaped organ. It is present on the ventral side of fourth abdominal segment. When it is released from the catch, it exerts a force against the substratum and the insect is propelled in the air.

(*ii*) Abdominal Appendages in Immature Insects

(1) *Tracheal gills*: Gills are lateral outgrowths of body wall which are richly supplied with tracheae to obtain oxygen from water in naiads (aquatic immature stages of hemimetabolous insects). Seven pairs of filamentous gills are present in the first seven abdominal segments of naiads of may fly and are called as lateral gills. Three or two leaf like gills (lamellate) are found at the end of abdomen of naiad of damselfly and are called as caudal gills. In dragonfly, the gills are retained within the abdomen in a pouch like rectum and are called as rectal gills.

(2) *Anal papillae*: A group of four papillae surrounds the anus in mosquito larvae. These papillae are concerned with salt regulation.

(3) *Dolichasters*: These structures are found on the abdomen of antlion grub. Each dolichaster is a segmental protuberance fringed with setae.

(4) *Proloegs*: These are present in the larvae of moth, butterfly and sawfly. Two to five pairs are normally present. They are unsegmented, thick and fleshy. The tip of the proleg is called planta upon which are borne heavily sclerotised hooks called crochets. They aid in crawling and clinging to surface.

(*iii*) Abdominal Appendages in Winged Adults

(1) *Cornicles*: Aphids have a pair of short tubes known as cornicles or siphonculi projecting from dorsum of fifth or sixth abdominal segment. They permit the escape of waxy fluid which perhaps serves for protection against predators.

(2) *Caudal breathing tube*: It consists of two grooved filaments closely applied to each other forming a hollow tube at the apex of abdomen. *e.g.* water scorpion.

(3) *Cerci*: (Cercus–Singular): They are the most conspicuous appendages associated normally with the eleventh abdominal segment. They are sensory in function. They exhibit wide diversity and form.

Long and many segmented: *e.g.* Mayfly

Long and unsegmented: *e.g.* Cricket

Short and many segmented: *e.g.* Cockroach

Short and unsegmented: *e.g.* Grasshopper

Sclerotised and forceps like: *e.g.* Earwig. Cerci are useful in defense, prey capture, unfolding wings and courtship.

Asymmetrical cerci: Male embiid. Left cercus is longer than right and functions as clasping organ during copulation.

(4) *Median caudal filament*: In mayfly (and also in a wingless insect silverfish), the epiproct is elongated into cercus like median caudal filament.

(5) *Pygostyles*: A pair of unsegmented cerci like structures are found in the last abdominal segment of scoliid wasp.

(6) *Anal styli*: A pair of short unsegmented structure found at the end of the abdomen of male cockroach. They are used to hold the female during copulation.

(7) *Ovipositor*: The egg laying organ found in female insect is called ovipositor. It is suited to lay eggs in precise microhabitats. It exhibits wide diversity and form.

Short and horny: *e.g.* Short horned grasshopper

Long and sword like: *e.g.* Katydid, long horned grasshopper

Needle like: *e.g.* Cricket

Ovipositor modified into sting: *e.g.* Worker honey bee.

Pseudoovipositor: An appendicular ovipositor is lacking in fruit flies and house flies. In fruit flies, the elongated abdomen terminates into a sharp point with which the fly pierces the rind of the fruit before depositing the eggs. In the house fly the terminal abdominal segments are telescopic and these telescopic segments aid in oviposition. The ovipositor of house fly is called pseudoovipositor or ovitubus or oviscapt.

Male genitalia: External sexual organs of male insects are confined to ninth abdominal segment. In damselfly, the functional copulatory organ is present on the ventral side of second abdominal segment

Internal Structures and Functions

Digestive System

The alimentary canal of insects is a long, muscular and tubular structure extending from mouth to anus. It is differentiated into three regions *viz.*, Foregut, midgut and hindgut.

1. Foregut

It is ectodermal in origin. Anterior invagination of ectoderm forms foregut (Stomodeum). Internal cuticular lining is present. Terminal mouth parts leads into a preoralcavity. Preoralcavity between epipharynx and hypopharynx is called as Cibarium. Preoralcavity between hypopharynx and salivary duct is Salivarium. Behind the mouth, a well musculated organ called Pharynx is present which pushes the food into oesophagous. Pharynx acts as a sucking pump in sap feeders. Oesophagous is a narrow tube which conduct food into crop. Crop is the dilated distal part of oesophagus acting as food reservoir. In bees, crop is called as honey stomach where nectar conversion occurs. Proventriculus or Gizzard is the posterior part of foregut and is musculated. It is found in solid feeders and absent in fluid feeders or sap feeders. Food flowing from foregut to midgut is regulated through cardial or oesophageal valve. The internal cuticle of gizzard is variously modified as follows.

(i) Teeth like in cockroach to grind and strain food.

(ii) Plate like in honey bee to separate pollen grains from nectar

(iii) Spine like in flea to break the blood corpuscles

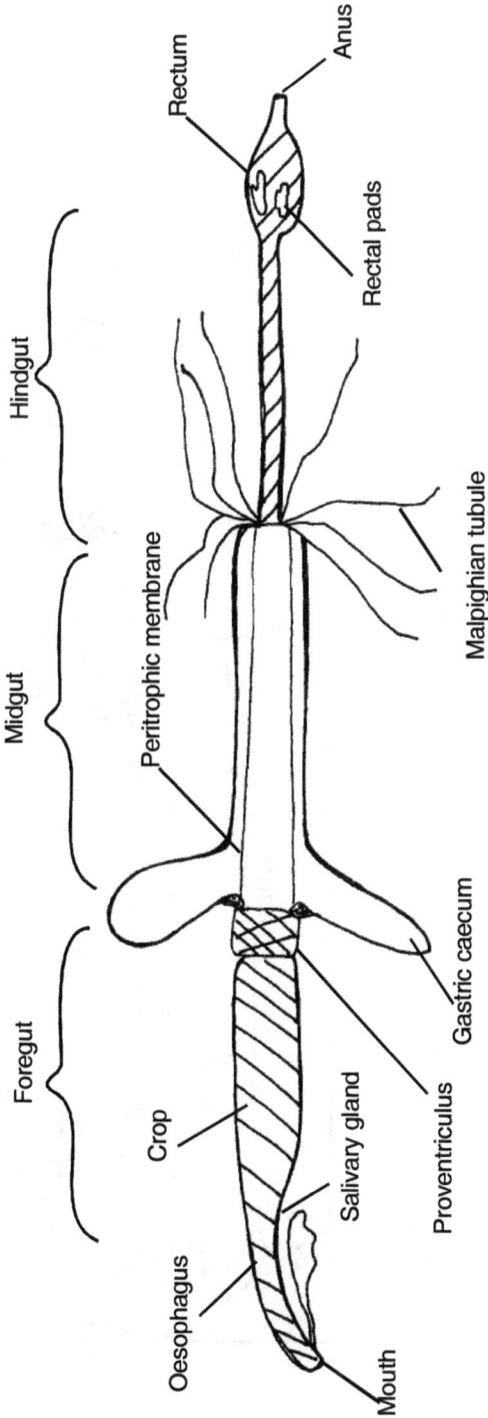

Figure 4.43(a): Digestive System of Insect Generalized Diagram (Typical of Grasshopper)

Oesophagus

First ventriculus

Second ventriculus

Third ventriculus

Malpighian tubules

Rectum

(2) Milkweed Bug Digestive System (Sucking Insect)

Oesophagus

Gastric caecum

Ventriculus

Longitudinal muscle bond (dorsal)

Silk (salivary) gland

Malpigian tubule

Small intestine

Bladder

Ventriculus

Large intestine

Rectum

Ventriculus

Ventriculus

Third ventriculus

Ileum

(1) Caterpillar Digestive System (Chewing Insect)

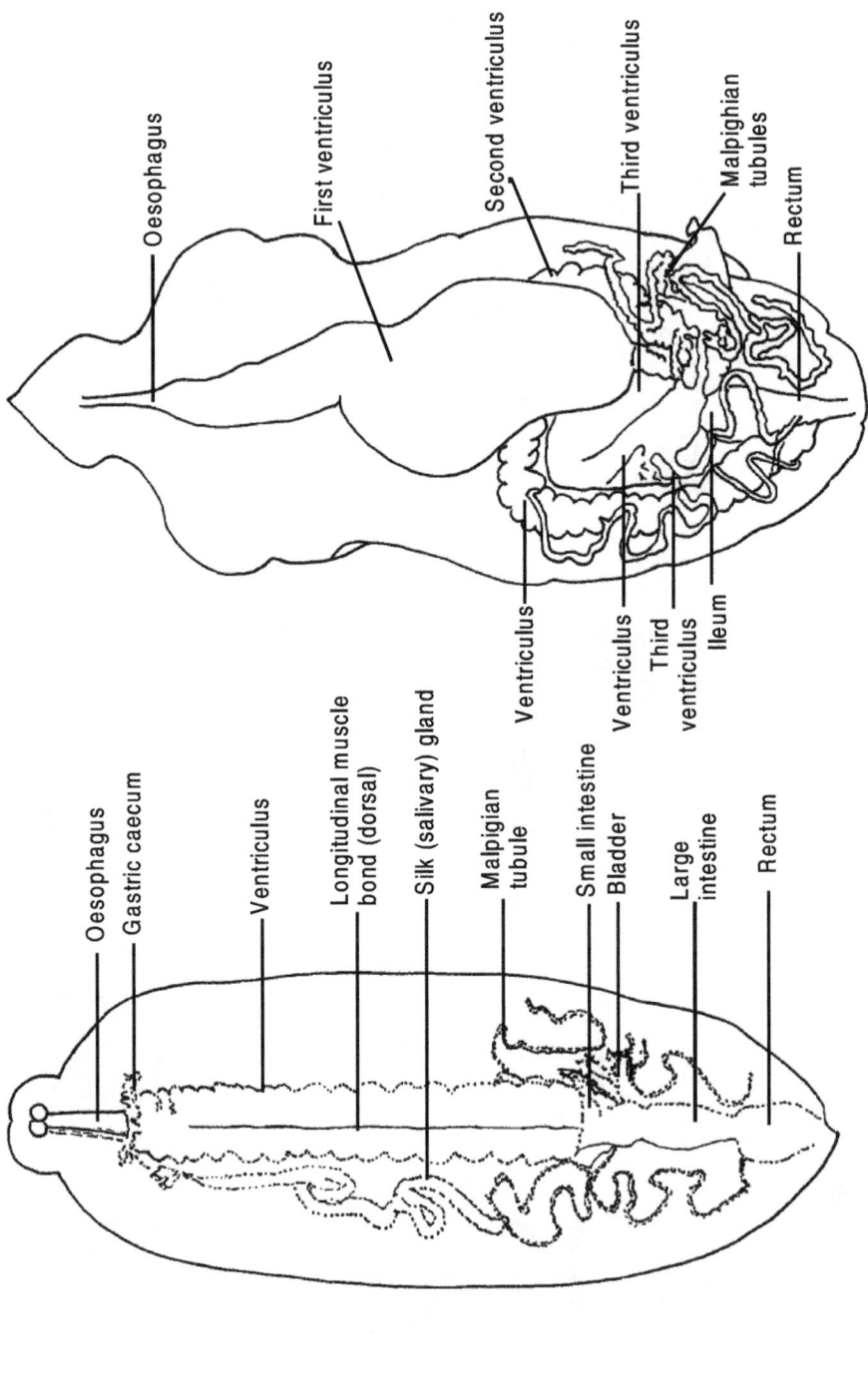

Figure 4.43(b): Insect Digestive System

2. Midgut

It is endodermal in origin and also called as mesentron. This part contains no cuticular lining. Midgut is made up of three types of epithelial cells.

(*i*) Secretory cells (Columnar cells) (*ii*) Goblet cells (aged secretory cells), (*iii*) Regenerative cells which replaces secretory cells. Important structures present in midgut are as follows:

(a) Peritrophic Membrane

It is the internal lining of midgut, secreted by anterior or entire layer of midgut epithelial cells. Present in solid feeders and absent in sap feeders. This layer is semipermeable in nature to digestive juices and digestion products. It lubricates and facilitates food movement. Envelops the food and protects the midgut epithelial cells against harder food particles.

(b) Gastric Caecae: (Enteric caecae or Hepatic caecae)

Finger like outgrowths found in anterior or posterior ends of midgut. This structure increases the functional area of midgut and shelter symbiotic bacteria in some insects.

(c) Pyloric valve: (Proctodeal Valve)

Midgut opens into hindgut through pyloric valve, which regulate food flow. In certain immature stages of insects, midgut is not connected to hindgut till pupation. *e.g.* Honey bee grub.

(d) Filter Chamber

It is a complex organ in which two ends of ventriculus and the beginning of hind gut are enclosed in a sac. This is useful to short circuit excess water found in liquid food in homopteran insects. This process avoids dilution of digestive enzymes and concentrates food for efficient digestion. Also helps in osmoregulation by preventing dilution of haemolymph.

3. Hindgut

It is ectodermal in origin and produced by the posterior invagination of ectoderm. Internal cuticular lining is present, which is permeable to salts, ions, aminoacids and water. The main functions of hindgut are the absorption of water, salt and other useful substances from the faeces and urine. Hindgut is differentiated into three regions *viz.*, ileum, colon and rectum. In the larva of scarabids and termites, illeum is pouch like for housing symbionts and acts as fermentation chamber. Rectum contains rectal pads helping in dehydration of faeces and it opens out through anus.

Gut Physiology

Primary functions of the gut is to digest the ingested food and to absorb the metabolites. Digestion process is enhanced with the help of enzymes produced by digestive glands and microbes housed in special cells.

Digestive Glands

(a) Salivary Glands

In Cockroach, a pair of labial glands acts as salivary gland, where the salivary ducts open into salivarium. In caterpillars, mandibular glands are modified to secrete saliva, where the salivary glands are modified for silk production. Functions of saliva:

1 To moisten and dissolve food.
2. To lubricate mouthparts.
3. To add flavour to gustatory receptors.
4. In cockroach, the saliva contains amylase for the digestion of starch.
5. In honey bee, saliva contains invertase for sucrose digestion.
6. In Jassid, saliva contains lipase and protease for lipids and protein digestion. Jassid saliva also contains toxins which produces tissue necrosis and phytotoxemia on the plant parts.
7. In plant bug, saliva contains pectinase which helps in stylet penetration and extra intestinal digestion.
8. In mosquito, saliva contains anticoagulin which prevents blood clotting.
9. In gall producing midges, saliva contains Indole Acetic Acid (IAA).
10. In disease transmitting vectors, the saliva paves way for the entry of pathogens.

(b) Hepatic Caecae and Midgut Epithelial Cells

It secretes most of the digestive juices. Two types of cells were involved in the enzyme secretion.

Holocrine: Epithelial cells disintegrate in the process of enzyme secretion.

Merocrine: Enzyme secretion occurs without cell break down.

Digestive Enzymes

Insect Group Enzyme Substrate

Phytophagous larvae: Amylase Starch, Maltase Maltose, Invertase Sucrose

Omnivorous insects: Protease Protein. Lipase Lipid

Nectar feeders: Invertase Sucrose

Wood boring insects and Termites: Cellulase Cellulose

Meat eating maggots: Collagenase Collagen and elastin

Bird lice: Keratinase Keratin

(c) Microbes in Digestion

In the insect body few cells were housing symbiotic microorganisms known as mycetocyte. These mycetocytes aggregate to form an organ called mycetome.

(i) *Flagellate Protozoa*: It produces cellulase for cellulose digestion in termites and wood cockroach.

(*ii*) *Bacteria*: It helps in wax digestion in wax moth.

(*iii*) Bed bug and cockroach obtain vitamin and aminoacids from microbes.

These microbes were transmitted between individuals through food exchange (mouth to mouth feeding) called *trophallaxis* and through egg called as *transovarial* transmission. In plant bug and ant lion grub, partial digestion occurs in the host body prior to food ingestion called as extra intestinal digestion. In most of the insects, digestion occurs in mid gut.

Absorption

In many insects, absorption of nutrients occurs through microvilli of midgut epithelial cells by diffusion. Absorption of water and ions occur through rectum. In cockroach, lipid absorption occurs through crop. In termites and scarabaeids (White grubs), absorption occurs through ileum. In solid feeders, resorption of water from the faeces occurs in the rectum and the faeces is expelled as pellets. In sap feeders (liquid feeders), the faeces is liquid like. The liquid faeces of homopteran bugs (aphids, mealy bugs, Scales and psyllids) with soluble sugars and amino acids is known as honey dew, which attracts ants for feeding.

Excretory System

Removal of waste products of metabolism, especially nitrogenous compounds from the body of insects is known as excretion. The excretion process helps the insect to maintain salt water balance and thereby establish physiological homeostasis. Following are the excretory organs:

1. *Malpighian tubules*: Thin, blind-ending tubules, originating near the junction of mid and hindgut, predominantly involved in regulation of salt, water

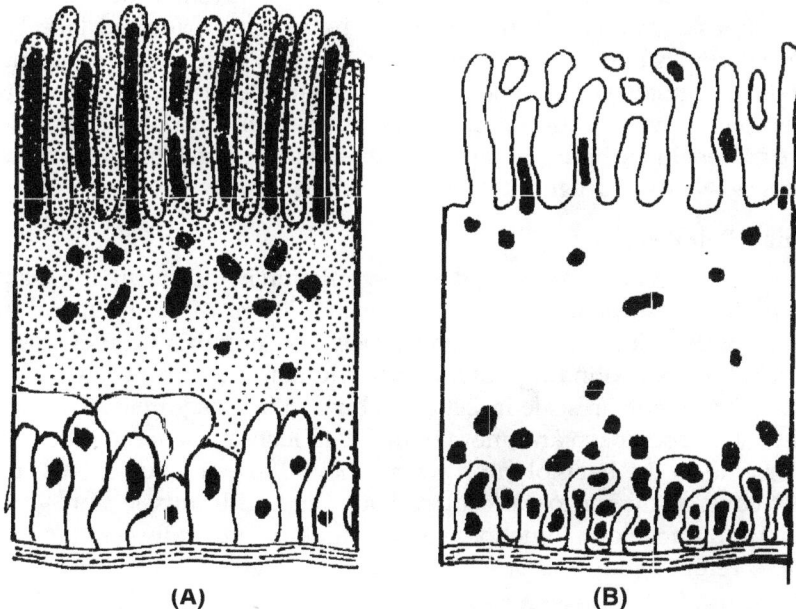

(A) **(B)**

Figure 4.44: Ultrastructure of (A) Honey-comb and (B) Brush-border epithelium of Malpighian tubules (After Wigglesworth and Salpeter, 1962)

and nitrogenous waste excretion. This structure was discovered by Marcello Malpighi.

2. *Nephrocytes*: Cells that sieve the haemolymph for products that they metabolize (pericardial cells).

3. *Fat bodies*: A loose or compact aggregation of cells, mostly trophocytes, suspended in the haemocoel, responsible for storage and excretion.

4. *Oenocytes*: These are specialised cells of haemocoel, epidermis or fat body with many functions. One of the function is excretion.

5. *Integument*: The outer covering of the living tissues of an insect.

6. *Tracheal system*: The insect gas exchange system, comprising tracheae and tracheoles.

7. *Rectum*: The posterior part of hind gut.

Among the above organs, malpighian tubules are the major organ of excretion.

Excretion and Osmoregulation

Insect faeces, either in liquid form or solid pellets, contains both undigested food and metabolic excretions. Aquatic insects excrete dilute wastes from their anus directly into water by flushing with water. But, terrestrial insects must conserve water. This requires efficient waste disposal in a concentrated or even dry form, simultaneously avoiding the toxic effects of nitrogen. Both terrestrial and aquatic insects must conserve ions, such as sodium (Na^+), potassium (K^+) and chloride (Cl^-), that may be limiting in their food or lost into the water by diffusion. Therefore, the production of insect excreta (urine or pellets) is a result of two related processes: excretion and osmoregulation (maintenance of favourable osmotic pressure and ionic concentration of body fluid). The system responsible for excretion and osmoregulation is referred to as excretory system and its activities are performed largely by the Malpighian tubules and hindgut. However, in fresh water insects, haemolymph composition is regulated in response to loss of ions to the surrounding water with the help of excretory system and special cells. Special cells are called Chloride cells which are present in the hindgut, capable of absorbing inorganic ions from the dilute solutions. (*e.g.* Naids of dragonflies and damselflies).

Malpighian Tubules

The main organ of excretion and osmoregulation in insects are the malpighian tubules, acting in association with rectum or ileum. Malpighian tubules are outgrowths of the alimentary canal and consists of long thin tubules formed of a single layer of cells surrounding a blind-ending lumen. They are absent in spring tail and aphids, 2 numbers in scale insects, 4 in bugs, 5 in mosquitoes, 6 in moths and butterflies, 60 in cockroach and more than 200 in locusts. Generally, they are free, waving around in the haemolymph where they filter out solutes. Each tubule is externally covered by peritonial coat and supplied with muscle fibres (aiding in peristalsis) and tracheloes. Functional differentiation of the tubules was seen, with the distal secretory region and proximal absorptive region.

Physiology

The malpighian tubules produce a filtrate (the primary urine) which is isosmotic but ionically dissimilar to the haemolymph and selectively reabsorbs water and certain solutes, but eliminates others. The malpighian tubules produces an iosmotic filtrate which is high in K+ and low in Na+ with Cl- as major anion. The active transport of ions especially K+ into the tubule lumen generates an osmotic pressure gradiant for the passive flow of water. Sugars and most amino acids are also passively filtered from the haemolymph via junctions between the tubule cells, whereas amino acids and nonmetabolizables and toxic organic compounds are actively transported into the tubule lumen. Sugar is resorbed from the lumen and returned to the haemolymph. The continuous secretory activity of each malpighian tubule leads to a flow of primary urine from its lumen towards and into the gut. In the rectum, the urine is modified by removal of solutes and water to maintain fluid and ionic homeostasis of the body.

Nitrogenous Excretion

Terrestrial insects excrete waste products as uric acid or certain of its salts called urates, which were water insoluble and requires less amount of water for waste product removal. This type of excretion is known as *Uricotelism*. In aquatic insects, ammonia is the excretory product, which is freely soluble in water and requires more amount of water for waste product removal. This type of excretion is known as *Ammonotelism*.

Cryptonephry: The distal ends of the Malpighian tubules are held in contact with the rectal wall by the perinephric membrane, which is concerned either with efficient dehydration of faeces before their elimination or ionic regulation, (*e.g.* adult Coleptera, larval Lepidoptera and larval symphyta).

Functions of Malphighian Tubule

Excretory in function, mainly concerned with removal of nitrogenous wastes. The other accessory functions are as follows:

1. Spittle secretion in spittle bug
2. Light production in Bolitophila
3. Silk production in larval neuroptera

Storage Excretion

The excretory waste materials are retained within the body in different sites.

(*i*) Uric acid is stored as urates in the cells of *fat body e.g.,* American cockroach.

(*ii*) Uric acid is stored in the *body wall*, giving white colour. *e.g.* Red cotton bug.

(*iii*) Uric acid is stored in the *male accessory glands* to produce the outer coat of spermatophore, which is excreted during copulation.

(*iv*) Uric acid is stored in the wing scales giving white colour. *e.g.,* Pierid butterflies.

(*v*) Waste products of pupal metabolism (Meconium) is stored and released during adult emergence.

Respiratory System

Similar to aerobic animals, insects must obtain oxygen from their environment and eliminate carbon dioxide respired by their cells. This is a gas exchange through series of gas filled tubes providing surface area for gaseous exchange (Respiration strictly refers to oxygen-consuming, cellular metabolic processes). Air supplied directly to the tissue and haemolymph (blood) is not involved in the respiratory role. Gas exchange occurs by means of internal airfilled tracheae. These tubes branch and ramify through the body. The finest branches called tracheole comes in contact with all internal organs and tissues and are numerous with high oxygen requirements. Air usually enters the tracheae via spiracular openings positioned laterally on the body. No insect has more than ten pairs (two thoracic and eight abdominal). Based on the number and location of functional spiracles respiratory system is classified as follows:

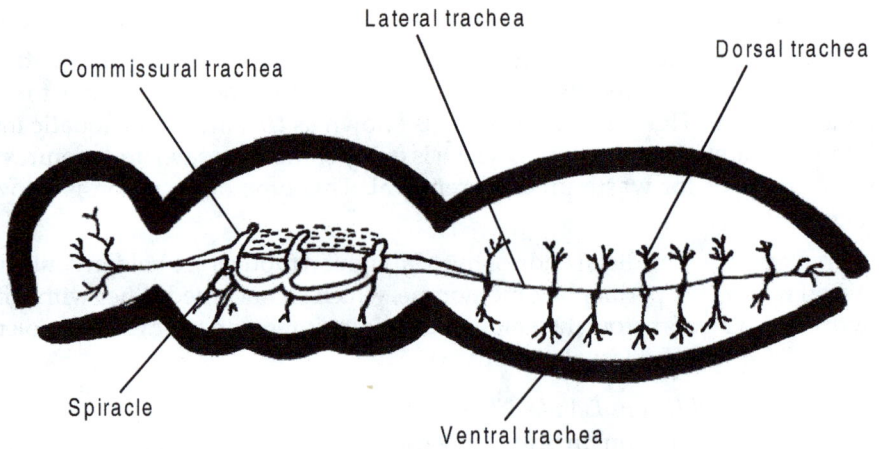

Figure 4.45: Insect Respiratory System

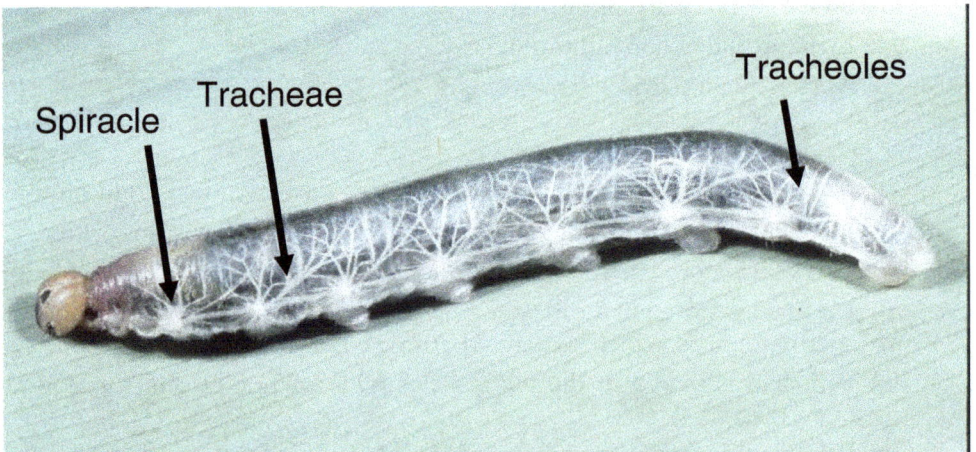

Figure 4.45(a): Insect Respiration

1. *Holopneustic*:10 pairs, 2 in thorax and 8 in abdomen. *e.g.* grasshopper
2. *Hemipneustic*: Out of 10 pairs, one or two non-functional
3. *Peripneustic*: 9 pairs–1 in thorax 8 in abdomen *e.g.* Caterpillar
4. *Amphipneustic*: 2 pairs–One anterior, one posterior, *e.g.* maggot.
5. *Propneustic*: 1 pair–anterior pair *e.g.* Puparium
6. *Metapneustic*: 1 pair–posterior pair *e.g.*Wriggler
7. *Hypopneustic*:10 pairs–7 functional (1 thorax + 6 abdominal), 3 non functional. *e.g.* head louse
8. *Apneustic*: All spiracles closed, closed tracheal system *e.g.* naiad of may fly.

Organs of Respiration

Spiracles: Spiracles have a chamber or atrium with a opening and closing mechanism called atrial valve. This regulates air passage and minimises water loss. Each spiracle is set in a sclerotized cuticular plate called a peritreme. Tracheae are invaginations of the epidermis and thus their lining is continuous with the body cuticle. The ringed appearance of the tracheae is due to the spiral ridges called taenidia. This allows the tracheae to be flexible but resists compression. The cuticular linings of the tracheae are shed during moulting. Tracheoles are less than 1 μm in diameter and they end blindly and closely contact the respiring tissues. Taenidia and waxlayer is absent. Cuticulin layer is permeable to gases. It is intracellular in nature, but enclosed only in the cytoplasm of tracheal and cell called tracheoblast. Gaseous exchange occurs across tracheoles. There are four tracheal trunks *viz*., lateral, dorsal, ventral and visceral, helping in the passage of air. In the trachea, thin walled-collapsable sac like dilations are present, called as airsacs where taenidia is absent. Airsacs acts as oxygen reservoir. Provide buoyancy to flying and aquatic insects. Provide space for growing organs. Acts as sound resonator and heat insulators.

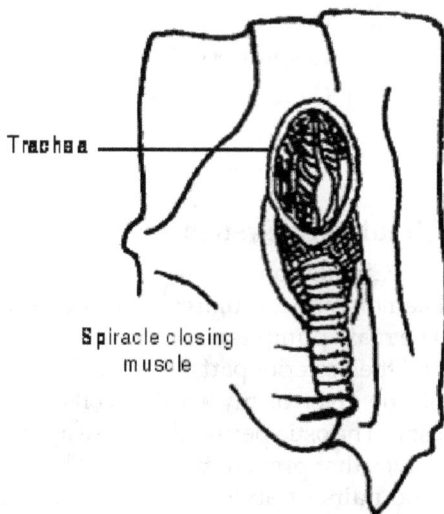

Figure 4.46: Spiracles

Mechanism of Respiration

Oxygen enters the spiracle and passes through the length of the tracheae to the tracheoles and into the target cells by a combination of ventilation and diffusion along a concentration gradient, from high in the external air to low in the tissue. When the net movement of oxygen molecules in the tracheal system is inward, its called Inspiration, whereas the net movement of CO_2 and water vapour molecules is termed as Expiration.

Respiration in Aquatic Insects

1. Closed Tracheal System

In some aquatic and many endoparasitic larvae, spiracles are absent and the tracheae divide peripherally to form a network. This covers the body surface, allowing cutaneous gas exchange, *e.g.*

Gills: Tracheated thin outgrowth of body wall.

Lamellate gills–mayfly naiad

Filamentous gills–damselfly naiad

Rectal gills–dragonfly naiad

2. Open Tracheal System

(*i*) Air store: Air bubble stored beneath wings acts as physical gill, *e.g.* water bug.

(*ii*) Respiratory siphon, *e.g.* Wriggler

(*iii*) Caudal breathing tube, *e.g.* Water scorpion

(*iv*) Plastron: Closely set hydrofuge hairs of epicuticle hold a thin film of air indefinitely.

Circulatory System in Insects

Circulation in insects is maintained by a system of muscular pumps moving haemolymph through compartments separated by fibromuscular septa or membranes. The main pump is the pulsatile dorsal vessel. The anterior part may be called aorta and the posterior part the heart. The dorsal vessel is a simple tube, generally composed of one layer of myocardial cells and with segmentally arranged openings called ostia. The ostia permit the one-way flow of haemolymph into the dorsal vessel due to valves that prevent backflow. There may be up to three pairs of thoracic ostia and nine pairs of abdominal ostia. The dorsal vessel lies in the pericardial sinus, a compartment above a dorsal diaphragm (a fibromuscular septum–a separating membrane) formed of connective tissue and segmental pairs of alary muscles. The alary muscles support the dorsal vessel but their contractions do not affect heartbeat. Haemolymph enters the periocardial sinus via segmental openings in the diaphragm and then moves into the dorsal vessel via ostia during a muscular relaxation phase. Waves of contraction start at the posterior end of the body, pump the haemolymph forward in the dorsal vessel and out via the aorta into the head. Thereafter, the appendages of the head and thorax are supplied with haemolymph as it circulates posteroventrally and finally returns to the pericardial sinus and dorsal vessel. Another important component of the insect circulatory system is the ventral diaphragm, a fibromuscular septum that lies in the floor of the body cavity associated with the ventral nerve cord. Circulation of the haemolymph is aided by active peristaltic contractions of the ventral diaphragm which directs the haemolymph backwards and laterally in the perineural sinus below the diaphragm. These movements are important in insects that use the circulation in thermoregulation. Ventral diaphragm also facilitates rapid exchange of chemicals between the ventral nerve cord and the

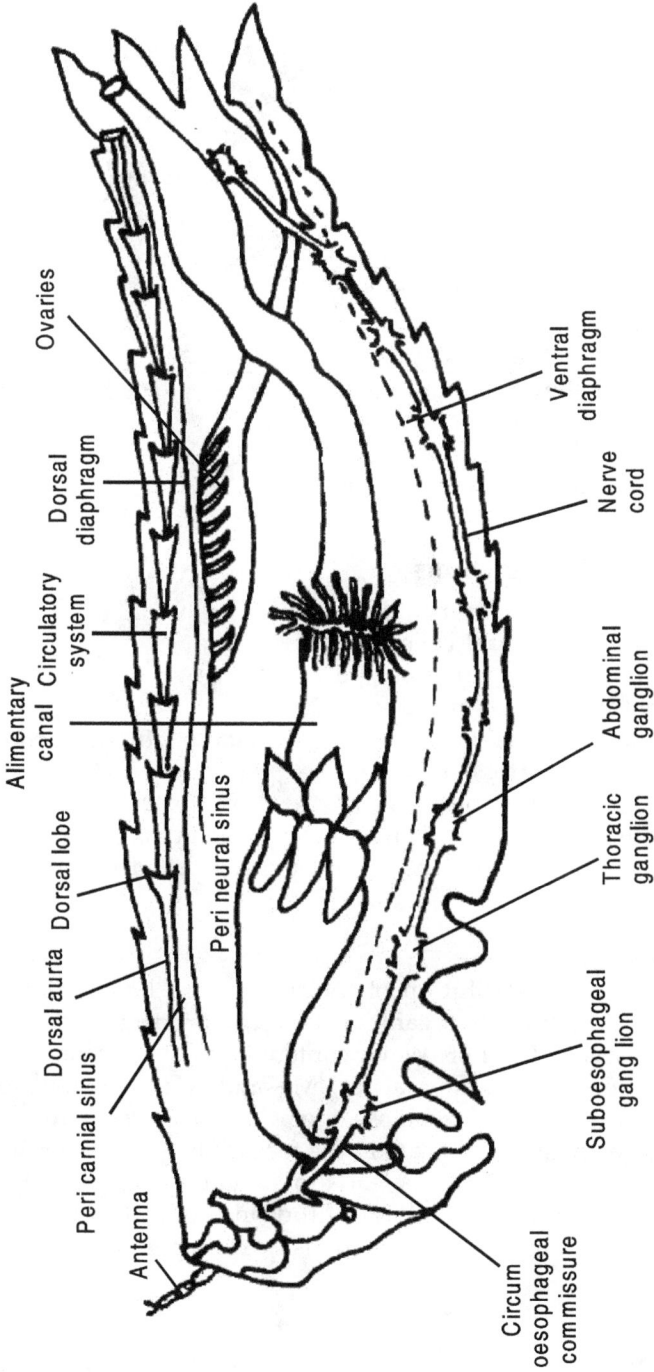

Figure 4.47: Insect Circulatory System

haemolymph. Haemolymph is generally circulated to appendages unidirectionaly by various tubes, septa, valves and pumps. The muscular pumps are termed accessory pulsatile organs and occur at the base of the antennae and legs. Antennal pulsatile organs releases neurohormones that are carried to the antennal lumen to influence the sensory neurones. Circulation occurs in the wings of young adult. In wing, circulation is sustained by influxes of air into the wing veins, rather than any pulsatile organs. Pulses of air in the fine tracheal tubes of the veins push the haemolymph through the enclosed space of the veins. The insect circulatory system shows high degree of coordination between dorsal vessel, fibro-muscular diaphragms and accessory pumps.

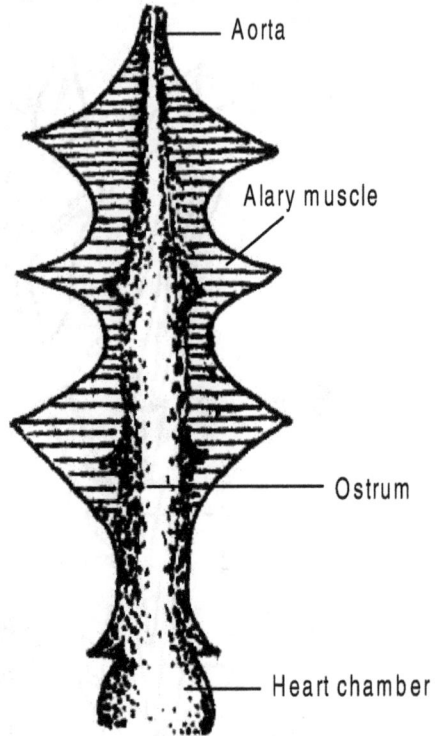

Figure 4.48: Heart and Aerota Muscles

Haemolymph and its Functions

Haemolymph is a watery fluid containing ions, molecules and cells. It is often clear and colourless but may be variously pigmented or rarely red due to haemoglobin in the immature stages of few aquatic and endoparasitic flies (*e.g.*, Chironomid larva). Haemolymph performs the function of both blood and lymph. It is not involved in gas transporting function (respiration). Haemolymph contains a fluid portion called plasma and cellular fractions called haemocytes.

1. Plasma

Plasma is an aqueous solution of inorganic ions, lipids, sugars (mainly trehalose), amino acids, proteins, organic acids and other compounds. pH is usually acidic (6.7). Density is 1.01 to 1.06. Water content is 84-92 per cent. Inorganic ions present are 'Na' in predators and parasites, 'Mg' and 'K' in phytophagous insects. Carbohydrate is in the form of trehalose sugar. Major proteins are lipoproteins, glycoproteins and enzymes. Lipids are in the form of fat particles or lipoproteins. Higher concentration of amino acids leads to a condition called aminoacidemia which effects the osmosis process. In high altitude insects, glycerol is present which acts as a anti freezing compound. Nitrogenous waste is present in the form of uric acid.

2. Haemocytes

The blood cells or haemocytes are of several types and all are nucleate. Different types of haemocytes are as follows:

(*a*) *Prohaemocyte*: Smallest of all cells with largest nucleus.

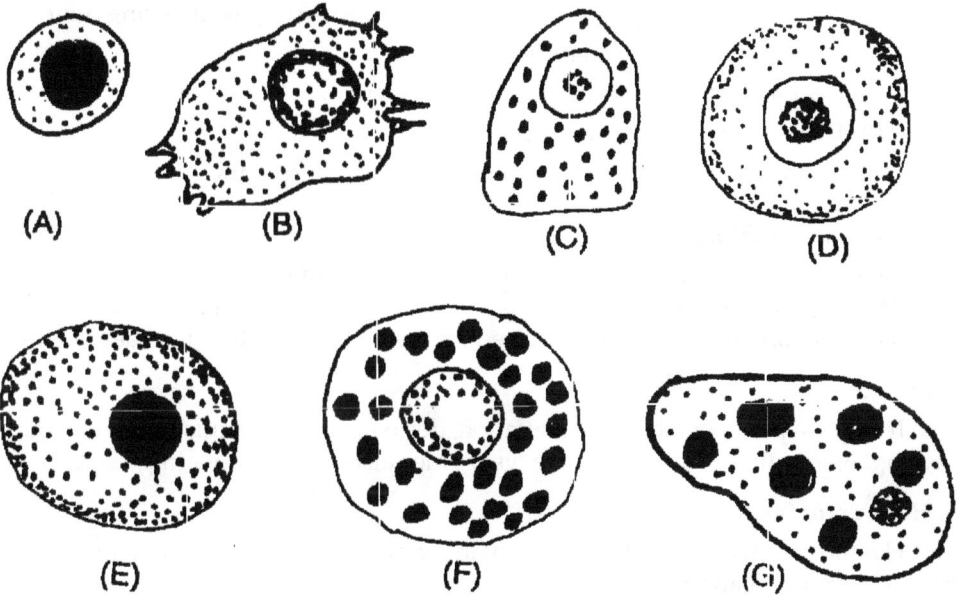

Figure 4.49: Types of Haemocytes (A) Prohaemocyte; (B) Plasmatocyte; (C) Granular; (D) Oenocytoids; (E) Cystocyte; (F) Spherule Cell; (G) Adipohaemocytes (After Jones, 1956)

(b) *Plasmatocyte* (Phagocyte) aids in phagocytocis

(c) *Granular heamocyte*: Contains large number of cytoplasmic inclusions

(d) *Spherule cell*: Cytoplasmic inclusions obscure the nucleus

(e) *Cystocyte* (Coagulocyte): Role in blood coagulation and plasma precipitation.

(f) *Oenocytoids*: Large cells with ecentric nucleus

(g) *Adipohaemocytes*: Round or avoid with distinct fat droplets

(h) *Podocyte*: Large flattened cells with number of protoplasmic projections.

(i) *Vermiform cells*: Rare type, long thread like.

Functions of Haemolymph

1. Lubricant

Haemolymph keeps the internal cells moist and the movement of internal organs is also made easy.

2. Hydraulic Medium

Hydrostatic pressure developed due to blood pumping is useful in the following processes.

(a) Ecdysis (moulting)

(b) Wing expansion in adults

(c) Ecolosion in diptera (adult emergence from the puparium using ptilinum)

(d) Eversion of penis in male insects

(e) Eversion of osmeteria in papilionid larvae

(f) Eversion of mask in naiad of dragonfly

(g) Maintenance of body shape in soft bodied caterpillars.

3. Transport and Storage

Digested nutrients, hormones and gases (chironomid larva) were transported with the help of haemolymph. It also removes the waste materials to the excretory organs. Water and raw materials required for histogenesis is stored in haemolymph.

4. Protection

It helps in phagocytocis, encapsulation, detoxification, coagulation, and wound healing. Non celluar component like lysozymes also kills the invading bacteria.

5. Heat Transfer

Haemolymph through its movement in the circulatory system regulate the body heat (Thermoregulation).

6. Maintenance of Osmotic Pressure

Ions, amino acids and organic acids present in the haemolymph helps in maintaining osmotic pressure required for normal physiological functions.

7. Reflex Bleeding

Exudation of heamolymph through slit, pore etc. repels natural enemies. *e.g.* Aphids.

8. Metabolic Medium

Haemolymph serves as a medium for ongoing metabolic reactions (trahalose is converted into glucose).

Nervous System

The basic component in the nervous system is the nerve cell or neuron, composed of a cell body with two projections (fibers) the dendrites that receive stimuli and the axon that transmits information, either to another neuron or to an effector organ such as a muscle. Axon may have lateral branches called Collateral and terminal arborization and synapse. Insect neurons release a variety of chemicals at synapses either to stimulate or to inhibit effector neurons or muscles. Acetylcholine and catecholamines such as dopamine are the important neurotransmitters involved in the impulse conduction. Neurons are of following types based on structure and function.

A. On Structural Basis

(i) *Monopolar*: neuron with a single axon

(ii) *Bipolar*: neuron with a proximal axon and a long distal dendrite.

(iii) *Multipolar*: neuron with a proximal axon and many distal dendrites.

Types of Neurons

Unipolar

Bipolar

Multi polar

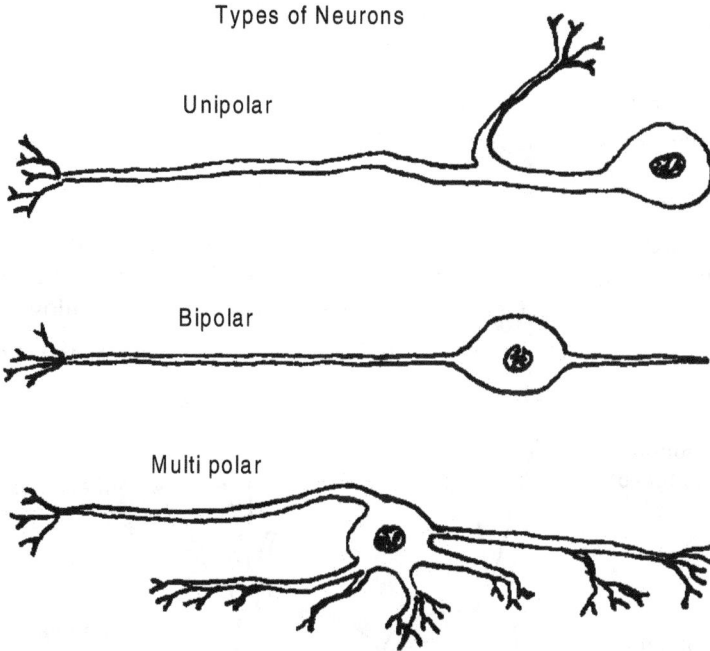

Figure 4.50: Structure of Neuron

B. Functional Basis

(i) *Sensory neuron*: It conducts impulse from sense organs to central nervous system (CNS).

(ii) *Motor neuron*: It conducts impulse from CNS to effector organs

(iii) *Inter neuron* (association neuron): It inter-links sensory and motor neurons.

The cell bodies of inter neurons and motor neurons are aggregated with the fibers inter connecting all types of nerve cells to form nerve centers called ganglia.

Mechanism of Impulse Conduction: Impulses are conducted by the neurons by two means.

Axonic Conduction: Ionic composition varies between inside and outside of axon resulting in excitable conditions, which leads to impulse conduction as electrical response.

Synaptic conduction: Neurochemical transmitters are involved in the impulse conduction through the synaptic gap. Neurotransmitters and the type of reactions helping in the impulse conduction are as follows.

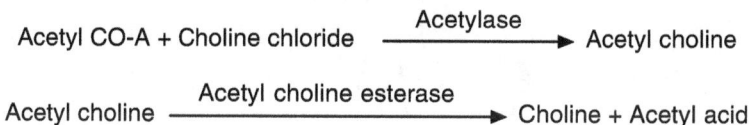

$$\text{Acetyl CO-A + Choline chloride} \xrightarrow{\text{Acetylase}} \text{Acetyl choline}$$

$$\text{Acetyl choline} \xrightarrow{\text{Acetyl choline esterase}} \text{Choline + Acetyl acid}$$

Nervous system can be divided in to three major sub-systems as:

(*i*) Central nervous system (CNS)

(*ii*) Visceral nervous system (VNS)

(*iii*) Peripheral nervous system (PNS)

(*i*) Central Nervous System

It contains double series of nerve centers (ganglia). These ganglia are connected by longitudinal tracts of nerve fibers called connectives and transverse tracts of nerve fibers called commissures. Central nervous system includes the following:

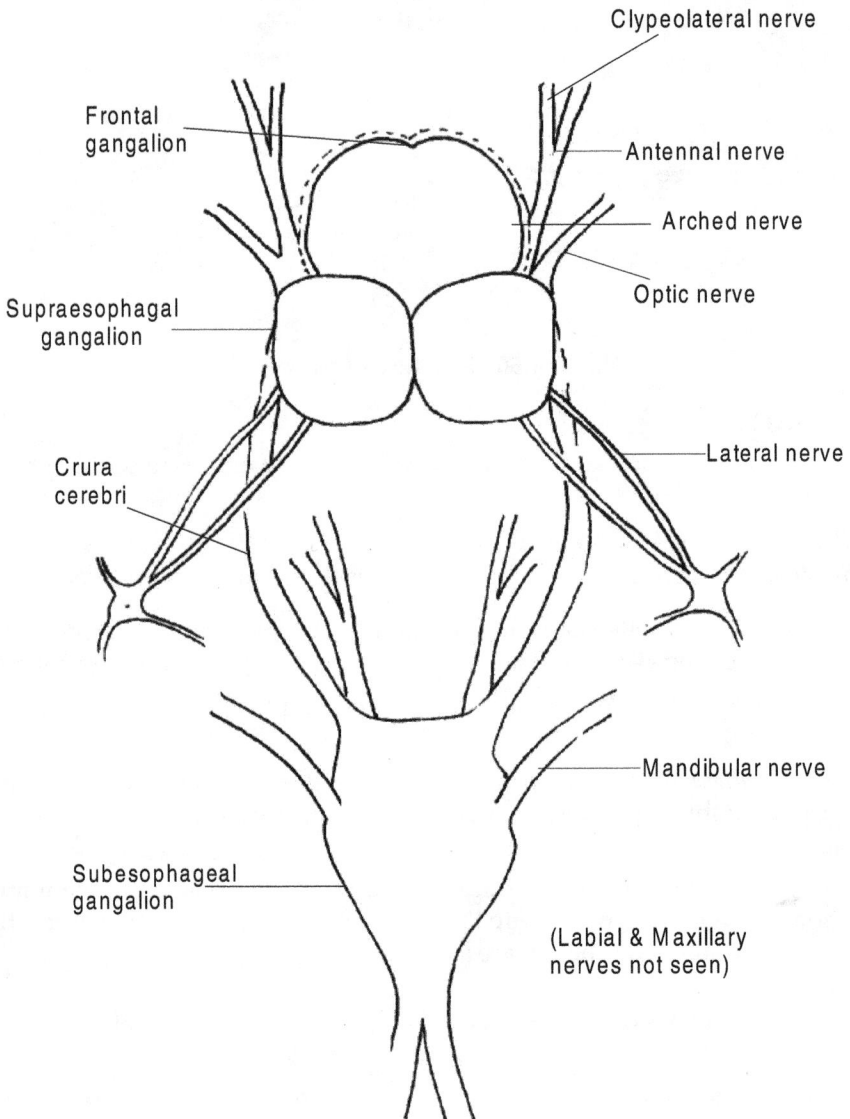

Figure 4.51: Insect Brain

(a) Brain

Formed by the fusion of first three cephalic neuromeres.

Protocerebrum: Large, innervate compound eyes and ocelli.

Deutocerebrum: Found beneath protocerebrum, innervate antennae.

Tritocerebrum: Bilobed, innervate labrum.

Brain is the main sensory centre controlling insect behaviour.

(b) Ventral Nerve Cord

Median chain of segmental ganglia beneath oesophagus.

(c) Sub Oesophageal Ganglia

Formed by the last three cephalic neuromeres which innervate mandible, maxillae and labium.

(d) Thoracic Ganglia

Three pairs found in the respective thoracic segments, largest ganglia, innervate legs and muscles.

(e) Abdominal Ganglia

Maximum eight pairs are present and number varies due to fusion of ganglia, innervate spiracles.

(f) Thoraco Abdominal Ganglia

Thoracic and abdominal ganglia are fused to form a single compound ganglia. Innervate genital organs and cerci.

(ii) Visceral Nervous System

The visceral (sympathetic) nervous system consists of three separate systems which are as follows:

1. The stomodeal/stomatogastric which includes the frontal ganglion and associated with the brain, aorta and foregut;
2. Ventral visceral, associated with the ventral nerve cord; and
3. Caudal visceral, associated with the posterior segments of abdomen.

Together the nerves and ganglia of these subsystems, innervate the anterior and posterior gut, several endocrine organs (Corpora cardiaca and Corpora allata), the reproductive organs, and the tracheal system including the spiracles.

(iii) Peripheral Nervous System

The peripheral nervous system consists of all the motor neuron axons that radiate to the muscles from the ganglia of the CNS and visceral nervous system plus the sensory neurons of the cuticular sensory structures (the sense organs) that receive mechanical, chemical, thermal or visual stimuli from an environment.

Sense Organs

Sensilla are the organs associated with sensory perception and develop from epidermal cells. The different types of sense organs are:

Figure 4.52b: Insect Nervous System

Ventral nerve cord

Segmental gangalion

Supra-esophageal ganglion

Sub-esophageal ganglion

Gut

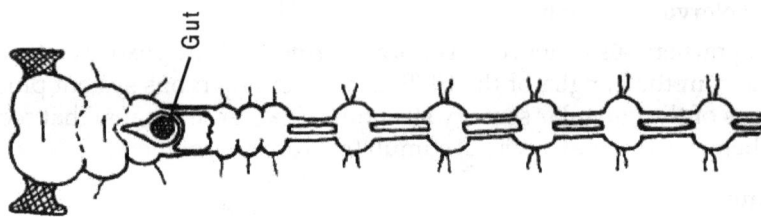

Figure 4.52a: Structure of Nervous System (General Form)

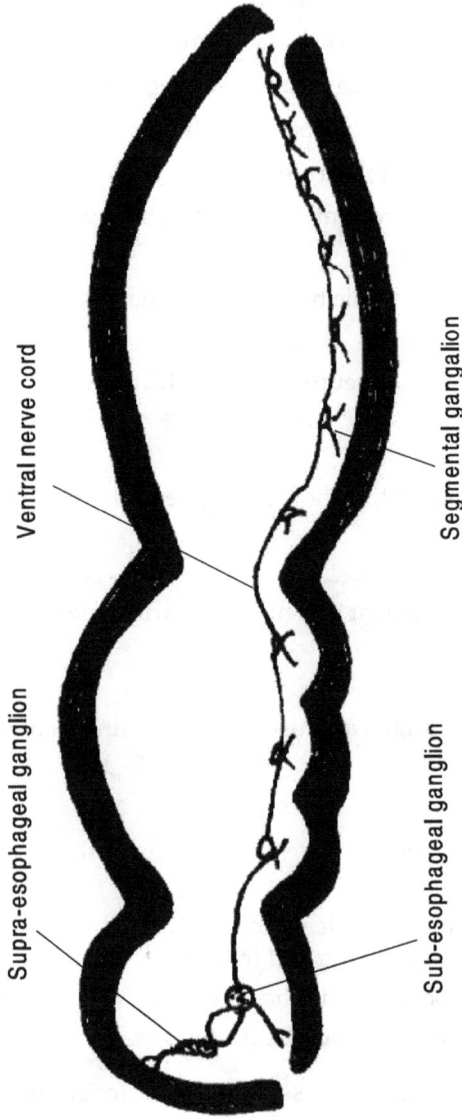

1. Mechano receptors
2. Auditory receptors
3. Chemoreceptors
4. Thermo receptors and
5. Photo receptors.

1. Mechano Receptors (Detect Mechanical Forces)

(i) Trichoid Sensilla

Hair like little sense organ. Sense cell associated with spur and seta. These cells are sensitive to touch and are located in antenna and trophy (mouth parts).

(ii) Campaniform Sensilla (Dome Sensilla)

Terminal end of these sensilla is rod like and inserted into dome shaped cuticula. These cells are sensitive to pressure and located in leg joints and wing bases.

(iii) Chordotonal Organ

The specialized sensory organs that receive vibrations are subcuticular whereas mechano receptors are called chordotonal organ. An organ consists of one to many scolopidia, each of which consists of cap cell, scolopale cell and dendrite. These organs are interoceptors attached to both ends of body wall.

Functions

(i) Proprioception (positioning of their body parts in relation to the gravity).

(ii) Sensitive to sound waves, vibration of substratum and pressure changes.

(iii) Johnston's organ: All adults insects and many larvae have a complex chordotonal organ called Johnston's organ lying within the second antennal segment (Pedicel). These organs sense movements of antennal flagellum. It also functions in hearing in some insects like male mosquitoes and midges.

(iv) Subgenual organ: Chordotonal organ located in the proximal tibia of each leg, used to detect substrate vibration. Subgenual organs are found in most insects, except the Coleoptera and Diptera.

2. Auditory Receptors (Detect Sound Waves)

(i) Delicate Tactile Hairs

Present in plumose antenna of male mosquito.

(ii) Tympanum

This is a membrane stretched across tympanic cavity and responds to sounds produced at some distance and transmitted by airborne vibration. Tympanal membranes are linked to chordotonal organs that enhance sound reception.

Tympanal organs are located:

1. Between the metathoracic legs of mantids.
2. The metathorax of many noctuid moths.

3. The prothoracic legs of many orthopterans.
4. The abdomen of short horned grasshopper, cicada.
5. The wings of certain moths and lacewings.

3. Chemoreceptors (Detect Smell and Taste)

Detect chemical energy. Insect chemoreceptors are sensilla with one pore (uniporous) or more pores (multiporous). Uniporous chemorceptors mostly detect chemicals of solid and liquid form by contact and are called as gustatory receptor. Many sensor neurons located in antenna are of this type. Multiporous chemoreceptors detect chemicals in vapour form, at distant by smell and are called as olfactory receptor. Few sensory neurons located in trophi and tarsi are of this type. Each pore forms a chamber known as pore kettle with more number of pore tubules that run inwards to meet multibranched dendrites.

4. Thermoreceptors (Detect Heat)

Present in poikilothermic insects and sensitive to temperature changes. In bed bug, it is useful to locate the host utilizing the temperature gradient of the host.

5. Photoreceptors (Detect Light Energy)

(a) Compound Eyes

The compound eye is based on many individual units called ommatidia. Each ommatidium is marked externally by a hexagonal area called facet. Compound eye is made up of two parts called optic part and sensory part. Optic part contains a cuticular lens called corneal lens secreted by corneagenous cells and crystalline cone covered by primary pigment cells. Function of the optic part is to gather light. Sensory part contains six to ten visual cells called retinular cells covered by secondary pigment cells which collectively secrete a light sensitive rod at the centre called rhabdom. Rhabdom contains light sensitive pigments called rhodopsin. Each ommatidium is covered by a ring of light absorbing pigmented cells, which isolates an ommatidium from other. Nerve cells are clustered around the longitudinal axis of each ommatidium.

Types of Ommatidia

(i) *Apposition type (light tight)*: Due to the presence of primary pigment cells light cannot enter the adjacent cells. The mosaic image formed is very distinct. The image formed by the compound eye is of a series of opposed points of light of different intensities. This functions well in diurnal insects.

(ii) *Super position type*: Primary pigment cells are absent allowing light to pass between adjacent ommatidia. Image formed in this way are indistinct, bright and blurred. This type is seen in nocturnal and crepuscular insects.

(b) Lateral Ocelli (Stemmata)

Visual organs of holometabolous larva. Structure is similar to ommatidium. It helps to detect form, colour and movement, and also to scan the environment.

(c) Dorsal Ocelli

Visual organs of nymph and it vary from 0-3 in numbers. It contains a single corneal lens with many visual cells individually secreting the rhabdomere. Dorsal ocelli perceive light to maintain diurnal rhythm and is not involved in image perception.

Reproductive System

In insects, male and female sexes are mostly separate. Sexual dimorphism is common where the male differ from the female morphologically as in bees, mosquito and cockroach. The other types are:

Gynandromorph (Sexual Mosaic)

Abnormal individual with secondary sexual characters of both male and female. *e.g.* mutant Drosophila.

Hermaphrodite

Male and female gonads are present in one organism. *e.g.* Cottony cushion scale.

Female Reproductive System

The main functions of the female reproductive system are egg production and storage of male's spermatozoa until the eggs are ready to be fertilized. The basic components of the female system are paired ovaries, which empty their mature oocytes (eggs) via the calyces (Calyx) into the lateral oviduct which unite to form the common (median) oviduct. The gonopore (opening) of the common oviduct is usually concealed in an inflection of the body wall that typically forms a cavity, the genital chamber. This chamber serves as a copulatory pouch during mating and thus is often known as the bursa copulatrix. Its external opening is the vulva. In many insects, the vulva is narrow and the genital chamber becomes an enclosed pouch or tube referred to as the vagina. Two types of ectodermal glands open into the genital chamber. The first is the spermatheca which stores spermatoza until they are needed for egg fertilization. The spermatheca is single and sac-like with a slender duct, and often has a diverticulum that forms a tubular spermathecal gland. The gland or glandular cells within the storage part of the spermatheca provide nourishment to the contained spermatozoa. The second type of ectodermal gland, known collectively as accessory glands, opens more posteriorly in the genital chamber. Each ovary is composed of a cluster of egg or ovarian tubes, the ovarioles, each consisting of a terminal filament, a germarium (in which mitosis gives rise to primary oocytes), a vitellarium (in which oocytes grow by deposition of yolk in a process known as vitellogenesis) and a pedicel. An ovariole contain a series of developing oocytes each surrounded by a layer of follicle cells forming an epithelium (the oocyte with its epithelium is termed a follicle), the youngest oocyte occur near the apical germarium and the most mature near the pedicel. There are different types of ovarioles based on the presence or absence of specialized nutritive cells called trophocytes/nurse cells for nourishment of oocytes.

Paniostic Ovariole

Lacks specialized nutritive cells so that it contains only a string of follicles, with the oocytes obtaining nutrients from the haemolymph via the follicular epithelium. *e.g.* Cockroach.

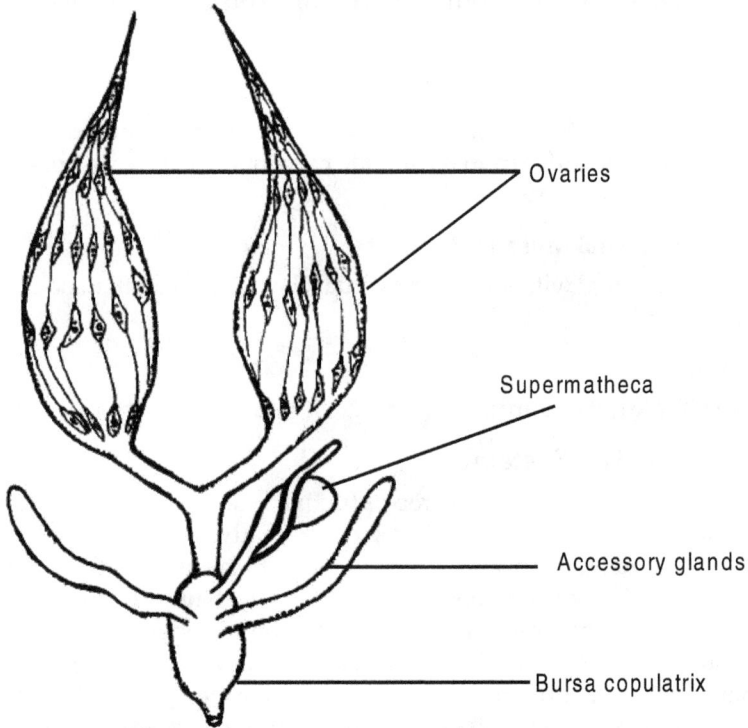

Figure 4.53: Female Reproductive System

Telotrophic Ovariole

(Acrotrophic) The trophocyte is present and its location is confined to the germarium and remain connected to the oocytes by cytoplasmic strands as the oocytes move down the ovariole, *e.g.* bugs.

Polytrophic Ovariole

A number of trophocytes are connected to each oocyte and trophocytes moves down along with the ovariole, providing nutrients until depleted. Thus, individual oocytes are alternated with groups of smaller trophocytes in the ovarioles, *e.g.* moths and flies.

Accessory glands of the female reproductive tract are often called as colleterial or cement glands, because their secretions surround and protect the eggs or cement them to the substrate, *e.g.* egg case production in mantis, ootheca formation in cockroach, venom production in bees.

Male Reproductive System

The main functions of the male reproductive system are the production and storage of spermatozoa and their transport in a viable state to the reproductive tract of the female. Morphologically, the male tract consists of paired testes.

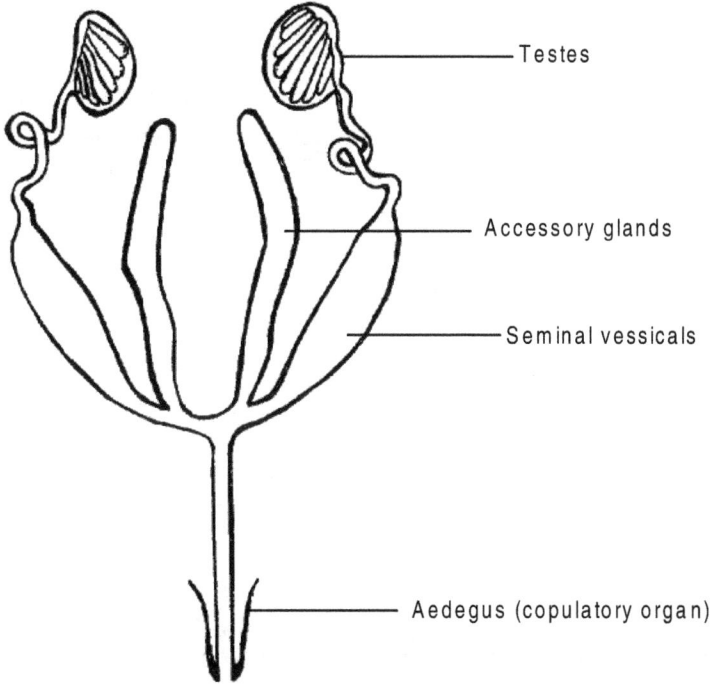

Figure 4.54: Male Reproductive System

Physiology of reproduction

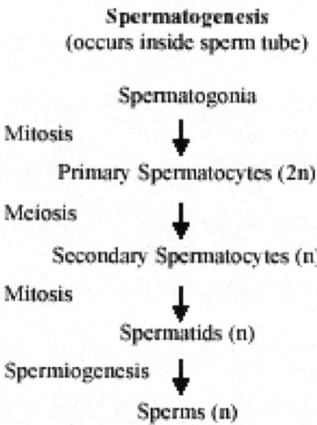

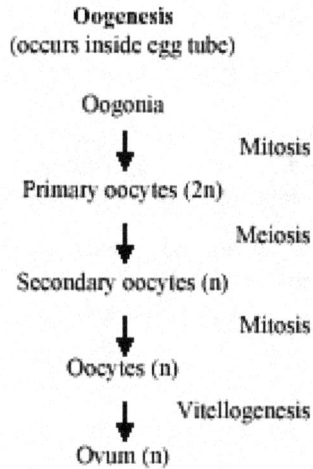

Spermatogenesis
(occurs inside sperm tube)

Spermatogonia

Mitosis $\quad\downarrow$

Primary Spermatocytes (2n)

Meiosis $\quad\downarrow$

Secondary Spermatocytes (n)

Mitosis $\quad\downarrow$

Spermatids (n)

Spermiogenesis $\downarrow$

Sperms (n)

Oogenesis
(occurs inside egg tube)

Oogonia

$\downarrow \quad$ Mitosis

Primary oocytes (2n)

$\downarrow \quad$ Meiosis

Secondary oocytes (n)

$\downarrow \quad$ Mitosis

Oocytes (n)

$\downarrow \quad$ Vitellogenesis

Ovum (n)

Sperm Transfer

(i) Intragenital

Common method, through aedeagus via vaginal orifice into female genital passage.

(*ii*) Haemocoelous

Sperms transferred into the body cavity, *e.g.* Bed bug.

(*iii*) External

Spermatophores are ejected out into open place by male while female walk over it and gets inseminated, *e.g.* Silver fish.

Fertilization

Sperm enter into egg to produce morphogenesis. Egg nucleus divides meiotically into female gamete nucleus and polar body. Then, the fertilization occurs by the fusion of male and female gamete nuclei.

Types of Reproduction

1. Oviparity

Majority of female insects, are oviparous, lay eggs. Embryonic development occurs after oviposition by utilizing the yolk, *e.g.*Head louse, moths.

2. Viviparity

Unlike oviparous, here initiation of egg development takes place within the mother. The life cycle is shortened by retention of eggs and even developing young within the mother. Four main types of viviparity are observed in different insect groups.

(i) Ovoviviparity

Fertilized eggs containing yolk are incubated inside the reproductive tract of the female and hatching of egg occur just prior to or soon after oviposition, *e.g.* Thrips, few cockroaches, beetles, and flesh fly. Fecundity of this group is low.

(ii) Pseudoplacental Viviparity

This occurs when a yolk deficient egg develops in the genital tract of the female. The mother provides a special placenta like tissue, through which nutrients are transferred to developing embryos. There is no oral feeding and larvae are laid upon hatching, *e.g.* aphids, some earwigs, psocids and polytenid bugs.

(iii) Haemocoelous Viviparity

This involves embryos developing free in the female's haemolymph with nutrients taken up by osmosis. This form of internal parasitism occurs only in strepsiptera and some gall midges.

(iv) Adenotrophic Viviparity

This occurs when a poorly developed larva hatches and feeds orally from accessory gland (milk gland) secretion within the uterus of the mother. The full grown larva is deposited and pupates immediately (*e.g.*) Tsetse flies, louse, ked, bat flies.

3. Parthenogenesis

Reproduction without fertilization is parthenogenesis. Different types of parthenogenesis are as follows:

(a) Based on Occurrence
  (i)   Facultative (not compulsory), *e.g.* bee.
  (ii) Obligatory or constant (compulsory), *e.g.* stick insect
  (iii) Cyclic/sporadic: alternation of gamic and agamic population, *e.g.*aphid.

(b) Based on Sex Produced
  (i)   Arrhenotoky: Produce male, *e.g.* bee
  (ii) Thelytoky: produce female, *e.g.* aphids
  (iii) Amphitoky/deuterotoky: produce both male and female, *e.g.* Cynipid wasp.

(c) Based on Meiosis
  (i) Apomictic: no meiosis occurs
  (ii) Automictic: meiosis occurs, but diploidy is maintained

4. Polyembryony
   This form of asexual reproduction involves the production of two or more embryos from one egg by subdivision. Mostly observed in parasitic insects (*e.g. Platygaster*). Nutrition for a large number of developing embryo cannot be supplied by the original egg and is acquired from the host's haemolymph through a specialized enveloping membrane called trophamnion.

5. Paedogenesis
   Some insects cut short their life cycles by loss of adult and pupal stages. In this precocious stage gonads develop and give birth to young one by parthenogenesis *i.e.* reproduction by immature insects.
  (i) Larval paedogenesis, *e.g.* Gall midges
  (ii) Pupal paedogenesis, *e.g. Miaster* sp.

Glandular System
   Glandular system is otherwise called as secretary system and is divided in to two major groups based on the presence or absence of ducts.

A. Exocrine Glands (Glands with Duct)
  1. *Salivary glands*: Salivary glands are modified labial glands which secrete saliva and open beneath hypopharynx.
  2. *Mandibular glands*: Secrete saliva in caterpillars when salivary glands are modified into silk glands. In queen bee, it secretes queen substance.
  3. *Maxillary glands*: Secretions are useful to lubricate mouth parts.
  4. *Pharyngeal glands*: Secrete bee milk or royal jelly in nurse bee.
  5. *Frontal glands*: Secrete sticky defensive fluid in nasute termites.
  6. *Setal glands*: Glandular seta (Scoli) secrete irritant fluid in hairy/slug caterpillar.

7. *Tenant hairs*: Secrete sticky fluid found in pulvilli of legs and helps in ceiling walking in house flies.

8. *Moulting glands*: Modified glandular epidermal cells, secrete moulting fluid necessary for moulting.

9. *Stink glands (Repugnatorial glands)*: Secrete bad smelling substance, *e.g.* Stink bugs, bed bugs.

10. *Osmeteria (Forked gland)*: Eversible gland in the thorax of papilionid larva with defense function, *e.g.* Citrus butterfly larva.

11. *Androconia (Scented scales)*: Secretions of glandular scales of male pierid butterflies to attract the opposite sex.

12. *Pheromone glands*: Found in abdominal terminalia of one sex and its secretions are released outside to attract opposite sex of the same species.

13. *Wax glands*: Dermal glands producing wax in bees and mealy bugs.

14. *Sting glands*: Modified accessory glands secreting venom in worker bees and wasps.

15. *Lac glands*: Dermal glands secreting resinous substances in lac insect.

16 *Milk glands*: Modified accessory gland nourishing larva developing in uterus. *e.g.* Sheep ked.

B. Endocrine Glands (Glands Without Duct)

1. *Neurosecretory cells*: A pair of median neuro-secretory cells and lateral neurosecretory cells are present. The axons of these neurosecretory cells form two pairs of nervi corpora cardiaci ending in carpora cardiaca. This structure influence, the functioning of other endocrine glands.

2. *Corpora cardiaca*: It consists of paired bodies fused in middle and have both nervous tissues and glandular tissues. It acts as a conventional storage and release organ for neurosecretory cells. It controls heart beat and regulate trehalose level in haemolymph.

3. *Corpora allata*: It is a paired gland attached to corpora cardiaca and secretes juvenile hormone (JH) thereby inhibit metamorphosis. It is needed for egg maturation and functioning of male accessory glands. Practically JH analogues interfere with insect development. Precocene is an anti JH which induce precocious metamorphosis and death in insects.

4. *Prothoracic glands*: Paired gland present in ventrolateral part of prothorax of larva and is degenerated in adults. It secretes the moulting hormone ecdysone. Neurosecretory cells activate prothoracic glands to secrete ecdysone.

5. *Weismann's ring*: Formed by the fusion of carpora cardiaca, carpora allata, prothoracic glands and hypocerebral ganglion to secrete puparium hardening hormone. It is present in maggots of Dipteran flies.

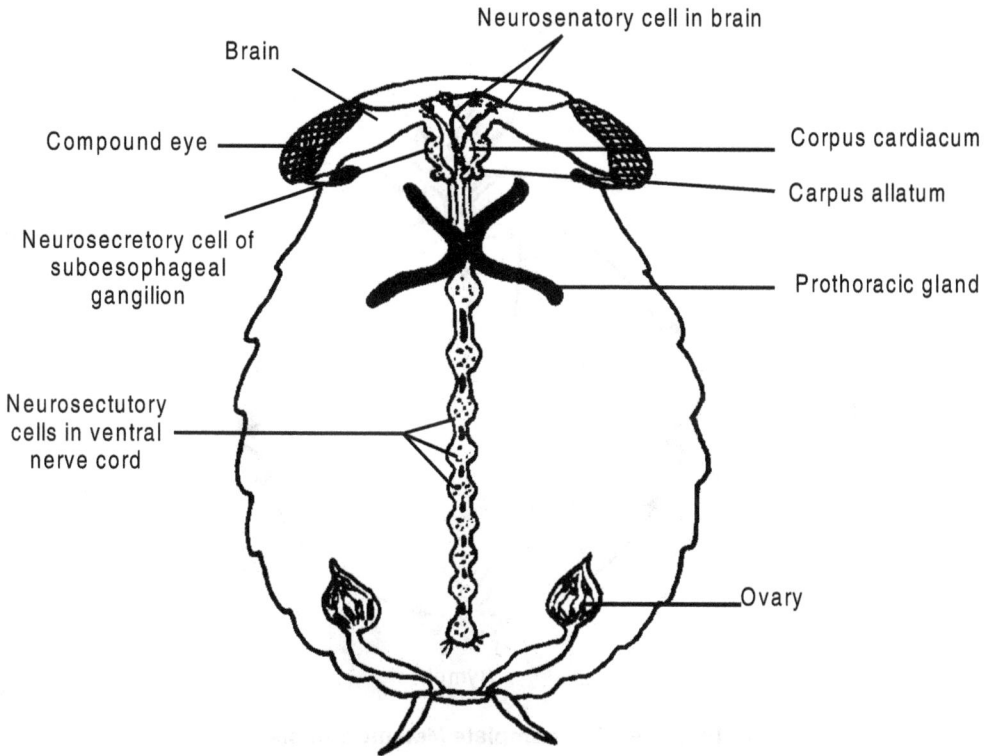

**Figure 4.55: The Location of Various Endocrine Glands and
Neurosecretory Cells in an Insect**

Metamorphosis and Immature Stages in Insects

Metamorphosis is the change in growth and development in insect which is undergone during its life cycle from birth to maturity. There are four basic types of metamorphosis in insects.

1. Ametabola

(No metamorphosis), *e.g.* Silver fish. These insects have only three stages in their life namely egg, young ones and adult. It is most primitive type of metamorphosis. The hatching insect resembles the adult in all respects except for the size and called as juveniles. Moulting continues throughout the life.

2. Hemimetabola

(Incomplete metamorphosis), *e.g.* Dragonfly, damselfly and may fly. These insects also have three stages in their life namely egg, young one and adult. The young ones which are aquatic are called as naiads. They are different from adults in habit and habitat. They breathe by means of tracheal gills. In dragonfly naiad, the lower lip (labium) is called mask which is hinged and provided with hooks for capturing prey. After final moult, the insects have fully developed wings suited to aerial life.

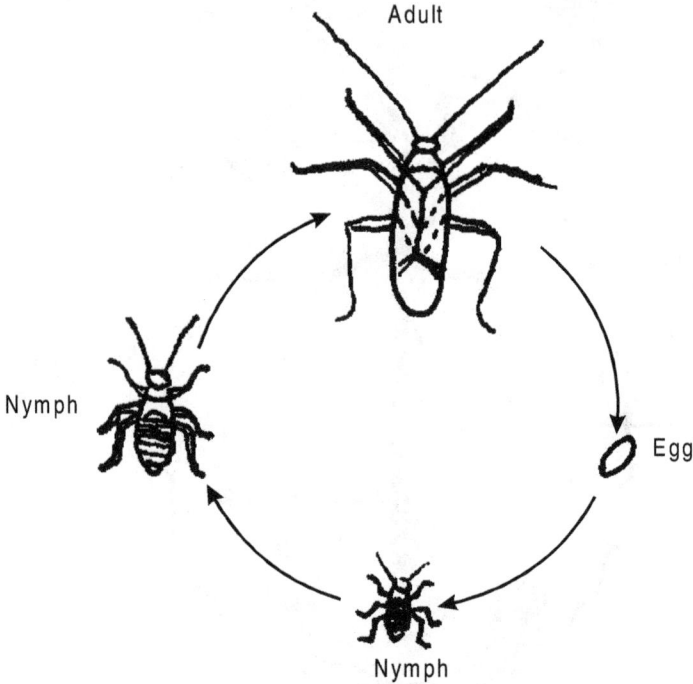

Figure 4.56: Incomplete Metamorphosis

3. Paurometabola

(Gradual metamorphosis), *e.g.* Cockroach, grasshopper, bugs. The young ones are called nymphs. They are terrestrial and resemble the adults in general body form except the wings and external genitalia. Their compound eyes and mouth parts are similar to that of adults. Both nymphs and adults share the same habitat. Wing buds externally appear in later instars. The genitalia development is gradual. Later instar nymphs closely resemble the adult with successive moults.

4. Holometabola

(Complete metamorphosis), *e.g.* Butterfly, moth, fly and bees. These insects have four life stages namely egg, larva, pupa and adult. Majority

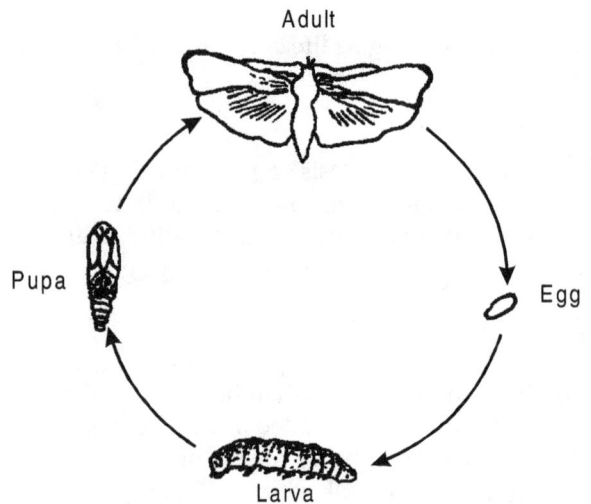

Figure 4.57: Complete Metamorphosis

of insects undergo complete metamorphosis. Larvae of butterflies are called caterpillar. Larva differs greatly in form from adult. Compound eyes are absent in larva. Lateral ocelli or stemmata are the visual organs. Their mouth parts and food habit differ from adults. Wing development is internal. When the larval growth is completed, it transforms into pupa. During the non-feeding pupal stage, the larval tissues disintegrate and adult organs are built up.

I. Eggs

The first stage of development in all insects is egg. Majority of insects are oviparous. Egg stage is inconspicuous, inexpensive and inactive. Yolk contained in the egg supports the embryonic development. Eggs are laid under conditions where the food is available for feeding of the future young ones. Eggs are laid either individually or in groups. The outer protective shell of the egg is called chorion. Near the anterior end of the egg, there is a small opening called micropyle which allows the sperm entry for fertilization. Chorion may have a variety of textures. Size and shape of the insect eggs vary widely.

Types of Eggs

(a) Singly Laid

1. *Sculptured egg*: Chorion with reticulate markings and ridges, *e.g.* Castor butterfly.
2. *Elongate egg*: Eggs are cigar shaped, *e.g.* Sorghum shoot fly.
3. *Round egg*: Eggs are either spherical or globular, *e.g.* Citrus butterfly
4. *Nit*: Egg of head louse is called nit. It is cemented to the base of the hair. There is an egg stigma at the posterior end, which assists in attachment. At the anterior end, there is an oval lid which is lifted at time of hatching.
5. *Egg with float*: Egg is boat shaped with a conspicuous float on either side. The lateral sides are expanded. The expansions serve as floats, *e.g. Anopheles* mosquito.

(b) Eggs Laid in Groups

1. *Pedicellate eggs*: Eggs are laid in silken stalks of about 1.25 mm length in one groups on plants, *e.g.* Green lacewing fly.
2. *Barrel shaped eggs*: Eggs are barrel shaped. They look like miniature batteries. They are deposited in compactly arranged masses, *e.g.* Stink bug.
3. *Ootheca (Pl. Oothecae)*: Eggs are deposited by cockroach in a brown bean like chitinous capsule. Each ootheca consists of a double layered wrapper protecting two parallel rows of eggs. Each ootheca has 16 eggs arranged in two rows. Oothecae are carried for several days protruding from the abdomen of female prior to oviposition in a secluded spot. Along the top,there is a crest, which has small pores which permit gaseous exchange without undue water loss. Chitinous egg case is produced out of the secretions of colleterial glands.

4. *Egg pod*: Grasshoppers secrete a frothy material that encases an egg mass which is deposited in the ground. The egg mass lacks a definite covering. On the top of the egg,the frothy substance hardens to form a plug which prevents the drying of eggs.

5. *Egg case*: Mantids deposit their eggs on twigs in a foamy secretion called spumaline which eventually hardens to produce an egg case or ootheca. Inside the egg case, eggs are aligned in rows inside the egg chambers.

6. *Egg mass*: Moths lay eggs in groups in a mass of its body hairs. Anal tuft of hairs found at the end of the abdomen is mainly used for this purpose, *e.g.* Rice stem borer. Female silk worm moth under captivity lays eggs on egg card. Each egg mass is called DFL (diseases free laying).

7. *Egg raft*: In *Culex* mosquitoes, the eggs are laid in a compact mass consisting of 200-300 eggs called egg raft in water.

II. Larvae

Larval stage is the active growing stage. It is the immature stage between the egg and pupal stage of an insect having complete metamorphosis. This stage differs radically from the adult.

Types of Larvae

There are three main types of insect larvae namely oligopod, polypod and apodous.

1. Oligopod

Thoracic legs are well developed. Abdominal legs are absent. There are two subtypes.

(a) Campodeiform

They are so called from their resemblance to the dipluran genus Campodea. Body is elongate, depressed dorsoventrally and well sclerotised. Head is prognathous. Thoracic legs are long. A pair of abdominal cerci or caudal processes is usually present. Larvae are generally predators and are very active, *e.g.* grub of antlion or grub of lady brid beetle.

(b) Scarabaeiform

Body is 'C' shaped, stout and subcylindrical. Head is well developed. Thoracic legs are short. Caudal processes are absent. Larva is sluggish, burrowing into wood or soil, *e.g.* grub of rhinoceros beetle.

2. Polypod or Eruciform

The body consists of an elongate trunk with large sclerotised head capsule. Head bears a pair of powerful mandibles which tear up vegetation. Two groups of single lensed eyes (Stemmata) found on either side of the head constitute the visual organs. The antenna is short. Three pairs of thoracic legs and upto five pairs of unjointed abdominal legs or prolegs are present. Thoracic legs are segmented and they end in claws which are used for holding on to the leaf. Bottom of the proleg is called planta which typically bears rows or circlet of short hooked spines or crochets which are useful in clinging to the exposed surface of vegetation and walking. Abdominal segments three to six and ten typically bear prolegs, *e.g.* Caterpillar (larvae of moths ad butterflies).

(a) Hairy Caterpillar

The body hairs may be dense, sparse or arranged in tufts. Hairs may cause irritation, when touched, *e.g.* Red hairy caterpillar.

(b) Slug Caterpillar

Larva is thick, short, stout and fleshy. Laval head is small and retractile. Thoracic legs are minute. Abdominal legs are absent. Abdominal segmentation is indistinct. Larva has poisonous spines called scoli distributed all over the body. Such larva is also called platyform larva.

(c) Semilooper

Either three or four pairs of prolegs are present. Prolegs are either wanting or rudimentary in either third or third and fourth abdominal segments, *e.g.* castor semilooper.

(d) Looper

They are also called measuring worm or earth measurer or inch worm. In this type, only two pairs of prolegs are present in sixth and tenth abdominal segments, *e.g.* Dhaincha looper.

3. Apodous

They are larvae without appendages for locomotion. Based on the degree of development and sclerotization of head capsule, there are three subtypes.

(a) Eucepalous

Larva with well developed head capsule are having functional mandibles, maxillae, stemmata and antennae. Mandibles act transversely, *e.g.* Wriggler (larva of mosquito) and grub of red palm weevil.

(b) Hemicephalous

Head capsule is reduced and can be withdrawn into thorax. Mandibles act vertically, *e.g.* Larva of horse fly and robber fly.

(c) Acephalous

Head capsule is absent. Mouthparts consist of a pair of protrusible curved mouth hooks and associated internal sclerites. They are also called vermiform larvae, *e.g.* Maggot (larva of house fly).

III. Pupa

It is the resting and inactive stage in all holometabolous insects. During this stage, the insect is incapable of feeding and is quiescent. During the transitional stage, the larval characters are destroyed and new adult characters are created. There are three main types of pupae.

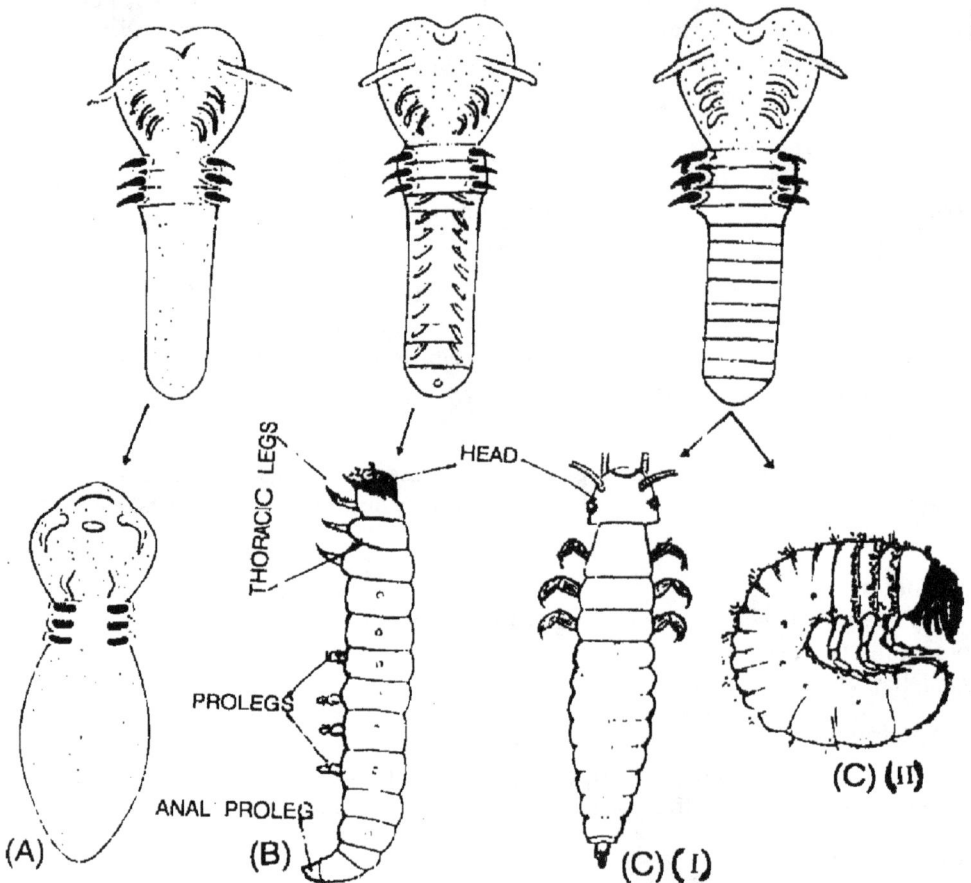

**Figure 4.58: Types of Larvae (A) Protopod; (B) Polypod eruciform;
(C) Oligopod–(i): Campodeiform; and (ii) Scarabaeform larvae**

1. Obtect

Various appendages of the pupa *viz.*, antennae, legs and wing pads are glued to the body by a secretion produced during the last larval moult. Exposed surfaces of the appendages are more heavily sclerotised than those adjacent to body, *e.g.* moth pupa.

(a) Chrysalis

It is the naked obtect pupa of butterfly. It is angular and attractively coloured. The pupa is attached to the substratum by hooks present at the terminal end of the abdomen called cremaster. The middle part of the chrysalis is attached to the substratum by two strong silken threads called girdle.

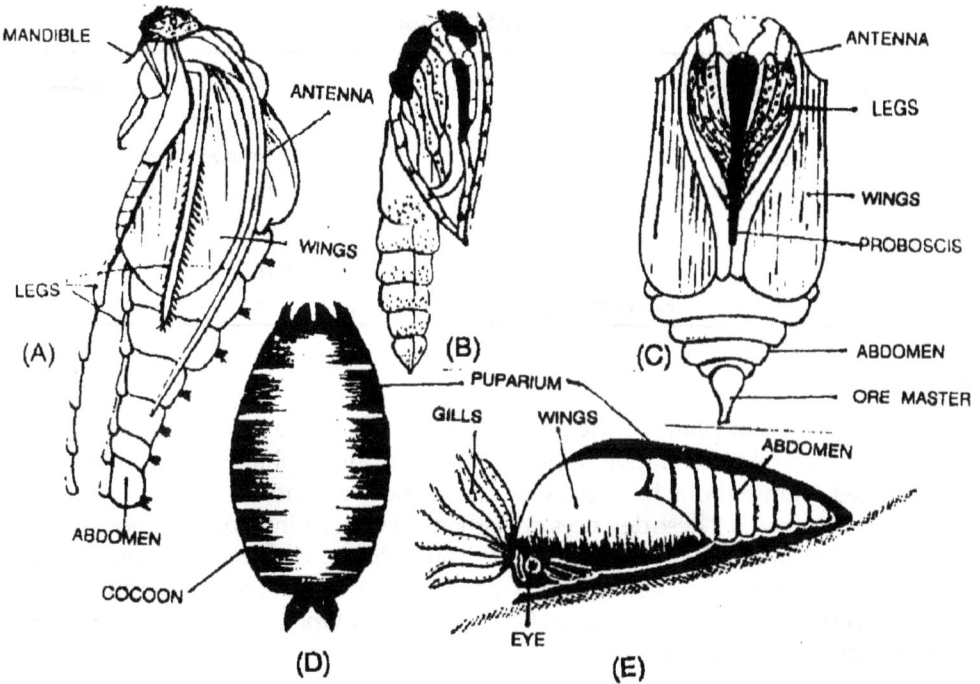

Figure 4.59: Types of Pupae (A) Decticous; (B) Exarate; (C) Obtect

(b) Tumbler

Pupa of mosquito is called tumbler. It is an obtect type of pupa. It is comma shaped with rudimentary appendages. Breathing trumpets are present in the cephalic end and anal paddles are present at the end of the abdomen. Abdomen is capable of jerky movements which are produced by the anal paddles. The pupa is very active.

2. Exarate

Various appendages *viz.*, antennae, legs and wing pads are not glued to the body. They are free. All oligopod larvae turns into exarate pupae. The pupa is soft and pale, *e.g.* Pupa of rhinoceros beetle.

3. Coarctate

The pupal case is barrel shaped, smooth with no apparent appendages. The last larval skin is changed into case containing the exarate pupa. The hardened dark brown pupal case is called puparium, *e.g.* Fly pupa.

Pupal Protection

In general, pupal stage lacks mobility. Hence, it is the most vulnerable stage. To get protection against adverse conditions and natural enemies, the pupa is enclosed in a protective cover called cocoon. Based on the nature and materials used for preparation of cocoons, there are several types shown as under.

Types of Cocoon

Cocoon Variability	Materials Used	Example
Silk cocoon	Silk	Silkworm
Earthern cocoon	Soil-saliva	Gram pod borer
Hairy cocoon	Body hairs	Woody bear
Fibrous cocoon	FIbres	Red plam weevil
Frassy cocoon	Frass+saliva	Coconut black headed caterpillar
Puparium	Hardened last larval skin	Housefly

Tropism in Insects

The majority of plant species support complex fauna of herbivores and are defined in relation to the range of plants used *viz.*, monophages, oligophages and polyphages. Many plants have developed defences against insects during the co-evolutionary process. These defences are either physical (spines, pubescence on stems and leaves, silica or sclerenchyma in leaf tissue, etc.) or chemical (secondary plant compounds-tannins, terpenoids, alkaloids, etc.) in nature. Secondary plant compounds may either repel an insect for feeding and oviposition or affect the biology of an insect due to antibiosis.

Insects are intimately associated with plant and are important to many in their reproduction, pollination or seed dispersal.

Entomophily–Insect pollination

  (*i*) Cantharophily–beetle pollinated

 (*ii*) Myophily–fly pollinated

(*iii*) Sphecophily- Wasp pollinated

 (*iv*) Myrmecophily- ant pollinated

  (*v*) Melittophily-bee pollinated

 (*vi*) Phalaenophily-moth pollinated

(*vii*) Psycophily- butterfly pollinated

Insect-plant interactions associated with pollination are clearly mutualistic. The plant is fertilized by appropriate pollen, while the insect obtains food, *e.g.* (*i*) Fig species pollinated by fig wasp. (*ii*) Correlation between moth proboscis length and flower depths–Madagascar star orchid with 30 cm length of flower parts is pollinated by giant hawk moth with long proboscis. *Figs and fig wasps*: The female wasp enters the fig syconium via ostiole (small hole), pollinates the female flowers, oviposit in some of them and dies. Each wasp larva develops within the ovary of a flower. Female flowers not oviposited by wasps will form seeds. Month after oviposition, emerging wingless male wasps mate with female. Shortly, after male phase of the flower starts and female wasp gather pollen and move out of this flower to locate another figtree in the female phase to continue this process. *Ants and seed dispersal*:

Many ants feed on seeds and also helps in dispersal of seed, known as myrmecochory. Ants plant interactions involve "domatia" (Domatia-little houses). Plants with true domatia are called myrmecophtes. These myrmecophytes receive some benefits from occupancy of their domatia, *e.g.* Acacia plant and Pseudomyrmex ants. Ants protect the plants from herbivores and the plant derives mineral nutrients and nitrogen from ant colony waste through the inner surface of domatia (Myrmecotrophy). *'Phytotelmata' (Plant held water containers)*: Many plants support insect communities in structures that retain water. In this process plant also derive some nutrition from insects *e.g.* Pitcher plants.

Insect Communication

Insects generally communicate to locate their food source and mate using:

(*i*) Semiochemicals

(*ii*) Light Production

(*iii*) Sound Production

(*iv*) Body language

I. Semiochemicals

a. Pheromones: "substances that are secreted to the outside by an individual and received by a second individual of the same species in which they release a specific reaction". Pheromones are mostly volatile but sometimes they are liquid. All pheromones are produced by exocrine glands, derived from epidermal cells. The scent organs may be located anywhere on the body:

(*i*) Female lepidoptera–eversible sacs or pouches between 8th and 9th abdominal segments.

(*ii*) Female honey bee–mandibular glands

(*iii*) Female aphids–hind tibia

(*iv*) Female cockroach–midgut.

Further classification of pheromones is based on categories of behaviour associated with sex, aggregation, spacing, trail forming and alarms.

(*i*) *Sex Phermones*: Used for mate location and courtship, *e.g.* Queen butterfly *Danaus gilippus*: Male insect produces courtship pheromone an alkaloid called danaidone.

(*ii*) *Aggregation pheromones*: Causes insects of both sexes to crowd around the source of the pheromone. This may lead to increase chances of mating.

The other benefits being security from predation, maximum utilization of scarce food resource, overcoming of host resistance and cohesion of social insects, *e.g.* Cockroaches and scolytid beetles.

(*iii*) *Spacing phermones*: Produces appropriate spacing on food resources, *e.g.* Many Tephritid flies lay eggs singly in fruit where a single maggot is to be developed. Here the female deposits an oviposition deterrent pheromone on the fruit to avoid subsequent oviposition.

(iv) *Trail-marking pheromones*: Many social insects use this pheromone to mark their trails to food and the nest. This pheromone is volatile and short-lived, *e.g.* in ants, the trail pheromone are commonly metabolic waste products excreted by the poison gland. These need not be species specific–several species share some common chemicals.

(v) *Alarm pheromones*: This causes alarm behaviour. Alarm is provoked by the presence of predator or in many social insects by the threat to the nest.

II. Light Production

The principal role of light emission is courtship signalling and prey finding. This involves species specific variation in duration, number and rate of flashes in a pattern and the frequency of repetition of the pattern, *e.g.* Lampyrid beetle (fire fly)

$$\text{Luciferin} \xrightarrow[\text{ATP} + O_2]{\text{Luciferaceae}} \text{Oxy - luciferin} + CO_2 + \text{light}$$

Variation in ATP release controls the rate of flashing and differences in pH, controls frequency and colour of light emitted.

III. Sound Production

It is useful in location of mate, *e.g.* Mosquitoes, Cicada.

IV. Body Language (The Dance Language of Bees)

(i) *Round dance*: Nearby food is communicated by a simple round dance, the incoming worker exchanges nectar and makes tight circles with frequent reversals for few seconds. The quality of nectar or pollen is communicated by the vigour of the dance.

(ii) *Waggle dance*: More distant sources are identified by this dance which involves abdomen shaking during a figure-of-eight circuit. Informative characteristics of the dance include the length of the straight part (measured by number of comb cells traversed), the dance tempo (number of dances per unit time) and the duration of waggling and noise production (buzzing) during straight line section. Messages conveyed are the energy required to get the source, quality of forage and direction relative to sun's position.

(iii) *Vibration dance*: It is useful in regulating the daily and seasonal foraging patterns in relation to fluctuating food supply. More vibration Indicates that colony needs more foraging.

Vibration with Queen:To increase queen rearing

No vibration:To swarm with queen or for a mating flight of new queens.

Offense and Defense Behaviour in Insects

Insects exhibit several types of defense mechanisms to escape from their natural enemies. Insects defense can be grouped into four types as follows.

    I. Behavioural defense

    II. Structural defense

    III. Chemical defense

    IV. Colourational defense

I. Behavioural Defense

Insects exhibit several kinds of escape reactions.

1. *Jumping*: It is an effective form of escape

2. *Reflex dropping*: Caterpillars often drop from their food on a strand of silk when disturbed and reel themselves back up when the danger has passed, *e.g.* Mango shoot webber.

3. *Thanatosis*: Many beetles and weevils feign dead which is a form of defense against a predator which prefers live prey.

4. *Threatening pose*: The insects frighten their enemies by threatening pose, *e.g.* gingelly sphingid and stag beetle.

5. *Protective constructions*:

    Eggs: Stalked eggs gain protection from predators, *e.g.* green lace wing fly.

    Nymph: Frothy secretions produced by the nymph of spittle bug protects it against predators and desiccation.

    Larva: Bag worms live in portable cases made out of plant materials along which they move, feed and finally pupate inside the cases.

    Pupa: Several materials like silk, frass, soil, body hairs and fibres are used for forming protective cocoons. In flies, the last larval skin hardens to form chitinous protective puparium. Mosquito pupa is unusually active resulting in its escape from aquatic predators.

    Adult: In mealy bug, waxy threads covering the body afford protection. Hard shell like covering protects the soft bodied insects, *e.g.* female scale insect.

II. Structural Defense

1. *Horny integument*: The integument is hard and highly sclerotised. Hence, it can resist beak penetration by insectivorous birds, *e.g.* jewel beetle.

2. *Sclerotised cerci*: In earwigs, the cerci are forecep like which are the organs of defense.

3. *Raptorial leg*: The grasping legs of preying mantis are also useful to attack its enemy

4. *Tentacles*: Varying number of paired movable processes are present on the body of the larvae of butterflies grouped in the family Danaidae. They are useful to brush off the enemies.

III. Chemical Defense

It is the primary mechanism of defense in many insects. Defensive chemicals may be of two types *viz.* venom and odoriferous compounds. The venom is injected into the body of the enemy, *e.g.* Honey bee. Bee venom is produced from acid glands which are the modified accessory glands. The ovipositor is modified into sting which is the organ of defense. Odoriferous or repugnatorial substances are not injected into the body of the enemy, but they may be acquired from the food source (exogenous) or produced inside the body (endogenous).

Exogenous

1. *Osmeteria*: (Osmeterium–Singular) Odoriferous plant components accumulated in thoracic pouches are expelled by the eversion of a pair of coloured protrusible structures called osmeteria releasing a disagreeable odour in response to a disturbance.

Endogenous

1. *Stink glands*: In stink bugs and rice earhead bug, a stink gland is present in metathorax emanates a bad odour when handled.
2. *Poisonous setae*: They are present in the body wall of certain caterpillar, *e.g.* castor slug. The setal tip breaks off issuing the poison from the poison gland cell.

IV. Colourational Defense

1. *Cryptic colouration*: By the deceptive look, the insect gains protection. The insect looks like a particular object that forms a common component of the environment or the colour of the insect blends with the background. There are three types.

   Homochromism: Colour is similar, *e.g.* Preying mantis.

   Homomorphism: Form is similar, *e.g.* cowbug.

   Homotypism: Both colour and form are similar, *e.g.* stick insect.
2. *Revealing colouration*: *e.g.* Giant silk worm. Forewings are crypticaly coloured. Hind wings are attractively coloured. Once the bird locates the insect, the prey insect exposes the bright hind wings with eye spots to startle the predator.
3. *Warning colouration*: Butterflies are usually attractively coloured. The bright colour serves as a warning to the predator. Larva of monarch butterfly while feeding on the milk weed plant, ingests cardiac glycosides. As a result, both the larva and adult become unpalatable.
4. *Mimicry*: One species of animal imitates the appearance of another better protected animal species, thereby sharing immunity against destruction. The former is called mimic and the latter is known as model. There are two types of mimicry. Batesian mimicry: Mimic is restricted to palatable species. Mimic alone gets protection because the predators are apparently misled.

e.g. (*i*) Viceroy butterfly: *Liminitis archippus*–Mimic Monarch butterfly: *Danaus plexippus*–Model; (*ii*) *Hypolimnas bolina*–Mimic *Euploea core*–Model; (*iii*) A fly mimicking wasp. Mullerian mimicry: Both model and mimic are unpalatable and the ingestion of either by a predator results in the avoidance of both the species. This form of mimicry is advantageous to both the mimic and model. *e.g. Danaus chrysippus, D. genetua.*

5

Insect Taxonomy

Taxonomy is the science of classification. It can be defined as placing biological organisms or forms in order. Simpson (1961) has defined taxonomy as the theoretical study of classification including basis, principles, procedures and rules. Taxonomy includes nomenclature and classification.

Systematics

The science of study of kind and diversity of organisms and any or all relations among them. Systematics includes taxonomy and evolution.

Stages of Taxonomy

The taxonomy of any group passes through several stages.

(i) *Alpha taxonomy* (α): It is concerned with naming and characterization of species.

(ii) *Beta taxonomy* (β): Concerned with classification

(iii) *Gama taxonomy* (γ): Concerned with evolutionary relations and phylogeny.

Basis for Classification

Classification is the ordering of a large group of organisms based on certain characters into small groups.

Classification is mainly based on evolutionary relationship and not based on superficial resemblance. Points considered while classifying are (*i*) external structure, (*ii*) internal characters, (*iii*) developmental history, (*iv*) physiological data and (*v*) cytogenetic data.

The biological system of classification is called hierarchial concept of classification. This was introduced by Carl Von Linnaeus (1758). A large group of organism is successively subdivided into small group. These groups are called taxa (taxon-singular). Each group is at a particular level in this system. This level is called the rank. Groups of the same rank are grouped together and that constitutes the taxonomic category, *e.g.* Class. Certain taxonomic categories are obligate, while others are optional. For describing and classifying any organism, the basic taxonomic category is species.

The lowermost category for classifying an organism is subspecies. Systematic position of Indian honey-bee

*1. Kingdom: Animalia
*2. Phylum: Arthropoda
*3. Class: Insecta
 4. Sub class: Pterygota
*5. Order: Hymenoptera
 6. Suborder: Apocrita
 7. Super family: Apoidea
*8. Family: Apidae
 9. Subfamily: Apinae
*10. Genus: *Apis*
*11. Species: *indica*

   (*: Obligate, while others optional)

Sub species is the geographic variety or race. Species is the natural reproductive unit among animals. It is a group of individuals having similar structure, development and behaviour which interbreed to produce viable offsprings. (Biological species concept says that a species is a natural, interbreeding population which is reproductively isolated from the individuals of other species).

Functions of Species

(*i*) Reproductive Community

The individuals of the species are able to recognise each other as potential mate and seek each other for the purpose of reproduction.

(*ii*) An Ecological Unit

Irrespective of the number of individuals, they act as a unit and interact with the individuals of other species with which they share the habitat.

(*iii*) A Genetic Unit

It consists of a large number of intercommunicating gene pool in which the individual is a temporary reservoir holding a portion of genes for a short period.

   1. Larval paedogenesis–*e.g.* gall midges.

Nomenclature and Identification

A name is required for identification of any organism. Generally, two types of names are used (*i*) common name (*ii*) scientific name.

I. Common Name

(*i*) They are inaccurate because it varies from region to region and country to country and there is no uniformity followed in naming the organisms, *e.g.* Locust is a bug referring cicada in European countries and normally locusts also refer to short horned grasshoppers living in groups.

(*ii*) Common name is not available for all organisms, *e.g.* Squash bugs present in cucurbitaceous plants are represented by many species, but no common name is available for each species. It is available only for a large group like order and family.

(*iii*) Same common name is used for insects of different orders, *e.g.* Flies. A true fly has only 2 wings, whereas other insects like mayfly, dragonfly, etc., are also mentioned as flies.

(*iv*) Homonym–Same name is used for describing two different type of insects, *e.g.* Boll worm is a common term used for more than five species of boll feeding insects.

(*v*) Synonym–More than one name denoting a single insect, *e.g.* Gram pod borer, American bollworm denotes *Helicoverpa armigera*.

II. Scientific Name

The system of naming organisms using two words is called Binomial nomenclature (Trinomial nomenclature if three words are used). This system of naming gives accurate information. It is universal and is accepted in all parts of the world. The rule regarding the naming of organisms is contained in International code for zoological nomenclature. Normally, there are two names, the first name is the generic name and the second name is the species name. The names that follow the generic name are called Trivial names. The trivial names may be either species or a subspecies name, *e.g.* Head louse: *Pediculus humanus captis*, Body louse: *Pediculus humanus corporis*. The first letter of the generic name is in capital and the first letter of species and subspecies are in small letter. All the words are Latinized and written in italics or it should be written and underlined separately. The authority name is written after the species name. It starts with capital letter. The author name is put in bracket if the taxa has been reclassified and placed in another group, *e.g.* Diamond black moth *Plutella xylostella* Linnaeus.

Identification

All insects present in the world are not yet identified and described (about 25 per cent of the insect species are unidentified). An already described species produce new race which requires further identification. Identification helps to understand about the organism and to take proper control measure. Insects can be identified through an expert by comparing the available collections, using photographs and pictures and by using taxonomic keys. Key is a tabular statement presenting alternatives, describing the features of an organism. Most of the keys are dichotomous *i.e.*, always dividing into two or they are always in the form of couplets and give a clear cut alternative. Keys can be constructed based on a single character (monothetic key) or many characters (polythetic key). Polythetic key is more advantageous. Monothetic key has three disadvantages (*i*) the organism may be an exception for a particular character (*ii*) chances of erring is more (*iii*) if the particular body part is broken on which the key is made, then the key cannot be used. Key can be classified based on evolutionary principles also as phylogenetic key and arbitrary key.

(i) *Phylogenetic key*: The key is based on the evolutionary relationship. The group appears only once in the phylogenetic key.

(*ii*) *Arbitrary key*: The taxa or group appears at several places in the key. It has more advantages.

Classification of Insects

Insects are the six legged flying arthropods coming under the phylum Arthropoda and class Insecta. The widely accepted classification of insects was proposed by the eminent insect taxonomist A.D.Imms.

Characters of Class Insecta

1. Body is divided into three regions
2. In head, a pair of antenna and a pair of compound eyes are usually present.
3. Thorax is the centre of locomotion with, 3 pairs of five jointed legs and two pairs of wings.
4. Excretion is mainly through malpighian tubules.
5. Tracheal system of respiration is well developed.
6. Brain is divided into protocerebrum, deutocerebrum and tritocerebrum.

The class Insecta has two subclasses *viz.*, Apterygota and Pterygota.

Apterygota	Pterygota
1. Primarily wingless-evolved from wingless ancestors.	1. Winged or secondarily wingless evolved from winged ancestors, *e.g.* Flea, head louse, bed bug.
2. Metamorphosis is totally absent or slight.	2. Present.
3. Mandibular articulation in head is monocondylic *i.e.* single	3. Dicondylic *i.e.*, double.
4. Pleural sulcus in thorax is absent.	4. Present.
5. Pregenital abdominal appendages present.	5. Absent

The subclass Apterygota has 4 orders namely:

1. Thysanura–Silverfish (Thysan-fringed, Ura-tail)

2. *Collembola-* Springtail or snowflea (coll-glue; embol-peg)

3. *Protura–*Proturans or Telsontail (Pro-first, Ura-tail)

4. *Diplura–*Diplurans or Japygids (Di-two; Ura-tail)

The sub class Pterygota has two divisions, namely exopterygota and endopterygota based on wing development.

Character	Exopterygota	Endopterygota
1. Wing development	External	Internal
2. Metamorphosis	Incomplete (Hemimetabola) or gradual (Paurametabola)	Complete (Holometabola)
3. Pupal stage	Absent	Present
4. Immature stage	Naiad or Nymph	Larva
5. No. of orders	16	9

The class Insecta is divided in to 29 orders (4 in Apterygota and 25 in Pterygota).

Exopterygota Groups

Group I. Paleopteran Orders (1,2)

01. Ephemeroptera–Mayflies

02. Odonata–Dragonfly, Damselfly

Group II. Orthopteroid Orders (3- 11)

03. Plecoptera–Stonefly

04. Grlloblatodia–Rock crawlers

05. Orthoptera–Grasshopper, locust, cricket, mole cricket

06. Phasmida–Stick insect, leaf insect

07. Dermaptera–Earwigs

08. Embioptera–Webspinners/Embids

09. Dictyoptera–Cockroach, preying mantis

10. Isoptera–Termites

11. Zoraptera–*Zorapterans*

Group III. Hemipteroid Orders (12-16)

12. Psocoptera–Book lice

13. Mallophaga–Bird lice

14. Siphunculata–Head and body louse

15. Hemiptera–Bugs

16. Thysanoptera–Thrips

Endopterygota

Group IV. Panorpoid Complex (1-6)

01. Neuroptera- Antilions, aphidlion, owl flies, mantispid flies.

02. Mecoptera–Scorpionflies.

03. Lepidoptera–Butterflies and moths.

04. Trichoptera–Caddisfly.

05. Diptera–True fly.

06. Siphonaptera–Fleas.

07. Hymenoptera–Bees, wasps, ants.

08. Coleoptera–Beetles and weevils.

09. Strepsiptera–Stylopids.

Paleopteran insects cannot flex the wings over the abdomen, *i.e.*, wing flexing mechanism is absent. The characters of insects belonging to Orthopteroid group include mandibulate mouth parts, anal area of hind wing is well developed, abdomen is always with cerci, many number of malpighian tubules and the ganglia of the ventral nerve cord are not fused. Hemipteroid insects have the haustellate mouth parts and mainly they feed on liquid food like plant sap. Panarpoid insects are holometabolous and they have larval and pupal stage.

Insect Orders

Order: Thysanura

Synonyms

Ectognatha, Ectotrophi

Etymology

Thysan–fringe; ura–tail.

Common Names

Silverfish, Fire brat, Bristle tail

Characters

Body is elongate and flattened. Body is glistening and clothed with scales. Compound eyes are present or absent. Antennae is long, filiform and multisegmented. Mouthparts are ecotognathous, biting type. They are primarily wingless insects. Abdomen is 11 segmented. Varying number of bilateral styli are present on abdominal sternites. Styli are believed to be reduced abdominal legs. Female has elongate jointed ovipositor. Abdomen at its tip carries a pair of elongate many segmented cerci and a median caudal filament. Insemination is indirect. Metamorphosis is absent. Moulting continues even after attaining sexual maturity.

Importance

It is often a pest in home and libraries. *Ctenolepisma* sp. is the common household silverfish. It feeds and destroys paper, book bindings and starched clothing. It can be collected from amongst old books, behind calendar, photo frames, etc.

Order: Collembola

Synonyms

Oligentoma, Oligoentomata

Etymology

Coll-glue; embol–wedge or peg.

Common Names

Spring tail, Snow flea

Characters

They are minute insects. Body is globose or tubular. Compound eyes are absent. One to several pairs of lateral ocelli form an eye patch. Antenna is four segmented. Mouthparts are entognathous biting type and found within a pouch. Tibia is fused with tarsus to form tibio-tarsus. They are primarily wingless. Abdomen is six segmented with there medially situated pregenital appendages.

1. *Ventral tube or Collophore or Glue peg*: It is a bilobed adhesive organ found on the first abdominal sternite. It is believed to be associated with respiration, adhesion and water absorption.

2. *Hamula or Tenaculum or Retinaculum*: It is present on the third abdominal sternite. It consists of a fused basal piece, corpus and free distal part called rami. It holds the furcula.

3. *Furcula or Springing organ*: It consists of a basal manubrium, paried dens and distal claws called mucro. It is held under tension beneath the abdomen by retinaculum when at rest. Malpighian tubules, tracheal system and metamorphosis are usually absent.

Importance

Sminthurus viridis is a pest on alfalfa. It can be collected from moist places in soil. They are also found in mushroom houses as a pest.

Order: Ephemeroptera

Synonyms

Ephemerida, Plectoptera

Etymology

Ephemero-living for a day; ptera-wing

Common Names

Mayflies, Shadflies, Dayflies

Characters

Small to medium sized soft bodied insects. Compound eyes are large. There are three ocelli. Antenna is short and setaceous. Mouthparts in adults are atrophied. Forewings are large and triangular. Hind wings are small and absent in some species. Numerous cross and intercalary veins are present. Wings are held vertically over the abdomen. Wing flexing mechanism is absent. Abdomen is slender with a pair of long cerci. Median caudal filament may be present or absent. Metamorphosis is incomplete with three stages *viz.*, egg, naiad and adult. Naiad is aquatic with biting mouthparts. It breaths through bilateral abdominal gills. At tip of the abdomen, a pair of long cerci and a median caudal filament are usually present. Immediately after the adult emergence, body of the insect is covered with closely set fine hairs called pellicle and this stage is called as subimago. It is dull in colour with opaque wings and legs and cerci are not well developed. In imago, wings are transparent. Legs and cerci are well developed. Body is shiny and not covered with pellicle. Adults are found near lakes and ponds and are also attracted by light.

Importance

Naiads are important fish food. Adults are short lived and hence the name mayfly. When they emerge in large numbers, they pose nuisance problem

Order: Odonata

Etymology

Odon–tooth

Common Names

Dragonflies and damselflies

Characters

Medium to large sized insects. They are attractively coloured. Head is globular and constricted behind into a petiolate neck. Compound eyes are large. Three ocelli are present. Mouthparts are adapted for biting. Mandibles are strongly toothed. Lacinia and galea are fused to form mala which is also toothed. Wings are either equal or sub equal, membraneous; venation is net work like with many cross veins. Wings have a dark pterostigma towards the costal apex. Sub costa ends in nodus. Wing flexing mechanism is absent. Legs are anteroventrally placed. They are suited for grasping, holding and conveying the prey to the mouth. Spinous femora and tibiae are useful for holding the prey. Forward shift of leg attachments allow easy transfer of prey items to mouth in flight. Legs are held in such a way that a basket is formed into which the food is scooped. Abdomen is long and slender. In male, gonopore is present on ninth abdominal segment. But the functional copulatory organ is present on the second abdominal sternite. Before mating, sperms are transferred to the functional penis. Cercus is one segmented. Metamorphosis is incomplete with three life stages. The naiad is aquatic. Labium is greatly elongated, jointed and bears two hooks at apex. It is called mask. It is useful to capture the prey.

Importance

Adults are aerial predators. They are able to catch, hold and devour the prey in flight. Naiads are aquatic predators. Dragonflies and damselflies can be collected with an aerial net near streams and ponds especially on a sunny day. Naiads can be collected from shallow fresh water ponds and rice fields.

Classification

There are two sub-orders. Dragonflies are classified under Anisoptera and damselflies are grouped under Zygoptera.

Order: Orthoptera

Synonyms

Saltatoria, Saltatoptera

Etymology

Ortho–straight; ptera-wings.

Common Names

Grasshoppers, Locust, Katydid, Cricket, Mole cricket

Characters

They are medium to large sized insects. Antenna is filiform. Mouthparts are mandibulate. Prothorax is large. Pronotum is curved, ventrally covering the pleural region. Hind legs are saltatorial. Forewings are leathery, thickened and known as tegmina. They are capable of bending without breaking. Hindwings are membranous with large anal area. They are folded by longitudinal or plates between veins and kept beneath the tegmina. Cerci are short and unsegmented. Ovipositor is well developed in female. Metamorphosis is gradual. In many Orthopterans the newly hatched first instar nymphs are covered by loose cuticle and are called pronymphs. Wing pads of nymphs undergo reversal during development. Specialized stridulatory (sound producing) and auditory (hearing) organs are present.

Classification

This order is sub divided into two suborders, *viz.*, Caelifera and Ensifera.

Caelifera	Ensifera
1. Antenna is short with less than than 30 segments.	1. Antenna is long with more than 30 segments.
2. Tympanum is found on the lateral side of the first abdominal segment.	2. Tympanum is fund on the foretibia.
3. Vision and hearing organs well developed.	3. Tactile response is develop.
4. Mandibles are specialized for feeding on monocot foliage.	4. Feed on dicot plants
5. Diurnal	5. Nocturnal
6. Rely on jumping to escape from predators.	6. Rely on crypsis
7. Eggs are laid in groups in soil inside shallow . burrows	7. Eggs are laid singly into plant tissue or Soil.

Some Important Families

Family: Acrididae

Short horned grasshoppers and locusts.

1. The antennae are smaller than body length.
2. The prothorax extends down on the sides reaching the base of forelegs.
3. Hind legs are well developed with enlarged femora which are adapted for jumping. There is row of peg like files on the femor used for sound production.
4. The tarsi are three segmented.
5. Tympanum is located on each side of the first abdominal segment.
6. The ovipositor is small and made of two pairs of valves.

Examples

(*i*) Kharif grasshopper (*Hieroglyphus nigrorepletus*) or *nigrorepictus*

(*ii*) Surface grasshopper (*Chrotagonus spp.*)

(*iii*) Rice grasshopper (*H. banian*)

(*iv*) Locust (*Schistocerca gregaria*)

Family: Tettigonidae

Long horned grasshoppers, Katydids and bush crickets.

1. Apterous forms are common.
2. When winged, the left tegmen usually overlaps the right one and in males the cubito-anal regions of the tegmina are modified asymmetrically for stridulation.
3. The eggs are not enclosed in pods, and the ovipositor frequently attains great length, even exceeding that of the body.

Examples

(*i*) Conocephalus

(*ii*) Tettigonia

Family: Gryllidae: Crickets.

1. The insects stridulate by friction of the tegmina and possess tibial auditory organs.
2. The stridulatory apparatus of the male tegmen occupies a larger area than in the Tettigoniidae and the two tegmina are similarly modified.

Examples

> (*i*) House cricket (*Brachytrypes megacephalus*)
>
> (*ii*) Field cricket (*Gryllus campestris*)

Family: Gryllotalpidae: Mole Crickets

1. Insects are subterrarean in habit.
2. The fore legs being greatly expanded and armed with strong teeth to assist digging.
3. The eyes are reduced and the ovipositor is vestigial.
4. The eggs are laid in a sort of nest, 10-15 cm below the surface.

Examples

> (*i*) Mole cricket (*Gryllotalpa gryllotalpa*)

Order: Phasmida

Synonyms

> Phasmodea, Phasmatodea

Etymology

> Phasma–an apparition

Common Names

> Stick insects, Leaf insects

Characters

Body is stick–like or leaf–like. Head is prognathous. Mouthparts are chewing type. Prothorax is short. Meso and metathorax are long. Metathorax is closely associated with the first abdominal segment. Legs are widely separated. They are long and slender resembling twigs in stick insect. Tibia and femur shows lamellate expansion in leaf insects. A line of weakness is found between the trochanter and rest of the leg. The legs get broken easily at this region and such legs get regenerated subsequently. Tarsus is five segmented. Wings may be present or absent. Forewings when present are small and modified into tegmina. In leaf insects, the wing venation mimics leaf venation. Cerci are short and unsegmented. They show protective resemblance. They are herbivorous.

Classification

There are two families. Stick insects are grouped under Phasmidae and leaf insects are classified under Phyllidae..

Order: Dermaptera

Synonyms

> Euplexoptera, Euplecoptera

Etymology

Derma–skin; ptera–wing

Common Names

Earwigs

Characters

They are generally elongate insects. Head is with a distinct 'Y' shaped epicranial suture. They have chewing mouthparts. Prothorax is large, well developed and mobile. Meso and Metathorax are fused with the first abdominal segment. Forewings are short, leathery and veinless. Both the wings meet along a mid dorsal line. They are called tegmina or elytra. They are protective in function and are not used for flight. Hindwings are large, membranous, semicircular and ear like. The anal area of the wings is large with a number of branches of anal veins which are radially arranged. They are folded fan like, longitudinally and twice transversely and kept beneath the forewings at rest. Wings do not cover the abdomen fully. Cerci are found at the end of the abdomen. They are unsegmented enlarged, highly sclerotised and forceps like. They are large and bowed in male and nearly straight in female. They are useful in defence, folding and unfolding of wings, prey capture and copulation. Parental care is shown by female earwigs. It literally, roost on the eggs until hatching occurs and also cares for the nymphs.

Importance

Euborellia annulipes bores into groundnut pods and feeds on the kernel.

Order: Embioptera

Synonyms

Embiodea, Embiidina

Etymology

Embia-lively; ptera-wings

Common Names

Embiids, Webspinners

Characters

They are small elongate soft bodied insects. Antenna is filiform. Mouthparts are chewing type. Basitarsus of the foreleg is greatly enlarged. Silk glands and spinnerets are found in the basitarsus. Hind femur is enlarged and helps in backward running. Male has well developed wings; while female is apterous. Wings are elongate, nearly equal, smoky brown with reduced wing venation. Radial vein is thick. Cerci are asymmetrical; left cercus is one segmented and it serves as clasper. Cerci are equal and two segmented in female. Embiids are gregarious and live inside tubular silken tunnels beneath stones, logs and bark of trees. Silken tunnels give protection against predators, prevent excessive water loss from the body and provide a humid atmosphere. Females show strong parental care and they nurse the eggs and nymphs. They feed on decaying plant matter.

Order: Dictyoptera

Synonyms

Oothecaria, Blattiformia

Etymology

Dictyon-net work; ptera-wings

Common Names

Cockroaches and preying mantids.

Characters

Head is hypognathous. Antenna is filiform. Mouthparts are chewing type. Tarsus is five segmented. Forewings are more on less thickened, leathery with a marginal costal vein. They are called tegmina. Hindwings are large, membranous and folded fanlike and kept beneath the forewings. Cerci are short and many segmented. Eggs are contained in an ootheca. Dictyoptera is divided into two suborders *viz.*, Blattaria (cockroaches) and Mantodea (preying mantids).

Some Important Families

Family: Blattidae

1. Pronotum covering head completely.
2. Cerci well developed
3. Styles distince, femora bearing strong spines.

Example

(i) Cockroach (*Blatta sp*)

Family: Mantidae

1. Pronotum does not cover head.
2. Ocelli 3 and well developed.
3. Inner ventral spines of fore femora are alternatively long and short, outer spines erect or oblique.

Example

(i) Praying mantis (*Mantis religa*)

Order: Isoptera

Synonyms

Termitina, Termitida,Socialia

Etymology

Iso–equal; ptera–wing.

Common Names

Termites, White ants

Characters

They are small greyish white, soft bodied insects. The body is pale yellow in colour because of weak sclerotization. Compound eyes are present in alate forms and usually absent in apterous forms. Antennae are short and moniliform. Mouthparts are adapted for biting and chewing. Two pairs of wings are present which are identical in size form and venation. Wings are membranous and semitransparent. Venation is not distinct. Veins near the costal and anal margin alone are distinct. Anterior veins are more sclerotised. Wings are flexed over the abdomen at rest. They are extended beyond the abdomen. Wings are present only in sexually mature forms during swarming season. Wing shedding takes place along the basal or humeral suture, after swarming. The remanant or the stump remaining behind is called 'scale'. Abdomen is broadly jointed to the thorax without constriction. External genital organs are lacking in both the sexes. Cerci are short.

Specialities

They are ancient polymorphic, social insects living in colonies.

Internal Characters

Salivary glands are well developed. Rectum is distended forming rectal pouch to accommodate large number of intestinal symbionts. Fat body development is extensive in male and female reproductives. Soil inhabiting termites construct earthern mounds called termitaria. They have evolved complex relationships with other organisms like bacteria, protozoa and fungi which help them in the digestion of wood. Incessant food sharing (trophallaxis) occurs between the members of the community by mouth-to-mouth and anus-to-mouth food transfer.

Importance

Termites are nature's scavengers. They convert logs, stumps, branches, etc., to humus. Many are injurious to crops, furniture and wood works of buildings.

Classification

These are the important families of economic importance.

Family: Mastotermitidae

1. Members of only this family bears hind wings with large anal lobes.
2. 5-segmented tarsi.
3. True workers are absent and pseudergates are present.

Example

Mastotermes darwiniensis

Family: Kalotermitidae

1. Pronotum is broad and flate.
2. True workers are absent, but pseudergates are present.
3. They are commonly called as dry-wood termites from their habits.

Example

Kalotermes, Neotermes, Cryptotermes

Family: Hodotermitidae

1. Ocelli absent.
2. Compound eyes are well developed in all casts.
3. Workers are present, they are commonly called harvester termites.

Example

Hodotermes spp.

Family: Rhinotermitidae

1. Pronotum is flat, scally wings.
2. Workers are present.
3. They are mostly found in warm and moist places.
4. Soldiers having curved mandibles.

Example

Rhinotermes spp, *Coptotermes* spp.

Family: Serritermitidae

1. Mandibles are falcate
2. It includes only a single Brazilian species *e.g. Serritermes serrifer.*

Family: Termitidae

1. This family is most destructive and agriculturally important.
2. Wings with reduced wing-venations but there are some small scales and hairs present.
3. Pronotum is narrow and ocelli present.
4. Workers present and generally called ground dwelling termites.
5. Solders with frontal restrum on head.
6. They constructs very high mounds.

Example

Termites (*Odontotermes* spp., *Macrotermes* spp., *Microtermes* spp.)

Order: Psocoptera

Synonyms

Corrodentia, Copeognatha

Etymology

Psoco-rub small; ptera—wings; Psochos-dust like.

Common Names

Book lice, Bark lice, Dust lice.

Characters

They are minute and soft bodied insects. Head has a distinct 'Y' shaped epicranial suture. Clypeus is swollen. Mouthparts are biting and chewing type. Mandibles are with well developed molar and incisor areas. Lacinia is rodlike ('pick') which is partially sunken into the head capsule. Legs are slender. Wings may be present or absent. Forewings are larger than hind wings. Wings are held roof like over the abdomen. Cerci are absent. Psocids are frequently gregarious. Some psocids have the ability to spin silk. Dorsal pair of labial glands are modified into silk glands.

Importance

The common book louse is *Liposcelis* sp. They feed on paper paste of book binding, fragments of animal and vegetable matter and stored products. They also damage dry preserved insects and herbarium specimens.

Order: Mallophaga

Synonyms

Phthirapters

Etymology

Mall-wool; phaga-eat.

Common Names

Chewing lice, Biting lice, Bird lice.

Characters

They are minute insects. Body is dorsoventrally flattened. Head is large triangular and broader than thorax. Compound eyes are reduced. Mouthparts are biting type with large dentate mandibles. Mandibles are useful to clip off the host's skin debris or feather and cling to the host. Prothorax is invariably free and not fused with pterothorax. Meso and metathorax may be free or fused. Legs terminate into a pair of claws usually which are adapted for clinging to feathers. The tarsus is either unsegmented or two segmented. Wings are absent and are secondarily wingless. Eggs are called nits and are cemented to the feathers.

Importance

They are obligate parasites on birds and less frequently on mammals. They severely infest the poultry bird. Affected birds will become restless and peck at one another continuously, leading to loss of plumage. Louse infestation results in reduced body weight and decline in productivity. Bird lice feed on feathers, hairs, skin scales, scabs and possible blood clots around wounds. They cause irritation while feeding and crawling. In order to obtain relief, birds have dust bath. *e.g. Menopon pallidum* and *M. gallinae* are the two common lice assoicated with poultry.

Order: Siphunculata

Synonyms

Anoplura

Etymology

Siphunculus–a little tube

Common Names

Sucking lice

Characters

They are minute insects. Body is dorsoventrally flattened. Head is small, conical and narrower than thorax. Mouth is surrounded by a row of hooks which are anchored in the host skin while feeding. Mouthparts are piercing and sucking type. There are three slender stylets which are withdrawn into a pouch in the head capsule at rest. This pouch is variously called stylet sac, buccal sac and trophic sac. Legs are scansorial, inwardly bent and adapted for clinging to mammalian hair. Tarsus is one segmented and ends in a single large claw which folds back on a thumb like projection of tibia forming an efficient organ of clinging. Wings are absent. They are secondarily wingless. Thoracic spiracles are dorsally located. Abdominal pleurites are highly sclerotised. Eggs are called nits and are strongly glued to the base of the hairs.

Importance

They are obligate blood sucking ecotoparasites on mammals. The presence of lice lesions on the skin is known as pediculosis. Louse infestation causes itching and anaemia. The following lice are associated with man.

1. *Head louse*: *Pediculus humanus captis*. Eggs are glued to hairs.

2. *Body louse*: *Pediculus humanus corporis*. It infests neck, armpits and crotch. It transmits epidemic typhus, relapsing fever and trench fever, which are serious and often fatal diseases to humans. Eggs are attached to clothing.

3. *Crab louse*: *Pthirus pubis*. It infests armpits, pubic and perianal regions. Pthiriasis causes intense itching.

Order: Hemiptera

Synonym

Rhynchota

Etymology

Hemi–half; ptera–wing

Common Name

True bugs

General Characters

Head is opisthognathous. Mouthparts are piercing and sucking type. Two pairs of bristle like stylets which are the modified mandibles and maxillae are present. Stylets rest in the grooved labium or rostrum. Both labial palps and maxillary palps are atrophied. Mesothorax is represented dorsally by scutellum. Forewings are either uniformly thickened throughout or basally coriaceous and distally membranous. Cerci are always absent. Metamorphosis usually gradual; rarely complete. Alimentary canal is suitably modified (filter chamber) to handle liquid food. Salivary glands are universally present. Extra-oral digestion is apparently widespread. Abdominal ganglia fused with thoracic ganglia.

Classification

There are two suboders *viz.*, Heteroptera and Homoptera.

Heteroptera	Homoptera
1. Generally two pairs of wings are present (except male scale insects). The anterior pair is often harder in consistency and throughout of the same texture. Some times wings may be absent in some individuals.	1. Two pairs of wings are always found in which forewing is thick at the base and membranous at apical portion.
2. Pronotum is small.	2. Mouth parts are piercing and sucking type, without palpi and arising from the front part of the underside of the head.
3. Tarsi are one to three segmented.	3. Body usually broad and flattened dorsoventrally.
4. Metamorphosis is simple excepting male scale insects and white flies.	4. The metamorphosis is incomplete.
5. The distal end of labium is attached with the head.	

Some Important Families of Heteroptera

Family: Reduviidae (Assassin Bugs, Kissing Bugs or Cone Nose Bugs)

1. Blood sucking predaceous insects, with long head, large eyes and ocelli.
2. Antennae geniculate and long.
3. Rostrum 3 segmented pointed.
4. Hemelytra without cuneus.
5. Body colour brown or black.

Example

> *Harpactor costalis*–predator of red cotton bug.

Family: Cimicidae: (Bed Bugs)

1. Oval, dorso-ventrally flattened body.
2. Very short hemelytra or apterous.
3. Ocelli absent.
4. Tarsi 3 segmented.
5. Ectoparasites of mammals and birds.

Example

> Bed bugs (*Cimex spp.*)

Family: Tingidae: (Lacewing Bugs)

1. Rostrum 4 segmented.
2. Crest-like pronotum covers scutellum and clavus posteriorly.
3. Ocelli absent.
4. Dorsal side of body and hemelytra densely reticulate.

Example

> Brinjal lace bug (*Urentius* sp.)

Family: Miridae: (Plant Bugs or Leaf Bugs)

1. Rostrum 4 segmented.
2. Antennae 3 segmented.
3. Short bodied slender insects.
4. Hemelytra with cuneus.
5. Tarsi 3 segmented.
6. Ovipositor well developed.

Example

> Sorghum earhead bug (*Calocoris angustatus*), Sunhemp bug (*Ragmus importunitas*).

Family: Lygaeidae: (Seed Bugs or Chinch Bugs)

1. Integument is strongly sclerotized, body oval-shaped, dark, small, brightly coloured.
2. Antennae 4 segmented.
3. Rostrum 4 segmented.
4. Winged and spterous forms.
5. Ocelli present.
6. Tarsi 3 segmented.
7. Coxae rotatory.

Example
Dusky cotton bug (*Oxycarenus spp.*), Predator *Geocoris tricolor*

Family: Pyrrhocoridae: (Red Bugs or Stainers)
1. Medium to large sized insects.
2. Both antennae and rostrum are four segmented.
3. The ocelli are absent.
4. The tersi are three segmented.
5. The prothorax is large and its distal ends are broader than the mesothorax.
6. Cubits is absent on the hemelytron.

Example
Red cotton bug (*Dysdercus cingulatus*)

Family: Coreidae: (Squash Bugs or Leaf Footed Bugs)
1. Minute to medium sized insects.
2. Both the rostrum and antennae are four segmented.
3. A pair of compound eyes and two ocelli are present.
4. Tarsi are three segmented.
5. Metathorax possesses a pair of odour glands which are capable of producing disagreeable odour.

Example
Rice gundhi bug (*Leptocorisa varicornis*)

Family: Pentatomidae: (Stink Bugs or Shield Bugs)
1. Medium to large sized bugs.
2. The head is small, triangular, having a longitudinal tyles on its centre.
3. A pair of compound eyes and two ocelli are present.
4. Scutellum is very large, reaching about one half or more distance from posterior margin of pronotum to the end of folded wings.
5. Antennae are generally five segmented.
6. The tarsi are two or three segmented.

Example
Painted bug (*Bagrada cruciferarum*)

Some Important Families of Homoptera
Family: Cicadidae: (Cicadas)
1. Head with 3 ocelli.
2. Tegmina transperant.

3. Wings with ambient vein.
4. Forelets with enlarged, femora empodia wanting.
5. Hind legs elongated slender.
6. Male with distinct stridulatory organs at the base of abdomen.
7. Long lived insects 13-15 years.

Example

Giant cicada (*Cyclochila Australasia*)

Family: Membracidae: (Tree Hoppers or Cowbugs)

1. Head modifies.
2. Pronotum enlarged with a spine or and prolonged backwards into a distinct elevated hood or horn lying over the abdomen.
3. Ocelli-2 in numbers.
4. Antennae concealed beneath eyes.
5. Stridulatory organs absent.
6. Tegmina transparent.

Example

Sugarcane hopper (*Tricenturs bicolor*)

Family: Cicadellidae: (Leaf Hoppers or Jassids)

1. Generally small, elongated, cylindrical insects having eight segmented setaceous type antennae.
2. The double row of spines are present on the hind tibia.
3. The compound eyes are large situated on the lateral surface of the head and two ocelli are also present.
4. Thorax is narrower than head and attached with the abdomen without any prominent division.
5. Tarsi are three segmented.
6. Mostly rest on the lower surface of the leaves and when disturbed they leao often several feet and fly.
7. The leaf hoppers are usually about ¼ inch I length and usually tapering posteriorly.

Example

Mango hopper (*Amritodus atkinsoni*), Cotton jassid (*Amrasca bigutulla bigutulla*)

Family: Fulgoridae

1. The head in the form of a snout protruted forward.
2. The antennae are three segmented.

3. Generally two ocelli are present.
4. The wings are membranous, small and not fully covered the abdomen.
5. Tarsi are five segmented.
6. Female with anal tuft.

Example
Sugarcane leaf hopper (*Pyrilla perpusilla*)

Family: Delphacidae: (Plant Hoppers)
1. Head without postclypeus, median ocelli wanting, antennal pedical with large, numerous sensilla.
2. Wings are mostly reduced.
3. Hind tibia with apical large movable spur.
4. Ovipositor fully developed.

Example
Indian rice hopper (*Perkinsiella sinensis*), Brown plant hopper (*Nilaparvata lugens*), White backed rice plant hopper (*Sogatella furcifera*).

Family: Aleyrodidae: (Whiteflies)
1. The adults are small, four winged insects about 3 mm in length.
2. Their body is generally covered with a white waxy powder.
3. The vasiform orifice opens on the dorsal surface of the last abdominal segment. The opening is covered with an operculum under which there is a tounge like organ termed ligula.
4. The anus opens at the base of the ligula and secretes honey-dew.
5. The compound eyes and two ocelli are present.
6. Tarsi are two segmented with paired claws in between them, there is a spur or pad.

Example
Sugarcane whitefly (*Aleurolobus barodensis*), Cotton white fly (*Bemisia tabaci*)

Family: Aphididae: (Aphids or Plant Lice or Greenflies)
1. They are small delicate and more or less pearshaped insects.
2. Both winged and wingles forms are found.
3. The head is hypognathous, small and fused with thorax.
4. The antennae are three to six segmented.
5. The compound eyes and three ocelli are present.
6. The tarsi are two segmented with paired claws.

7. The cornicles or honey-tubes are the most characterstic organs located on the 5th abdominal segment.

8. In the 9th tergum of abdomen, anus is situated which secrete honey dew.

Example

Musard aphis (*Lipaphis erysimi*)

Family: Coccidae: (Scale Insects or Soft Scales)

1. The insects are quite small, body is compressed and cobered with whitish meal.

2. The female is wingless, flat and round in structure.

3. The males are generally winged and the fore wings are only well developed.

4. The legs are highly reduced and tarsi one segmented.

5. The metamorphosis is complete in male and incomplete in female.

6. The mouth parts are absent in male insects hence they do not feed in adult stage.

Example

Mango mealy bug (*Drosicha mangiferae*)

Family: Lacciferidae: (Lac Insect)

1. The insect of he family are mainly confined to the tropics and sub-tropics.

2. They are the extremely small insects with irregular globular body.

3. The legs and abdominal spiracles are absent.

4. The antennae are vestigial and three or four segmented.

5. Both winged and wingless males are found having four ocelli.

Example

Lac insect *(Laccifer lacca)*

Order: Thysanoptera

Synonyms

Physopoda

Etymology

Thysano–fringe; ptera–wings

Common Name

Thrips.

Characters

They are minute, slender, soft bodied insects. Mouthparts are rasping and sucking. Mouth cone is formed by the labrum and labium together with basal segments of maxillae. There are three stylets derived from two maxillae and left mandibles. Right mandible is absent. Hence, mouthparts are asymmetrical. Wings are either absent or long, narrow and fringed with hairs which increase the surface area. They are weak fliers and passive flier in wind is common. Tarsus is with one or two segmented. At the apex of each tarsus, a protrusible vesicle is present. Abdomen is often pointed. An appendicular ovipositor may be present or absent. Nymphal stage is followed by prepupal and pupal stages which are analogous to the pupae of endopterygote insects.

Importance

Most of the thrips species belong to the family Thripidae and are phytophagous. They suck the plant sap. Some are vectors of plant diseases. Few are predators. *e.g.* Rice thrips: *Stenchaetothrips biformis* is a pest in rice nursery.

Classification

This order is subdivided into two suborders.

Terebrantia	Tubulifera
1. Female with an appendicular ovipositor.	1. Ovipositor is absent.
2. Abdomen end is not tube like.	2. The abdomen end is tubular.
3. Wing venation is present.	3. Wing venation is absent.
4. Wings with hairs (Microtrichia)	4. Wings without hairs (Microtrichia)

Some Important Families

Family: Thripidae

1. Ovipositor curved downwards.
2. Wings narrow and pointed.
3. Antennae 6-9 segmented, 6th antennal segment is larger one.
4. Last abdominal segment of female is conical

Example

Onion thrips (*Thrips tabacci*)

Family: Phlaeothripidae

1. Antennae with 7-8 segments, 3rd antennal segment is larger one.
2. Head rounded.
3. Maxillary palp 2 segmented.

Example

Phlaeothrips, Liothrips, Haplothrips, etc.

Order: Neuroptera

Etymology

Neuro-nerve; ptera–wings.

Common Names

Lace wings, Ant lions, Mantispidflies, Owlflies.

Characters

They are soft bodied insects. Antenna is filiform, with or without a terminal club. Mouthparts are chewing type in adults. Wings are equal, membranous with many cross veins. They are held in a roof-like manner over the abdomen. They are weak fliers. Larva is campodeiform with mandibulosuctorial mouthparts. Pupa is exarate. Pupation takes place in a silken cocoon. Six out of eight Malpighian tubules are modified as silk glands. They spin the cocoons through anal spinnerets.

Classification

This order is subdivided into two suborders *viz.*, Megaloptera and Planipennia.

Megaloptera	Planipennia
1. Body size is large.	1. Small insects.
2. Short and filiform antennae.	2. Very long many segmented filiform, monoliform antennae.
3. Wings without pterostigma.	3. Wings with pterostigma and net like venation.
4. Larvae having biting mouth parts.	4. Larvae having piercing, suctorial mouthparts.

Important Family

Chrysopidae: (Green Lacewings, Goldeneyes, Stinkflies, Aphid Lions)

1. Insects with bright green coloured body, wing veins and iridescent eyes.
2. Antennae filiform, longer than the body.
3. Fore wings with Rs arising as one stem.
4. Larvae with long hooked hairs.
5. Ocelli absent.
6. Predators of aphids and mites.

Example

Green lace wing (*Chrysoperla* spp.)

Order: Diptera

Etymology

Di-two; ptera-wing

Common Names

True flies, Mosquitoes, Gnats, Midges,

Characters

They are small to medium sized, soft bodied insects. The body regions are distinct. Head is often hemispherical and attached to the thorax by a slender neck. Mouthparts are of sucking type, but may be modified. All thoracic segments are fused together. The thoracic mass is largely made up of mesothorax. A small lobe of the mesonotum (scutellum) overhangs the base of the abdomen. They have a single pair of wings. Forewings are larger, membranous and used for flight. Hind wings are highly reduced, knobbed at the end and are called haltere. They are rapidly vibrated during flight. They function as organs of equilibrium. Flies are the swiftest among all insects. Metamorphosis is complete. Larvae of more common forms are known as maggots. They are apodous and acephalous. Mouthparts are represented as mouth hooks which are attached to internal sclerites. Pupa is generally with free appendages, often enclosed in the hardened last larval skin called puparium. Pupa belongs to the coarctate type.

Classification

This order is sub divided in to three suborders.

Nematocera (Thread-horn)	Brachycera (Short-horn)	Cyclorrhapha: (Circular-crack)
Antenna is long and many segmented in adult. Larval head is well developed. Larval mandibles act horizontally. Pupa is weakly obtect. Adult emergence is through a straight split in the thoracic region.	Antenna is short and few segmented in adult. Larval head is retractile into the thorax. Larval mandibles act vertically. Pupa is exarate. Adult emergence is through a straight split in the thoracic region.	Antenna is aristate in adult. Larval head is vestigial with mouth hooks. Larval mouth hooks act vertically. Pupa is coarctate. The coarctate pupa has a circular line of weakness along which the pupal case splits during the emergence of adult. The split results due to the pressure applied by an eversible bladder ptilinum in the head.

Some Important Families

Family: Culicidae: (Mosquitoes)

1. Body slender, long.
2. Ocelli are absent.
3. Mouth parts are piercing type, antennae plumose in male and pilose in female.
4. Wings are fringed, legs are long and covered with scales.
5. Larval stages are aquatic with respiratory siphon, metapneustic.
6. Female are blood sucker and male nector feeder.

Example

Mosquitoes (*Anopheles* spp., *Aedes* spp., *Culex* spp.)

Family: Cecidomyiidae: (Gall Midges)

1. Minute flies, antennae are long with moniliform.
2. Ocelli are absent.
3. Legs long with short coax.

Example

Sorghum midge (*Cobtarina sorghicola*), Mango inflorescence midge (*Erosomyia indica*)

Family: Asilidae: (Robber Flies)

1. Long, bristly flies, with a horny piercing proboscis.
2. Adults are predaceous on dragonflies, hoppers, beetles and wasps.
3. Larvae predaceous or scavenger living in soil.

Example

Robber fly (*Hyperechia* spp.)

Family: Tabanidae: (Horse flies)

1. Body without bristles, flattened, metallic brown.
2. Antennae without style.
3. Eyes are large (holoptic in male) and dichoptic in female.
4. Female blood suckers and male nector feeder.
5. Larvae 12 segmented.
6. The adult flies are active blood suckers of cats, horses and man.

Example

Horse fly (*Tabanus rubidus*–causes *"surra"* disease of horse)

Family: Syrphidae: (Hover Flies, Flower Flies)

1. Wasp or bee like brightly blue, blackish flies.
2. Antennae with dorsal arista.

Example

Family: Tephritidae: (Fruit Flies)

1. Commonly occurring yellow or orange coloured fruit flies.
2. Small flies with conspicuously marked, a horny flattened oviscapt of visible length.
3. Wings with spots or bands.

4. Tarsi with spurs.

5. Arista dorsoventral.

6. Larvae muscoid type, phytophagous, developing in leaf, stem or fruit.

Example

Cucurbit fruit fly [*Bactrocera* (syn. Dacus) *cucurbitae*], Oriental fruit fly (*B. dorsalis*), Ber fruit fly (*Carpomyia vesuviana*)

Family: Drosophilidae: (Vinegar Gnats, Pomace Flies)

1. Small flies occurring about flowing sap, decaying fruits, wine, vat, vinegar and fermented products.

2. Eyes are red, antennae arista and 2 segmented.

3. Larvae 11 segmented body and each segment is covered with girlde or hook like structure.

4. Larvae feeds on decaying fruits and other products.

Example

Vinegar fruit fly (*Drosophila melanogaster*)

Family: Tachinidae: (Tachinid Flies)

1. Small or medium sized black or brown flies with hypopleural row of bristles.

2. They are endoparasites (larvae).

Example

Tachinid flies (*Tachina* spp)

Family: Muscidae: (True Fly)

1. Small or medium sized flies.

2. Eyes large, holoptic in male.

3. Mouth parts are sponging or piercing type.

4. Large labella with pseudotrachaea.

5. Larvae apodus and with distinct head.

Example

House fly (*Musca domestica*)

Family: Agromyzidae (Leaf Miner Flies)

1. Minute black or yellow flies.

2. Subcoasta vestigial, vibrissa present.

3. Larvae are leaf miner, tunnel into stem, roots, developing seeds, pods, etc.

Example

Pea leaf miner (*Phytomyza* spp.), Tur pod borer (*Melanagromyza phaseoli*).

Order: Coleoptera

Synonym

Elytroptera

Etymology

Coleo–Sheath; ptera-wing

Common Names

Beetles, Weevils

Characters

They are minute to large sized insects. Antenna is usually 11 segmented. Mouthparts are chewing type. Mandibles are short with blunt teeth at the mesal face in phytophagous group. In predators the mandibles are long, sharply pointed with blade like inner ridge. In pollen feeders, teeth are absent and the mandibles are covered with stiff hairs. Prothorax is large, distinct and mobile. Mesothorax and metathorax are fused with the first abdominal segment. Forewings are heavily sclerotised, veinless and hardened. They are called elytra. Forewings do not overlap and meet mid-dorsally to form a mid-dorsal line. It is not used for flight. They serve as a pair of convex shields to cover the hind wings and delicate tergites of abdomen. Hind wings are membranous with few veins and are useful in flight. At rest they are folded transversely and kept beneath the elytra. In some weevils and ground beetles, the forewings are fused and hind wings are atrophied. A small part of the mesothorax known as scutellum remains exposed as a little triangle between the bases of elytra. Cerci and a distinct ovipositor are absent. Metamorphosis is complete. Larvae are often called grubs. Pupae are usually exarate and rarely found in cocoons.

Importance

It is the largest order. It includes predators, scavengers and many crop pests. They also damage stored products.

Classification

This order is divided into two suborders.

Adephaga (Devourers)	*Polyphaga (Eaters of Many Things)*
Hind coxae immovably fixed to the meta sternum, completely dividing the 1st visible abdominal sternite which is more or less fused with the 2nd and 3rd wings usually with 2 *m-cu* cross veins defining an oblongum. Larva with a tarsus and 2 (rarely1) claws developed in the legs, mandibles without a molar area.	Wings with distal parts not spiratly coiled, never with 2 *m-cu* cross veins. Larval legs without a distinct tarsus and mearly always with 1 claw.

Important Families

Family: Cicindelidae: (Tiger Beetles)

1. Large, active predaceous beetles.
2. Eyes large and prominent.

3. Mandibles large, acutely tooted.
4. Legs long adapted for running.
5. Larvae with 6 ocelli.

Example

Cicindela cancellata

Family: Carabidae: (Ground Beetles)

1. Live under stones, ground, barks, etc.
2. Oval, broad and dorso-ventrally flattened beetles with bright colouration.
3. Legs adapted for running and digging.
4. Antennae conspicuous and anteriorly projected.

Example

Anthia spp., *Calosoma* spp., *Amara* spp.

Family: Dytiscidae: (True Water Beetles, Predaceous Diving Beetles)

1. Large, dorsoventrally flattened, oval, beetle.
2. Antennae filiform.
3. Hind legs natatorial (for swimming).
4. Forelegs of males with adhesive peds and suckers.
5. Elytra store air beneath them.
6. Larvae with long sickle shaped mandibles.

Example

Cybister tripunctatus

Family: Gyrinidae: (Whirligig Beetles)

1. Body oval, elyptical, flattened.
2. Antennae very short, stout and inserted beneath the frons.
3. Forelegs long prehensile and dilated tarsi with suckers in males.

Example

Gyrinus marinus

Family: Coccinellidae: (Lady Bird Beetles)

1. Adult small, oval or spherical convex, brightly coloured with red, black, yellow, brown spots.
2. Mostly carnivorous feeding on aphids, coccids, etc.
3. Head concealed in pronotum.
4. Tarsi 4 segmented, 3rd segment lodged into a cavity of 2nd segment.

Example

Lady bird beetle (*Coccinella septempunctata*)

Family: Scarabaeidae: (Scarabs, Dung Beetles)

1. Small or large, oval or spherical or convex body.
2. Head and prothorax with horns.
3. Antennae 8-10 segmented.
4. Hind tibia with spur.
5. Mandibles membranous.
6. They feed upon organic matter.

Example

Rhinocerous beetle (*Oryctes rhinocerous*)

Family: Melolonthidae

1. Body black, brown, green, blue or metallic coloured.
2. Labrum highly sclerotized.
3. Male with horny processes on head and with long antennal club.
4. Abdomen with 6 tergites.
5. Adults and larvae feed on roots and underground parts of plants.

Example

White grub (*Holotrichia spp.*)

Family: Anobiidae: (Wood Worms, Wood Borers)

1. Small, cylindrical beetles.
2. Head concealed in prothorax.
3. Antennae very short in larvae and long in adult and serrate, clavate or pectinate type.
4. Coxal cavities excavated.
5. Hind coxal groove receives femora.
6. They mostly destroy furniture.

Example

Cigarette beetle (*Lasioderma serricorne*), or dung store beetle (*Stegobium panaceum*).

Family: Bostrychidae: (Grain Borers)

1. Insects are small or large black or brown cylindrical body.
2. Head is hypognathous.
3. Antennae clubbed and less than 11 segments.
4. Fore coxae with cavities facing back.

Example

Rice grain borer (*Rhizopertha dominica*)

Family: Bruchidae: (Pulse Beetles, Seed Beetles)

1. Adult beetles are small, short, stout body covered with small setae or scales and are pale grey or brown in colour.
2. Antennae short clavate, serrate or pectinate.
3. Elytra striated, short and do not cover caudal end of abdomen.
4. Head hypognathous and snout like proboscis.
5. Tarsi 5 segmented.

Example

Pulse beetle (*Callosobruchus spp.*)

Family: Tenebrionidae: (Meal Worms)

1. Small, elongate. Flattened, black, brown coloured beetles.
2. Ground beetles are apterous or with immovable elytra and vestigial wings.
3. First three abdominal segments are fused.

Example

Red flour beetle (*Tribolium castaneum*).

Family: Dermestidae

1. Small or medium, oval, red or black coloured insects, body covered with fine hairs or scales.
2. Adults with dorsal ocelli.
3. Antennae short and clubbed.

Example

Khapra beetle (*Trogoderma granarium*)

Family: Buprestidae

1. Large or small mostly brilliantly metallic coloured, flattened or cylindrical bodied insects.
2. Elytra with scutellar stria.
3. Metasternum with a distinct transverse suture.
4. Antennae 11 segmented, sharp, serrated.
5. Thorax and abdomen firmly fused.
6. Beetles are active fliers.

Example

Jute stem buprestid (*Agrilus spp.*)

Family: Cerambycidae: (Longicorn Beetles)
1. Body long, cylindrical and with attractive colouration.
2. Antennae very long at least 2/3 of body length.
3. Tibia with 2 spurs.
4. Tarsi 5 segmented.
5. Stridulatory organs on hind margin of prothorax.
6. Grubs are apodus.

Example

Orange trunk borer (*Anoplophora versteegi*), Mango stem borer (*Bactocera rufomaculata*).

Family: Curculionidae: (Weevils, Snout Beetles)
1. Head modified into a long rostrum.
2. Antennae often clubbed, geniculate or clavate.
3. Mouthparts at the tip of rostrum.
4. Labrum wanting, palps are reduced.
5. Rostrum is used by female as a drilling instrument to bore hole for placing the eggs. It is more developed in female then male.
6. Larvae apodous.

Example

Rice weevil (*Sitophilus oryzae*)

Family: Chrysomelidae
1. The beetles are of moderate size, oval, convex, black, blue, green, etc.
2. Antennae widely separated.
3. Dorsal surface of body is greatly plane and shining with metallic colouration.
4. Eyes small.
5. Prothorax laterally, margined and narrow.
6. Elytra cover entire body.
7. Third tarsal segment is bilobed and median groove is occupied by tibiae and swollen hind femora.
8. Legs well adapted for jumping and running.

Example

Red pumpkin beetle (*Raphidophpalpa foveicollis*)

Order: Hymenoptera

Etymology

Hymen–membrane; ptera–wings.

Hymeno–god of marriage; ptera–wings,

(Marriage/union of fore and hind wings by hamuli)

Common Names

Ichneumonflies, Ants, Bees, Wasps, Parasitoids.

Characters

Mouthparts are primarily adapted for chewing. Mandibles are very well developed. In bees, both labium and maxillae are integrated to form the lapping tongue. Thorax is modified for efficient flight. Pronotum is collar like. Mesothorax is enlarged. Metathorax is small. Both prothorax and metathorax are fused with mesothorax. Wings are stiff and membranous. Forewings are larger than hindwings. Wing venation is reduced. Both forwings and hindwings are coupled by a row of hooklets (hamuli) present on the leading edge of the hindwing. Abdomen is basally constricted. The first abdominal segment is called propodeum. It is fused with metathorax. The first pair of abdominal spiracles is located in the propodeum. The second segment is known as pedicel which connects the thorax and abdomen. Abdomen beyond the pedicel is called gaster or metasoma. Ovipositor is always present in females. It is variously modified for oviposition or stinging or sawing or piercing plant tissue. Metamorphois is complete. Often the grub is apodous and eucephalous. Larva is rarely eruciform. Pupa is exarate and frequently enclosed in a silken cocoon secreted from labial glands. Sex is determined by the fertilization of the eggs. Fertilized eggs develop into females and males are produced from unfertilized eggs. Males are haploid and females diploid.

Classification

This order is subdivided into two suborders.

Symphyta	*Apocrita*
1. Abdomen is broadly joined to the thorax.	1. Abdomen is petiolated.
2. Larva is a caterpillar and belongs to eruciform type	2. Larva is a grub and it belongs to apodous eucephalous type.
3. Stemmata are present.	3. Stemmata are absent.
4. Both thoracic and abdominal legs are present.	4. Legs are absent.
5. Ovipositor is saw like.	5. Ovipositor is not saw like
6. Behavioural sophistication is less.	6. Behavioural sophistication is more.
7. They are phytophagous	7. They are generally parasitic

Some Important Families

Family: Tenthredinidae: (Sawflies)

1. Body less than 1.2 cm long.
2. Antennae filiform and 9 segmented.
3. Ovipositor with serrated plates and both plates move alternately.

4. Larvae is caterpillar like feed on leaves or galls and having 6 to 8 pro legs.
5. Tarsi 5 segmented.
6. Pupation inside cocoon under soil.

Example

Mustard saw fly (*Athalia lugens proxima*)

Family: Ichneumonidae: (Ichneumon flies)

This is the largest family of the order Hymenoptera.

1. Body of insects is large, slender, black or yellow brown.
2. Antennae is filiform, never geneculate but long and having 16 or more segments.
3. Larva is a parasite.
4. Ovipositor is long.
5. Larvae apodus.

They are parasites on larvae of coleopteran, Lepidoptera, diptera and hymenopterous insects.

Example

Isotima spp., on rice yellow borer, *Scirpophaga incertulas*, *Xanthopimpla punctata* on *Chilo partellus*.

Family: Braconidae: (Braconid Wasps)

1. Small, stout insects.
2. Hind wings with cross vein *r-m* and forewings without *2m-cu* cross vein.
3. Larva body is 13 segmented and endoparasitic, pupation within cocoon inside or outside of the body of host.
4. They are generally parasite of lepidopteran caterpillars.

Example

Apanteles, *Bracon* are common parasite of *Chilo partellus* and hairy caterpillar. *Bracon brevicornis* is a parasite of coconut pest *Nephantis serinopa*.

Family: Chalcididae: (Chalcid Wasps)

1. Insects are minute to small, metallic coloured, primary or secondary parasite of eggs, larvae and pupae of lepidopteran, hemipteran and dipteran insects.
2. Mandibles stout with 3-4 teeth.
3. Antennae geniculate 2-13 segmented.
4. Wings not folded and longitudinally.

Example

Meyeriella indica, *Brachymeria* sp.

Family: Eulophidae: (Pupal Parasitoids)

1. Tarsi with 4 or 5 segmented.
2. Forewings narrower with pubescence not in rows or line.
3. The insects are pupal parasitoids.

Example

Tetrastichus sokolowskii is a pupal parasitoid of *Plutella xylostella*.

Family: Trichogrammatidae: (Egg Parasitoids)

1. Minute insects.
2. Tarsi 3 segmented.
3. Forewings broad, pubescence in rows or lines.

Example

Trichogramma chilonis is a common egg parasitoid of lepidopteran pests.

Family: Vespidae: (Yellow Jackets, Hornets)

1. Red and yellow body with black markings.
2. Antennae generally geniculate and 12 segmented in female and 13 segmented in male.
3. Forewings are longitudinal, folded.
4. Hindwings without anal lobes.
5. Mandibles are strongly dentate.
6. Abdomen is sessile, conical with well developed terminal ovipositor or well developed stinfs.
7. Trimorphic-queen, workers and male, males are parthenogenetically produced. Workers are sterile females.

Example

Vespa orientalis, Vespa rufa, etc.

Family: Formicidae: (Ants)

1. Male and female (queen) are primary winged cast and workers are wingless caste.
2. Mouthparts are mandibulate, antennae geniculate 4-13 segmented.
3. Labrum vestigial.
4. Wings with 1-2 cubital and one dorsal cells.
5. Polymorphism-29 different forms in a colony of a single species. These forms are (*i*) workers-they are sterile apterous female (*ii*) soldiers- they are modified workers with large head and mandibles (*iii*) queen-these are fertile females and, (*iv*) king-these are fertile males.

Example

Ants (*Formica* sp.).

Family: Apidae: (Honey Bees)

1. Insects are small, black or brown, densely hairy, golden brown coloured.
2. Hind tibia without spurs.
3. Forewings with 3 submarginal and 1 redical cell.
4. Hind tibia flattened and fringed with pollen collecting apparatus.
5. Stings present on last abdominal segment.
6. They prepare honey comb on trees and secreate wax by dermal glands of worlers.
7. Castes are well developed-queen, workers and drone.

Example

Rock bee (*Apis dorsata*), Indian bee (*Apis indica*, Stingless bee (*Mellipona spp.*)

Family: Megachilidae: (Leaf Butter Bees)

1. Moderate, densely hairy solitary bees with black coloured body.
2. Labrum long and narrow.
3. Labial palpi 4 segmented.
4. Forewings with alike 2 submarginal cells.
5. Mandibles stout, long and dentate.

Example

Megachile spp.

Order: Lepidoptera

Synonym

Glossata

Etymology

Lepido–scale; ptera–wings.

Common Names

Moths, Butterflies, Skippers

Characters

Body, wings, appendages, are densely clothed with overlapping scales, which give colour, rigidity and strength. They insulate the body and smoothen air flow over the body. Mouthparts in adults are of siphoning type. Mandibles are absent. The galeae of maxillae are greatly elongated and are held together by interlocking hooks and spines. The suctorial proboscis is coiled up like a watch spring and kept beneath the head when not in use. Wings are membranous and are covered with overlapping pigmented scales. Forewings are larger than hindwings. Cross veins are few. Wings are coupled by either frenate or amplexiform type of wing coupling. Larvae are polypod-eruciform type. Mouthparts are adapted for chewing with strong mandibles. A group of lateral ocelli is found on either side of the head. The antenna is short and three segmented. There are three pairs of five segmented thoracic legs ending in claws. Two to five pairs of fleshy unsegmented prolegs are found in the abdomen. At the bottom of the proleg, crochets are present. Pupa is generally obtect. It is either naked or enclosed in a cocoon made out of soil, frass, silk or larval hairs.

Classification

This order is generally divided into Rhopalocera (butterflies) and Heterocera (moths) simply because of the difference in general appearance of the two groups. Some have divided the order into Macro and Micro Lepidoptera simply on the basis of size. According to Imms (1961) the order may be divided into following three suborders.

(*i*) Zeugloptera

(*ii*) Monotrysia

(*iii*) Ditrysia

Zeugloptera	Monotrysia	Ditrysia
1. Adults with functional wings	1. Mandibles are vestigial.	1. The female has a copulatory pore on the 8th sternite and an egg pod on the 9th sternite.
2. Lacinia is well developed and galea is not forming a proboscis.	2. Lacinia is absent.	2. The forewings are without jugam lobe.
3. The female genital chamber opens in the main oviduct.	3. Female with one or two genital opening behind the 9th abdominal sternite.	3. The hindwings are often with frenulum and reduced venation. * This is the largest suborder and includes all the families of agricultural importance.

Some Important Families

Family: Nymphalidae: (Brush Footed or Four Footed Butterflies)

1. Butterflies brightly coloured with clubbed shaped antennae.
2. Forelegs reduced and vestigial, folded on the thorax.
3. Tibia short, covered with long hairs.
4. Larvae with spines and a pair of anal processes developed back wordly.

Example

Painted lady butterfly (*Vanessa cardui*)

Family: Lycaenidae: (Blues, Coppers, Hair Streaks)

1. Fore legs normal but without claws.

2. Hindwings with tail lobes, but without humeral vein.
3. Antennae encircled by a ring of white scale.
4. Larvae with both ends tapering with scales.

Example

Pomegranate fruit fly (*Virachola isocrates*), Guava fruit borer (*Rapala varuna*)

Family: Papilionidae: (Swallow Tails)

1. Iridescent black, blue, orange, spotted, yellow orange coloured butterflies.
2. Antennae clubbed.
3. Large butterflies with tail like prolongation of hindwings.
4. R vein of forewing with 5 branches.
5. Larvae without setae.

Example

Lemon butterfly (*Papilio demoleus*)

Family: Pieridae: (Whites and Sulphurs)

1. Forewings normal with bifid claws.
2. Fire tibia without pads.
3. Antennae clubbed.

Example

Cabbage butterfly (*Pieris brassicae*)

Family: Arctiidae: (Tiger Moths)

1. Body stout, broad, hairy white orange or yellow coloured.
2. Hindwing with two anal veins.
3. Forewing with tufts of scales.
4. Tympanum present on the metepimeron.
5. Larvae large covered with long hairs densely.

Example

Red hairy caterpillae (*Amsacta albistriga*)

Family: Bombycidae: (Silk Worm Moths)

1. Male and female with pectinate antennae.
2. Moths robust and large.
3. Probosis vestigial or absent.
4. Tympanum absent.
5. Larvae large, wrinkled, white silk worm.

Example

Mulberry silk worm (*Bombyx mori*)

Family: Cochlididae: (Slug Caterpillar)

1. Medium sized moth with stout body.
2. Wings heavily and loosely scaled.
3. Antennae large, bipectinate in male, smaller, simple in female.

Example

Tea slug caterpillar (*Thosea cerrina*)

Family: Pyralidae: (Grass Moths)

1. Small moths of cryptical colouration.
2. Forewing narrow and elongated.
3. Anal area larger.
4. Labial palp porect.

Example

Sorghum stem borer (*Chilo partellus*), Sugarcane shoot borer (*C. infuscatellus*)

Family: Gelechiidae: (Paddy Moth)

1. Small, cryptically coloured moth.
2. Forewings narrower than hind wings.

Example

Groundnut leaf miner (*Stomopteryx subsecivella*), Potato tuber moth (PTM) (*Phthorimaea operculella*), Cotton pink bollworm (*Platyedra gossypiela*), Rice grain moth (*Sitotroga cerealella*).

Family: Geometridae: (Loopers)

1. Body slender shaped.
2. Wings large, keeping horizontally at rest.
3. Well developed proboscis.
4. Frenulum present.
5. In larvae, prolegs are absent and exhibit looping movement.

Example

Hyposidra sesbania, a pest of castor

Family: Lymantridae: (Tussock Moths)

1. Moderate sized moths.
2. Antennae bipectinate in male.

3. Ocelli absent.

4. A tuft of hairs at anal end of female is present.

5. Female mostly apterous.

6. Larvae densely haired with hairy tufts.

Example

Tussock moth (*Hemerocampa* spp.).

Family: Noctuidae: (Noctua Moths)

1. Wings with areola, cryptic and dull colouration.

2. *Sc* free at base.

3. Proboscis present.

4. Hind tarsi slender, tibial spurs long and frenulun present.

Example

Gram pod borer (*Helicoverpa armigera*), Tobacco caterpillar (*Spodoptera litura*). Castor semilooper (*Achaea janata*) and Cut worm (*Agrotis* spp.)

Family: Pyralidae: (Grass Borers)

1. These are large pyralid moths.

2. Proboscis vestigial.

3. Female moths larger than male moth.

4. Hindwings fringed.

Example

Rice yellow stem borer (*Scirpophagous incertulas*), Sugarcane top borer (*Scirpophaga nivella*).

Family: Plutellidae

1. These are small moths.

2. Maxillary palpi short and correct.

3. The larvae feed in a slight web in leaves.

Example

Diamondback moth (*Plutella xylostella*).

Family: Phycitidae

1. Small, delicate moths.

2. Forewings elongated.

3. Hindwings broad with a pectin of hairs dorsally.

4. Male with swollen antennal pedicel.

Example

Indian meal moth (*Plodia interpunctella*), Indian flour moth (*Cadra cautella*), Pulse pod borer (*Etiella zinckenella*), Brinjal stem borer (*Euzophera perticella*).

Family: Saturniidae: (Moon Months, Giant Silk Worm Moths)

1. Very large coloured moths.
2. Wings with transparent fenestrae or eye spots.
3. Probocis aborted (vestigial or absent).
4. Antennae pectinate in male and female, long in male.
5. Hindwings with tail like processes.
6. Larve large, smooth with scoli, with spiny tubercules.

Example

Oak silk moth (*Antheraea pernyi*), Eri silk worm (*Philosamia ricini*).

Family: Sphingidae: (Hawk Moths, Sphinx Moths, Horn Worms)

1. Large, robusts, flying moths.
2. Antennae hooked apically.
3. Proboscis extremely long.
4. Forewings larger than hindwings.
5. Larvae with segmented body.

Example

Death's head moth (*Acherontia* spp.)

6

Insect Ecology

The term ecology is derived from the Greek term "oikos" meaning "house" combined with "logus" meaning "the science of" or "the study of". Thus literally ecology is the study of earth's household comprising of the plants, animals, microorganisms and people that live together as interdependent components. The term ecology was coined by a German biologist Ernst Haekel (1869).

Definition of Ecology

☆ Ecology can be defined as the science of plants and animals in relation to their environment.

☆ Webster's dictionary defines ecology as "totality of pattern of relation between organisms and their environment."

☆ Eugene P. Odum defined ecology as "the study of organisms at home"

☆ Insect Ecology may be defined as the understanding of physiology and behaviour of insects as affected by their environment.

Division of Ecology

(a) Autecology

This is known as ecology of individuals, where we study the relation of individual species to its environment. Thus, in autecological studies, the assessment of the intrinsic capacities of organisms which enable them to cope with environmental conditions is important.

(b) Synecology

Under the natural conditions the organisms live together as a natural group affecting each others' life in several ways. Thus, more complex situations exist where the units of study, instead of individual organism, are groups of organisms known as community. Such an approach, where units are groups of organisms is called synecological approach. The synecology deals with population ecology, community ecology, biome ecology and ecosystem ecology.

Branches of Ecology

1. *Habitat ecology*–deals with living places of organisms, *e.g.* Marine ecology, freshwater ecology. Estuarine ecology and terrestrial ecology–a. forest ecology b. grassland, desert, and crop plants ecology
2. *Community ecology*–study of distribution of the animals and plants
3. *Population ecology*–deals with the study of the manner of growth, structure and regulation of population of organisms (Dem ecology)
4. *Evolutionary ecology*–study of evolution and speciation in animals
5. *Taxonomic ecology*–Ecology of different taxonomic groups of living organisms, *e.g.* Insect ecology, Parasitology, Human ecology, etc.
6. *Applied ecology*–study of the application of the ecological concepts to the human needs. This can be classified as:
   (*i*) Wildlife management
   (*ii*) Insect control
   (*iii*) Forest conservation
   (*iv*) Agriculture
   (*v*) Horticulture, etc.
7. *Ecosystem dynamics*–Ecological study of the soil formation, nutrient cycle, energy flow and productivity.
8. *Production ecology*–study of gross and net production of different ecosystem.
9. *Physiological ecology (Ecophysiology)*–study of the survival of the population.
10. *Ecological energetics*–study of the energy conservation and its flow in the organism with in the ecosystem.
11. *Chemical ecology*–study of the adaptations of animals to particular chemical substances.
12. *Pedology*–Soil properties and influence on organisms.
13. *Ethology*–Interpretation of animal behavior under natural condition.
14. *Sociology*–Ecology and Ethology of mankind.
15. *Systems ecology*–modern branch concerned with the analysis and understanding of the function and structure of ecosystems employing mathematical model, computer science, etc.

Ecology Related Terminology

1. Habitat is the place where the organism lives.
2. Population denotes groups of individuals of any kind of organism. Insect populations are groups of individuals set in a frame that is limited in time and space.

3. Community in the ecological sense includes all the populations of a given area. Community can also be defined as interacting 'web' of populations where individuals in a population feed upon and in turn are fed upon by individuals of other populations.

4. Ecosystem

   ☆ Ecosystem or ecological system is the functioning together of community and the nonliving environment where continuous exchange of matter and energy takes place.

   ☆ In other words, ecosystem is the assemblage of elements, communities and physical environment.

   ☆ Ecosystem is the ultimate unit of study in ecology as, they are composed of living organisms and the nonliving environment.

5. Biosphere is the term used for all of the earth's ecosystems functioning together on the global scale.

Agroecosystem is largely created and maintained to satisfy human wants or needs. It is not a natural ecosystem but is man made. Agroecosystem is the basic unit of pest management–a branch of applied ecology.

Balance of Nature

Balance of Nature is defined as the natural tendency of plant and animal population resulting from natural regulative processes in an undisturbed ecosystem (environment) to neither decline in numbers to extinction nor increase to indefinite density.

In unmanaged ecosystems, a state of balance exists or will be reached, *i.e.* species interact with each other and with their physical environment in such a way that on average, individuals are able to replace themselves. Each species in the community achieves a certain status that becomes fixed for a period of time and is resistant to change which is termed as balance of nature.

When man begins to manage creating new ecosystem (agroecosystem) where natural ecosystem existed previously, the balance is altered. The exceptionally strong forces react in opposition to our imposed change toward a return to the original system (*e.g.* outbreak of a pest is one of the forces). So, insect pests are not ecological aberrations. Their activities counter wants and needs of human populations.

Factors Influencing Pest Populations

The pest populations have a tendency to fluctuate as a result of their inherent characteristics as influenced by the environmental factors. The degree of influence of various environmental factors determines the magnitude of increase or decrease in numbers of a pest population. The rate of change in a pest population is determined by the fecundity, speed of development and survival among its numbers. The environmental factors that favour fecundity or speed of development and are not inimical to survival promote increase, and those having a reverse influence cause a decline in numbers. The same factor may be favourable in case of one population but may become unfavourable for another. It is thus necessary, to consider the influence of various factors with respect to a particular pest population.

The environmental factors may be grouped into two main categories, *i.e.* abiotic and biotic. Among the abiotic factors, it is primarily the physical factors such as temperature, moisture and light that have a direct influence on the populations of insect pests. These factors also influence the pest populations indirectly by modifying the biotic factors. The biotic factors include food and populations, primarily the natural enemies of pest populations.

Abiotic Factors (Density Independent Factors)

The climatic factors exercise a dominating influence on the development, longevity, reproduction and fecundity of insect pests. It is well known that densities of pest populations fluctuate with the prevailing weather conditions such as temperature, moisture, light and wind. Extremes of temperature, humidity or rainfall cause mortality among the pest and its natural enemies. The chances of an insect population to survive and reproduce first, increase and then decrease, as the population is exposed to unfavourable low range through the optimum into the unfavourable high range.

Temperature

The insects are all *poikilothermic, i.e.* they have no precise mechanism for regulating the temperature of their bodies. Their body temperature, therefore, follows more or less closely that of the surrounding medium. There is a fairly well defined favourable range of temperature for every insect species within which it is able to survive. This temperature range is determined by the prevailing temperature at which the normal physiological activities of the insect takes place. This narrow band of temperature has been temperatures beyond the favourable range, whether low or high, may retard growth and development of the insect or may even cause its death. The upper lethal limits are usually between 40° and 50°C, but some insects, such as stored product and desert insects, can withstand temperatures in the neighborhood of 60°C. The lower lethal limits vary widely and may lie below the freezing point of water.

The departure from the optimal range on both sides is tolerated to some extent, depending upon the physiological adaptations of the conceived populations. The reaction to changed temperature depends upon the suddenness of the change. In case of a gradual change, the insects become conditioned or acclimatized. The rate of acclimatization is dependent on the duration for which they are conditioned. The total heat required for the completion of physiological processes in the life-history of a species is considered as thermal constant. Within the favourable range, the thermal constant is not affected by the level of temperature.

Exposure to lethal low or high temperatures may result in instant killing and even the survivors may fail to grow and reproduce normally. The duration of exposure to such a condition is important and the harmful effects of exposure to sub-lethal temperatures may be manifested at some later critical stage, *i.e.* ecdysis and pupation. Some of the insect species when exposed to extremes of temperatures beyond the favourable range may become dormant, *i.e.* undergo hibernation or aestivation, which are reversible processes as the individuals may resume activity on being exposed to favourable temperature.

The insect populations can be grouped into three categories according to their responses to low temperature:

1. Those which cannot survive for any considerable time if the temperature falls below the lower limit of the favourable range. They cannot become dormant and hence must either develop or die. Such species have originated from the tropical or subtropical climates, *e.g.* locust.

2. Species which have a stage in the life-cycle adapted to survive exposure to low temperature and the other stages resemble those in the above category in lacking the capability to become dormant at low temperature. Such species have originated from the temperate climates, *e.g. Helicoverpa* sp. and *Agrotis* sp.

3. This group has a diapause stage and they have also originated from the temperate climates, *e.g.* many lepidopterous borers.

The species adapted to live in temperate climates, hibernate in a particular developmental stage and frequent this very stage and none other is capable of undergoing diapause. Some other species depend for their survival during winter on the insulating protection of the hibernacula such as debris, soil, plant remnants, snow and ice, and also on their ability to withstand exposure to low temperature through under-cooling. The combination of these two protective mechanisms prevents mortality, but in the case of incomplete protection, exposure to severe cold period may lead to eradication of the insect population. This may be more relevant in the case of introduced natural enemies.

Moisture

A constant supply of moisture is essential for metabolic reactions as well as for the dissolution and transport of salts. The water content in insects varies from less than 50 per cent to more than 90 per cent of the total body weight. Variation occurs between different species and even between different stages in the life-cycle of the same species. Soft bodied insects such as caterpillars tend to have comparatively large amount of water in their tissues, whereas many insects with hard bodies tend to have somewhat lesser amounts. Active stages commonly have higher water content than dormant stages. The range of moisture required is not so broad as in the case of temperature. Most of the insect species are capable of maintaining their body water at fairly constant levels while living under varying conditions. For most of the species, food in the shape of plant or its products is the source of water, and they have the adaptation to cope with conditions of excessive moisture and shortage of water. The other sources of gaining moisture are direct drinking of water or absorption through the integument. The loss of water from the body is prevented by insect cuticle having a waxy layer. A number of adaptations-morphological, biological and physiological-in nature help insect populations in overcoming unfavourable conditions of excessive moisture or acidity. As in temperature, the phenomenon of humidity preferendum also operates in insects, and it helps insects to congregate in suitable places. The humidity preference is influenced by the prevailing temperature. Most adverse effects of moisture are due to its scarcity or absence. Exceptionally dry air may prove lethal because some insect species may not survive the loss of even a small percentage of body water for a long time. Those in aestivation or diapause lose a large proportion of body water without adverse effects. Larva of an African chironomid midge, *Polypedilum vandesplanki*, can tolerate dehydration and suspension of metabolism for several years.

The exposure to excessive moisture can prove harmful to insect populations in the following ways:

1. By adversely affecting the normal development and feeding activity of insects.
2. By encouraging disease-causing micro-organisms such as fungi, bacteria, viruses and mycoplasma, and thereby causing mortality among insects.
3. Excessive moisture in insect body during winter reduces its capability to withstand exposure to low temperature and thus leads to an adverse effect on its cold hardiness.

Moisture also influences the speed of development and fecundity of most insect species. In some species, these activities are accelerated by excessive moisture, while in others they get retarded. A Study shows that sugarcane black bug *Cavelerius excavatus* (Distant), multiplied more rapidly at high humidity (90 per cent R.H.), whereas relative humidity above 70 per cent was harmful for multiplication of cotton jassid, *Amrasca biguttula biguttula* (Ishida). For eggs and pupae of spotted bollworm, *Earias insulana* (Boisduval), both very high (around 100 per cent) and low (<40 per cent) relative humidity were not conducive for development of these stages.

Light

Light is an essential ecological factor for many biological processes such as orientation or rhythmic behaviour of insects, bioluminescence, periodicities of occurrence and periods of inactivity. Light acts as a token stimulus by enabling insects to regulate and synchronize their life-cycles with the change in seasons. Unlike temperature, light is a non-lethal factor and it has specific direction in its flow. A characteristic feature of light is its quantitative shift from a minimum to a maximum and vice-versa in a short time. The properties of light that influence insect life are intensity or illumination, quality or wavelength and duration of light hours or photoperiod.

The insects orientate to the source of light and thus reach the right place at the right time. Such phototactic behaviour of insects is altered and modified by a number of factors such as temperature, humidity or moisture and food. Several species of moths, leafhoppers and beetles are attracted to light at dusk or during the night, and this behaviour has been used extensively through light traps for observing brood-emergence or fluctuations in their populations. Direct sunshine may injure or kill an exposed insect largely because of heat and desiccation.

Photoperiodism influences the motor activity rhythms of insects such as locomotion, feeding, adult emergence, mating and oviposition, and also moulting and growth in some species. The reproductive cycle in most of the temperate-zone insects is so tuned that they reproduce only during favourable periods and the remaining period is passed in diapause state. The induction of diapause, a genetically determined state of suppressed development and the manifestation of which is induced by environmental factors, is influenced by photoperiod in most of the lepidopterous insects. Photoperiodic responses in insects also influence polymorphism, ecological adaptations and phenological synchronization with the sources of food.

Oxygen and Carbon Dioxide

Insects can tolerate a wide range of oxygen and carbon dioxide. Some insects can survive several days in the absence of oxygen by reducing their metabolic rates and utilizing the oxygen in their tissues. Excessive carbon dioxide causes varied reactions in different insects. Some insects can live in an atmosphere with high carbon dioxide for several days. However, an excess of this gas in the atmosphere causes growth retardation in many insects. If the environment is high in carbon dioxide, the spiracles of insects tend to remain open, which may lead to excessive water loss.

Air Currents

Air currents are of great value to insect displacement and, therefore, affect population changes by influencing the numbers into or out of an area. Most insects will not undertake flight when the speed of wind exceeds the normal flight speed. However, insects are rarely blown about at random but have evolved such patterns of behaviour which enable them to exploit the wind to achieve their migratory needs.

Strong flying insects tend to fly with the winds during migrations and are displaced long distances as in case of spruce budworm moths. Many insects which are weak fliers also have specific behaviour patterns enhancing their opportunity to migrate to specific areas with the help of wind. The air currents carry aphids, leafhoppers and scale insects to far-off places and thus are instrumental in their dispersal. Air currents may be directly responsible for the death of insects. Firstly, severe wind coupled with heavy rains may cause mortality. Secondly, movement of air above a surface where evaporation is occurring (*e.g.* insect cuticle) increases the gradient of water vapour concentration and hence tends to increase the rate of evaporation.

Water Currents

Water currents often determine which species of insects would inhabit a particular area. The various genera of mayflies may be classified into still- and rapid-water forms. The legs and bodies of these insects are appropriately adapted. Black fly larvae fasten themselves to stones or other stationary material in the water. Caddisflies attach their cases to submerged objects. The mosquito larvae are unable to survive in moving water. Another important feature of currents in the aquatic environment involves the circulation of dissolved gases, salts and nutrients. For example, caddisfly and mayfly larvae may be found under conditions of relatively high oxygen concentration, and midge and black fly larvae at somewhat lower concentration, while certain mosquito and other fly larvae at very low concentrations.

Edaphic Factors

Edaphic factors include the structure, texture and composition of soil along with its physical and chemical characteristics. Each soil has a distinctive flora as well as fauna of fungi, bacteria, algae, protozoa, rotifers, nematodes, mollusks, arthropods, etc. Some of these organisms help in maintenance of soil fertility through nitrogen fixation while others are responsible for return of the essential elements back to the soil by decomposition of dead organic matter. In humus formation, earthworms, millipedes, dipterous larvae, slugs and snails play important role in breaking up and division of litter. Several properties of the soil like texture, moisture, drainage, chemical composition and physiography (topography) affect the distribution and abundance of insects. Soil texture varies from hard-packed clays to loose sands. Few insects dwell in hard-packed types, because they are unable to push or dig their way through them. The loams allow digging and burrowing operation and are usually favourable in their characteristics like moisture, drainage and organic matter. The cutworm, *Agrotis ipsilon* (Hufnagel), larvae live in soil of fairly light texture in which they move around freely in response to daily or seasonal temperature and moisture changes. Drainage and texture together exert considerable influence on the distribution of insects which pass part of their life in soil. The wireworms in wet arid land become important pests of potato, onion, lettuce and many other crops grown in irrigated fields.

Chemicals naturally present in the soil affect both the abundance and distribution of phytophagous insects. Deficiencies of mineral elements, resulting in similar plant deficiencies, inhibit the growth of some insects. Nitrogen deficiency lowers the productivity of some species of insects but results in outbreak of others.

Major topographic factors which affect biogeography of insects include height of mountain chains, steepness of the slope, directions of mountains and valleys, and exposure of the slope. These features also affect the climate of an area, thus influencing the distribution of certain insects. The mountain range and large bodies of water such as seas act as physical barriers to the spread of insects.

Biotic Factors (Density Dependent Factors)

Under natural conditions, organisms live together influencing each other's life directly. Such vital processes such as growth, nutrition and reproduction depend upon the interaction between the individuals of the same species (intraspecific) or between those of different species (interspecific). The interdependencies between insects themselves as well as between insects, other animals and plants exist throughout the life, or are casual and temporary. However, interdependency may exist between species which are taxonomically widely different such as between insects and bacteria, screw-worms and cattle, etc. The relationship between species may be beneficial to both, harmful to both or beneficial or harmful to one and neutral to the other. The most important biotic components of the insect life-system are natural enemies and food.

Natural Enemies

The natural enemies of insect pest populations include predators, parasites/parasitoids and disease causing micro-organisms such as fungi, bacteria, viruses and rickettsiae. The natural enemies are also influenced by various environmental factors such as weather and hyper-parasitism. The degree of influence of various natural enemies on pest population would thus vary. The abundance of predators influences the abundance of their prey in field conditions. Predators respond to an increase in prey population density through numerical response (increase in the density of predators in a given area) and functional response (increase in consumption by individual predators).

Weather is known to determine the effectiveness of a natural enemy. For example, in the control of Apple woolly aphid, with *Aphelinus mali* (Haldeman), the prevalent temperature in spring would determine the effectiveness of the parasitoid. Both the host and the parasitoid remain dormant during winter but the activity of host is resumed at a slightly lower temperature, and thus it is able to breed in the absence of the parasitoid in the early spring. In such a situation, the parasitoid may fail to press heavily on the host population as compared with the one when relatively higher temperature prevailed during early spring. Similarly, the larval parasitoid, *Apanteles flavipes* (Cameron), of maize borer after successful overwintering in the host larvae, resumes its activity in the spring earlier than its host but is unable to parasitize this particular host during dry and hot summer. This is because of lethal effects of high temperature (above 35°C) and low humidity (below 60 per cent R.H.) on the parasitoid.

Food

Insects being heterotrophic depend directly or indirectly on plants for food. The quantity and quality of food plays an important role in insects' survival, longevity, distribution, reproduction, speed of development, etc. The quantitative aspects of food may be expressed as absolute food shortage (when food is completely destroyed within a localized region) and effective food shortage (when food shortages may occur in patches throughout the distribution of a given insect population). The causes of absolute and effective food shortages are numerous:

There may be large number of the same individuals per unit quantity of food (intraspecific competition).

There may be more than one species consuming the same food materials (interspecific competition).

There might be other species that influence the food of a particular species without actually consuming it. For example, an epizootic of the fungus, *Entomophthora*, so reduced the number of the caterpillars of *Plutella xylostella* (Linnaeus), that the predator, *Angitis* sp. suffered severely from a shortage of food when it had been abundant before the outbreak of the disease.

Any environmental factor causing reduction in populations of particular plants or insects will probably cause reduction of not only the insect populations that utilize these plants or animals for food but also of parasitoids or predators of these animals.

Monophagous insects are more likely to be affected by such reductions in food supply than polyphagous species.

The quality of available food greatly influences egg production, larval development, longevity and size of insects. Many insect species store sufficient nutrients during the larval stage to accomplish adult activities. The adult longevity of these species is of comparatively short duration and commonly do not feed at all. Mayflies which live long enough to copulate and egg laying, depend entirely on these reserves. In other species, larvae may store nutrients sufficient for egg production, but the adults must ingest water and carbohydrates to survive.

There are differences with regard to suitability of a host at the species or even at the varietal level within the same crop. The differences in varieties/species with regards to their suitability for a particular insect population are governed by their acceptability as host, nutritional adequacy and absence of metabolic inhibitors or agents toxic to pest species.

Factors that Determine Insect Abundance

(i) Biotic Potential

It is the innate ability of the population to reproduce and survive. It depends on the inherited properties of the insect *i.e.*, reproduction and survival. Potential natality is the reproductive rate of the individuals in an optimal environment. Survival rate depends on the feeding habits and protection to young ones (*e.g.* Viviparity). Generally, insects with high reproductive rate tend to have low survival rate and vice versa.

Insect pests with high reproductive rate and low survival rate are called *r* strategists named after the statistical parameter *r*, the symbol for growth rate coefficient. Such pests succeed because of sheer numbers. *e.g.* Aphids.

K strategists reproduce slowly but effectively compete for environmental resources and so their survival rate is high. (K letter denotes flattened portion of growth curve) *e.g.* Codling moth of apple.

Birth rate or natality is measured as the total number of eggs laid per female per unit time. Factors determining birth rate are fecundity, fertility and sex ratio.

Death rate or mortality denotes the number of insects dying over a period. Environmental resistance is the physical and biological restraints that prevent a species from realizing its Biotic potential. Environmental resistance may be of 2 types.

1. Biotic factors–includes
   (*a*) Competition (interspecific and intraspecific)
   (*b*) Natural enemies (predators, parasites and pathogens).
2. Abiotic factors–
   (*a*) Temperature
   (*b*) Light
   (*c*) Moisture and water
   (*d*) Substratum and medium

Bioresources in Ecosystem

Ecosystem comprises of biological communities and non-living environment, *e.g.* Agro ecosystem, pond ecosystem, etc. Bioresources refers to the biodiversity of various organisms living in that ecosystem, *e.g.* pests of vegetables, their natural enemies, hyperparasitoids, microbes, etc. are referred to the bioresources in vegetable ecosystem.

The ecosystem should have more bioresources. Such ecosystem will be more stable. Insecticides will deplete the bioresources in ecosystem and make it less stable and prone to pest outbreak.

Natural control will be high when bioresources (*e.g.* Parasitoids and Predators) are more.

Population Dynamics and Role of Biotic Factors

Characteristics of a Population

1. *Density*: Population size per unit area
2. *Birth rate (Natality)*: Rate at which new individuals are added to the population by reproduction
3. *Death rate (Mortality)*: The rate at which individuals are lost by death.
4. *Dispersal*: The rate at which individuals immigrate into and emigrate out of the population.
5. *Dispersion*: The way in which individuals are distributed in space. It may be of 3 types.
   (*a*) Random distribution
   (*b*) Uniform distribution
   (*c*) Clumped distribution
6. *Age distribution*: The population of individuals of different ages in the group.
7. *Genetic characteristics*: Adaptiveness, reproductive fitness, persistence.
8. *Population growth form*: The way in which population changes/grows as a result of natality, mortality, and dispersal.

Population Dispersal

This is the movement of individuals into or out of the population. This plays an important role in the geographical distribution of insects to areas that were not previously occupied by them. The process of dispersal takes place due to the following reasons-

☆ in search of food
☆ to avoid predators
☆ preventing overcrowding
☆ due to abiotic factors *viz.*, wind, water, light, etc.
☆ breeding behaviour
☆ physiological reasons

The dispersal of population occurs in three ways:

1. Emigration
2. Immigration
3. Migration

1. Emigration

This is outward movement of an organism from one place to another for permanent settlement. Emigration results in depopulation.

2. Immigration

This is inward movement of the organism to any pace. Immigration leads to rise in population.

3. Migration

It involves mass movement of entire population, where some individual return again to the area from which they have moved.

Population Dynamics

Population dynamics is the branch of ecology which investigates the causation of population changes and the mechanism of natural control of a population.

Populations grow in two contrasting ways. They are:

1. J- shaped growth form
2. S- Shaped or sigmoid growth form

In the J–shaped growth form, the population density increases in exponential or geometric fashion; for example 2,4,8,16,32 ... and so on until the population runs out of some resource or encounters some limitation. Growth then comes to a more or less abrupt halt and density declines rapidly. Populations with this kind of growth form are unstable. Their reproductive rate is high and survival rate is less and so they are r strategists. Factors other than density regulates the population. (eg. Aphids).

In the S-shaped growth pattern the rate of increase of density decreases as the population increases and levels off at an upper asymptote level K, called the carrying

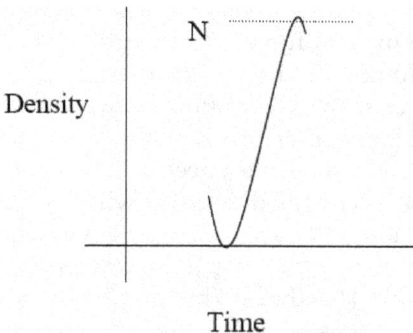

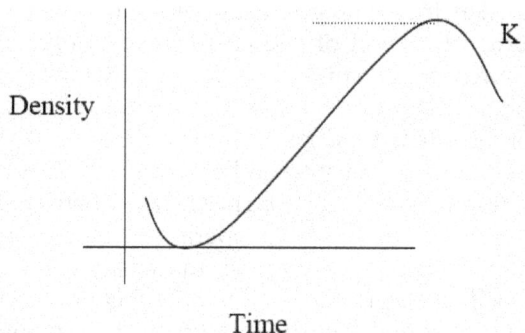

Figure 6.1: J-Shaped Growth Form   **Figure 6.2: S-Shaped or Sigmoid Growth Form**

capacity, or maximum sustainable density. Their reproductive rate is less and survival rate is more. So they are K strategists. This pattern has more stability since the population regulates itself. (*e.g.* Hymenopterans).

The population growth rate or change is worked out using the formula,

$$N_t = N_0 e^{(b-d)t} - E_t + I_t$$

where,

   N_t: Number at the end of a short time period

   N_0: Number at the beginning of a short time period

   e: Base of natural logarithm = 2.7183

   b: Birth rate

   d: Death rate

   t: Time period

   E: Emigration

   I: Immigration.

Population Theories

1. Nicholson's Theory

According to this theory, populations are self-governing system. They regulate their densities in relation to their own properties and those of their environments. This they do by depleting and impairing essential things to the threshold of favorability, or by maintaining reactive inimical factors, such as the attack of natural enemies, at the limit of tolerance. The mechanism of density governance is almost always intraspecific competition, either amongst the animals for a critically important requisite, or amongst natural enemies for which the animals concerned are requisites.

Governing reaction induced by density change holds population in a state of balance in their environments. The characteristics of balance is sustainable and effective compensatory reaction which maintains populations in spite of even violent changes in the environment, and which adjusts their densities in general conformity with prevailing conditions.

Far from being a stationary state, balance is commonly a state of oscillation about the level of the equilibrium density which is for ever changing with environmental conditions. Destructive factors do not add to mortality when they continue to operate over long periods, but merely cause a redistribution of mortality, for the intensity of competition automatically relaxes sufficiently to make room for the destruction they cause. Such compensatory reaction causes the effect of destructive factors upon density to be much less when balance is reattained than that which they produce when they first operate. Although population densities can be governed only by factors which react to density change factors which are uninfluenced by density may produce profound effect upon density. This they do by modifying the properties of the animals, or those of their environments, so influencing the level at which governing reaction adjusts population densities.

2. The Theory of Andrewartha and Birch

The theory of Andrewartha and Birch (1954) states that the number of animals in a natural population may be limited in three ways (i) by shortage of material resources, such as food, place in which to make nests, etc (ii) by inaccessibility of these material resources relative to the animals capacities for dispersal and searching and (c) by shortage of time when the rate of increase r is positives. Of the three ways, the first is probably the least, and the last is probably the most, important in nature. Concerning c, the fluctuations in the value of r may be caused by weather, predators, or any other component of environment influences the rate of increase.

3. Milne's Theory

Milne's theory is based on the natural control of insect populations. According to this, control of increase is due to the combined action of (a) density independent and (b) imperfectly density dependent environmental factors. In comparatively rare cases where this combined action occasionally fails, increase to the points of collective suicide is prevented by competition between individuals of the population. Decrease of numbers to zero is prevented ultimately by density independent factors alone. For, unless the letter is begin at the appropriate time to favour increase instead of decrease, the remnant of individuals left by the imperfectly density dependent factors must perish.

Milne's environmental factors are of three types:

A. Density-independent Factors

(I) Physical circumstances, mainly weather (II) actions of other species, such as their indiscriminate browsing, grazing, fouling and treading on vegetation, or casual predation and parasitism.

B. Imperfectly Density-dependent Factors

Action of other species competing for the same resources (interspecific competition), and of predators, parasites and pathogens in general.

C. Density Dependent Factor

Competition (for food, space, etc.) within the population itself, *i.e.* intraspecific competition.

4. Chitty's Theory

Chitty emphasizes the inherent properties of species, and suggests that, in general, there may be an inverse relationship between the average vitality of the individuals in a population and population density, which can act as a mechanism or part thereof for stabilizing numbers.

5. Pimentel Theory

Pimentel (1961) revived the application of evolutionary principles to the problems of population dynamics by proposing a genetic mechanism for the determination of numbers. He summarized his views as: a genetic feed-back mechanism functions to regulate population of the herbivorous, parasites and predators is supported by evidence from the biomathematics of population dynamics and study of natural population. The mechanism functions as a feed-back system through the dynamics of density pressure, selective pressure and genetic changes in interacting populations. In a herbivorous plant system, animal density influences selective pressure on plants, this selection influences genetic make-up of plant and in turn, the genetic make-up of plant influences animal density. The action and reactions of interacting populations in the food chain cycling in the genetic feed-back mechanism results in the evolution and regulation of animal population.

Life Table

Life tables are tabular statements showing the number of insects dying over a period of time and accounting for their deaths.

Ecosystem

The term ecosystem was coined by the British ecologist A. G. Tensely in 1935. Ecosystem may be defined as "A natural unit of living community (biotic factor-plants and animals) and non living environment (abiotic factors-light, temperature, humidity, soil condition, etc.) interacting and exchanging materials between themselves.

The Ecosystem Concept

Living organisms and their non living (abiotic) environment are inseparably interrelated and interact upon each other. Any unit that includes all the organisms (*i.e* community) in a given area interacting with the physical environment so that a flow of energy leads to clearly defined tropic structure, biotic diversity and materials cycles (*i.e.* exchange of materials between living and non living parts) within the system is an ecological system or ecosystem. For descriptive purposes, it is convenient to recognize the following components comprising the ecosystem.

1. Inorganic substances–C,N, Carbon di-oxide, water, etc.
2. Organic compounds–Proteins, carbohydrates, lipids, humic substances, etc.
3. Climate regime–Temperature and other physical factors.
4. Producers–autotrophic organisms, largely green plants, which are able to manufacture food from simple inorganic substances.
5. Macroconsumers or Phagotrophs (Phago= to eat)–heterotrophic organisms, chiefly animals, which ingest other organisms or particulate animal matter.
6. Microconsumers or Saprotrophs (sapro= to decompose) or Osmotrophs (osmo= to pass through a membrane)–heterotrophic organisms, chiefly bacteria and fungi, which break down the complex compounds of dead protoplasm, absorb some of the decomposing products, and release inorganic nutrients that are usable by the producers together with organic substances, which may provide energy sources or which may be inhibitory or stimulatory to other biotic components of the ecosystem. Just as ponds and forest are systems of interacting elements, so are cropped areas. An

ecosystem largely created and maintained to satisfy human wants or need is called an agroecosystem. It is not a natural ecosystem but is a man made.

A typical agroecosysyetm is composed of:

1. more or less uniform crop-plant population
2. weed communities
3. animal communities (including insects)
4. microbiotic communities
5. and the physical environment in which they interact.

Unique Features of Agroecosystem

☆ Agroecosystems often lack of temporal continuity-the existence may be of limited duration and may undergo immense, abrupt changes in microclimate because of cutting, ploughing, disking, burning, chemical application and other cultural or agronomic practices. So, agriculture is interfering with nature.

☆ Dominated by plants selected by humans, many consisting of imported genetic material. Other crop plants not from imported material have been in protective cultivation so long that they hardly resemble the parent stock from which they were derived.

☆ Most agroecosystems have little species diversity and the crop species has little intraspecific diversity. In other words, the crop tends to be genetically uniform. Usually, a single species dominates an agroecosystem, and elimination of weeds further simplifies the environment.

☆ With crop plants of similar type and age in the system, the vegetative structure is uniform, and a given phenological event (*e.g.* flowering, poding) occurs in all the plants at the same time.

☆ Nutrients are added to agroecosystems, which result in crop plants with uniformly succulent, nutrient rich tissues.

☆ Agroecosystems often have frequently occurring insect, weed and disease outbreaks due to the preceding 5 features.

Horticultural Ecosystem

It is an ecosystem similar to agroecosystem with some differences (Annual and perennial crops). Compared to agroecosystems, the orchard ecosystems are more permanent ecosystems with greater habitat continuity for phytophagous and entomophagous arthropods. If undisturbed by pesticides or management practices that diminish diversity, certain deciduous tree fruit ecosystem components remain relatively stable and some pests are maintained below tolerance level by diseases, predators and parasitoids. The cultivated varieties (cultivars) of most agricultural and horticultural crops bear little resemblance to their original wild ancestors. In the process of breeding for higher yield and quality of produce, the possible resistance to pests has often been ignored or considered to be of secondary importance.

Forest Ecosystem (Silviculture)

It is a natural ecosystem where plant communities normally consist of a complex group of species.

☆ Natural communities are complex whereas agricultural communities are simple.

☆ Complex communities tend to be stable because they incorporate many checks and balances and therefore no one species is likely to explode in numbers. Since it is an undisturbed ecosystem, usually pest outbreak is not there; most of the pests occur at a level below economic threshold level by the action of natural enemies and diseases.

☆ The activity of natural enemies will check the build up of the pest population.

☆ Mixed strands are much safer from pest attack than pure strands. i.e if other things being equal, the degree of environmental stability is in direct proportion to the number of species living together in an environment. The diversity, both of insects living on the trees and of parasites and predators feeding upon them, results in a multiplicity of interactions that tend to limit the increase of all species in the complex environment. As examples of mixed types in which disastrous outbreaks seldom or never occur. Forest ecosystem consists of Forest trees–Pests–parasitoids and predators– secondary consumers–Mutualists and Decomposers (Termites).

☆ There is no manmade intervention.

☆ No fertilizer or pesticides (or) minimum.

Food Chain, Food Web and Energy Flow in Ecological Systems

Organisms at or near the surface of the earth are immersed in a radiation environment consisting of solar radiation and long wave thermal radiation flux from nearby surfaces. Both contribute to climatic environment (temperature, evaporation of water, movement of air and water, etc), but only a fraction of the solar radiation can be converted by photosynthesis to provide energy for the biotic components of the ecosystem.

Food is an important or even central issue in ecology because food shortage will limit population increase if nothing else prevents population growth before this ultimate factor.

Elton (1927) was the first to coin the phrase food chain signifying the feeding links between organisms as food passes from the plant to the herbivore to the carnivore, and perhaps to the higher carnivores. This concept can be illustrated with one of Elton's examples.

Pine trees → aphids → spiders → cicadas → hawks

This example also illustrates where insects and other arthropods commonly fit into the food chain, and how they are important links in the movement of energy in the community/For less wooded areas the food chain may be based on corn as follows:

Sun → corn → corn earworm → ichneumon parasite → warbler → hawk

Sun → corn seed → corn seed beetle (carabid carnivore) → carabid beetle predator (eats eggs and larvae of corn seed beetle) → Vole → owl

The trophic level concept categorizes these links in the food chain into the following hierarchy or levels:

Producers–Plants produce food (some bacteria produce food also)

Consumers–All other organisms consume food

Primary consumers–Herbivores

Secondary consumers–Carnivores

Tertiary consumers–Secondary carnivores.

The transfer of food energy from the source in plants through a series of organisms with repeated eating and being eaten is referred to as the *food chain*. At each transfer, a large proportion, 80-90 per cent of potential energy is lost as heat. Therefore, the number of steps or links in a sequence is limited, usually 4 or 5. The shorter the food chain, the greater the available energy. Food chains are of 2 types.

1. *The grazing food chain*, starting from a green plant base, goes to grazing herbivores and onto carnivores.
2. *Detritus food chain*, which goes from dead organic matter into microorganisms and then to detritus feeding organisms (detrivores) and their predators.

Food chains are not isolated sequences, but are interconnected with each other. This interlocking pattern is often spoken as the food web. In complex natural communities, organisms whose food is obtained from plants by the same number of steps are said to belong to the same trophic level.

☆ Producers (green plants) occupy the trophic level I

☆ Primary consumer (plant eaters) occupy trophic level II

☆ Secondary consumer (carnivores or plant eaters) occupy trophic level III

☆ Tertiary consumer (secondary carnivores) occupy trophic level IV

The energy flow through a trophic level, equals the total assimilation (A) at that level, which inturn equals the production (P) of the biomass plus respiration (R).

Biogeochemical Cycles

The chemical elements, including all the essential elements of protoplasm tend to circulate in the biosphere in characteristic paths from the environment to organisms and back to the environment. These more or less circular paths are known as biogeochemical cycles. The movement of those elements and inorganic compounds that are essential to life can be conveniently designated as nutrient cycling. For each cycle, it is also convenient to designate two components or pools. (*i*) the reservoir pool, the large slow moving, generally non biological component and (*ii*) the exchange or cycling pool, a small but more active portion that is exchanging (*i.e.* moving back and forth) rapidly between organisms and their immediate environment. From the stand point of the biosphere as a whole, biogeochemical cycles fall into 2 basic groups. (*i*) gaseous types, in which the reservoir is in the atmosphere or hydrosphere (ocean) and (*ii*) sedimentary types, in which the reservoir is in the earth's crust.

(A) Hydrological Cycle

Water is the most essential inorganic nutrient for living organism. A continuous cyclic flow of water is maintained between living organism, atmosphere and earth via precipitation, evaporation and transpiration. This cycle is called as hydrological or water cycle. Water covers about 73 per cent of the earth's surface. It occurs in rivers, ponds, lacks, etc. water keeps on evaporating in the atmosphere from each of the water bodies. These water vapors condense at higher altitude of atmosphere and form clouds. These clouds are moved by winds and pass over land where further cooling takes place to precipitate water as rain or snowfall. Water on land surface eventually percolates into the soil water. Some of the ground water is absorbed by plant through their roots. This water is returned to the atmosphere through evaporation and transpiration. Animals drink water and part of it returns to the atmosphere through evaporation and excretion. Water is returned to the biosphere after the death of the organism through a process of decomposition.

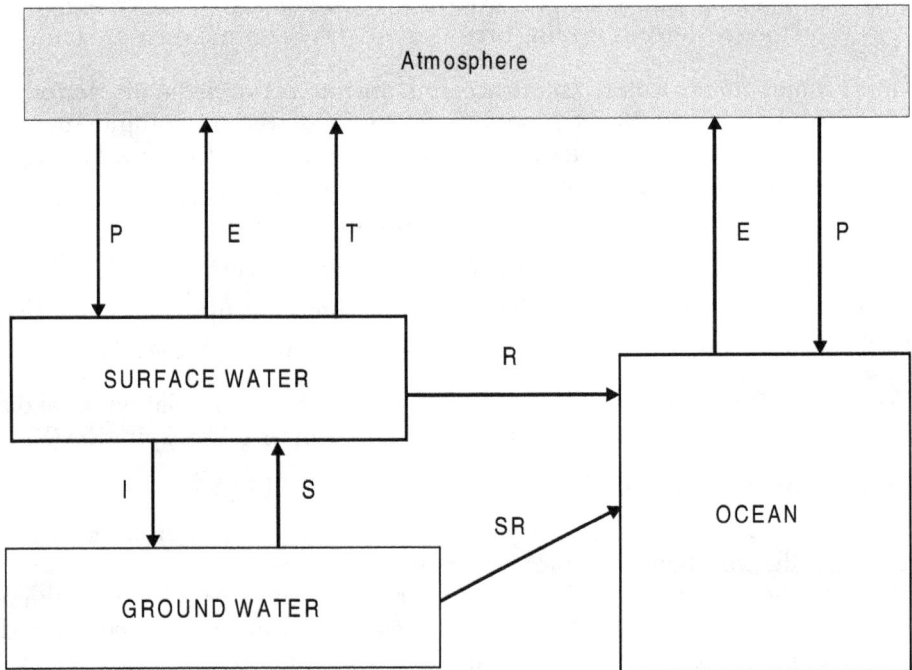

Figure 6.3: Hydrological Cycle-I

The hydrologic cycle, showing the transfer of water from the oceans to the atmosphere to the continents and back to the oceans again.

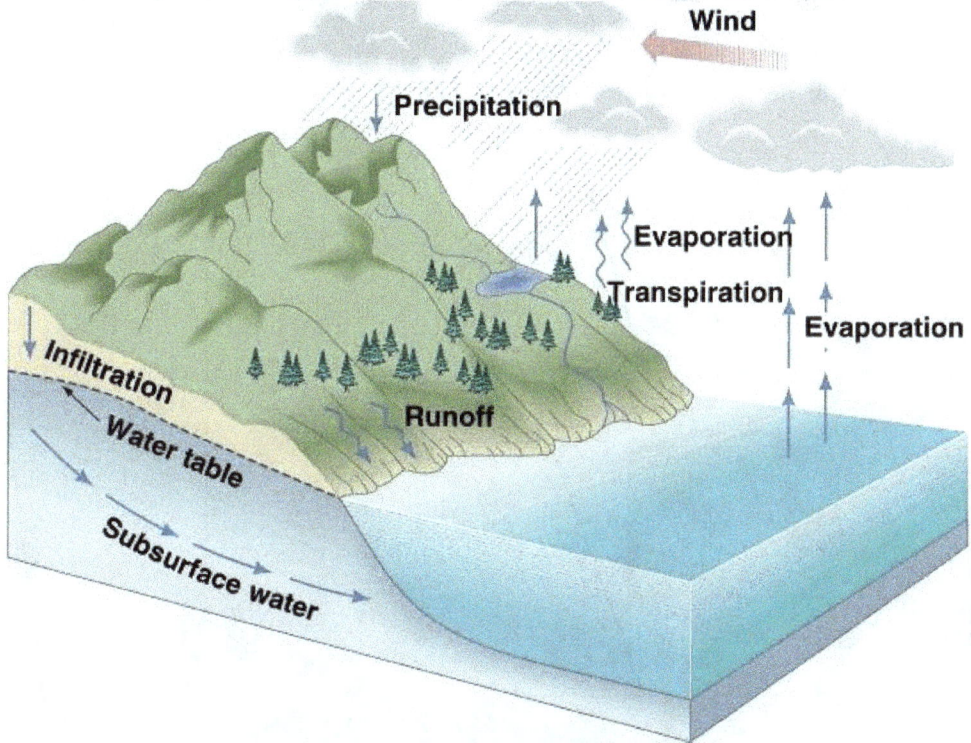

Figure 6.4: Hydrological Cycle-II

(B) Carbon Cycle

Carbon is the basic building element of all the living organisms. By volume, carbon di oxide in atmosphere is very low *i.e.*, about 0.03 per cent. This atmospheric carbon is used by green plants to manufacture carbohydrate by the process of photosynthesis. In photosynthesis, carbon from atmospheric carbon di oxide is incorporated into different consumers. From both the producers and consumers carbon is returned back to the atmosphere during respiration. After death organisms like bacteria, fungi and viruses decompose and degrade the complete organic compound into their simplest form.

(C) Nitrogen Cycle

Nitrogen is one of the most essential elements required for the synthesis of amino acids and protein in living organisms,. The quantity of nitrogen is very high *i.e.*, 79 per cent in the atmosphere. But free nitrogen is not used by plants and animals. They need it in the form of soluble nitrogen compounds such as ammonia (NH_3), nitrite (NO_2), nitrate (NO_3) and amino acids, respectively. Nitrogen cycle is completed in various steps (a) Nitrogen fixation- first step is the conversion of gas into nitrates

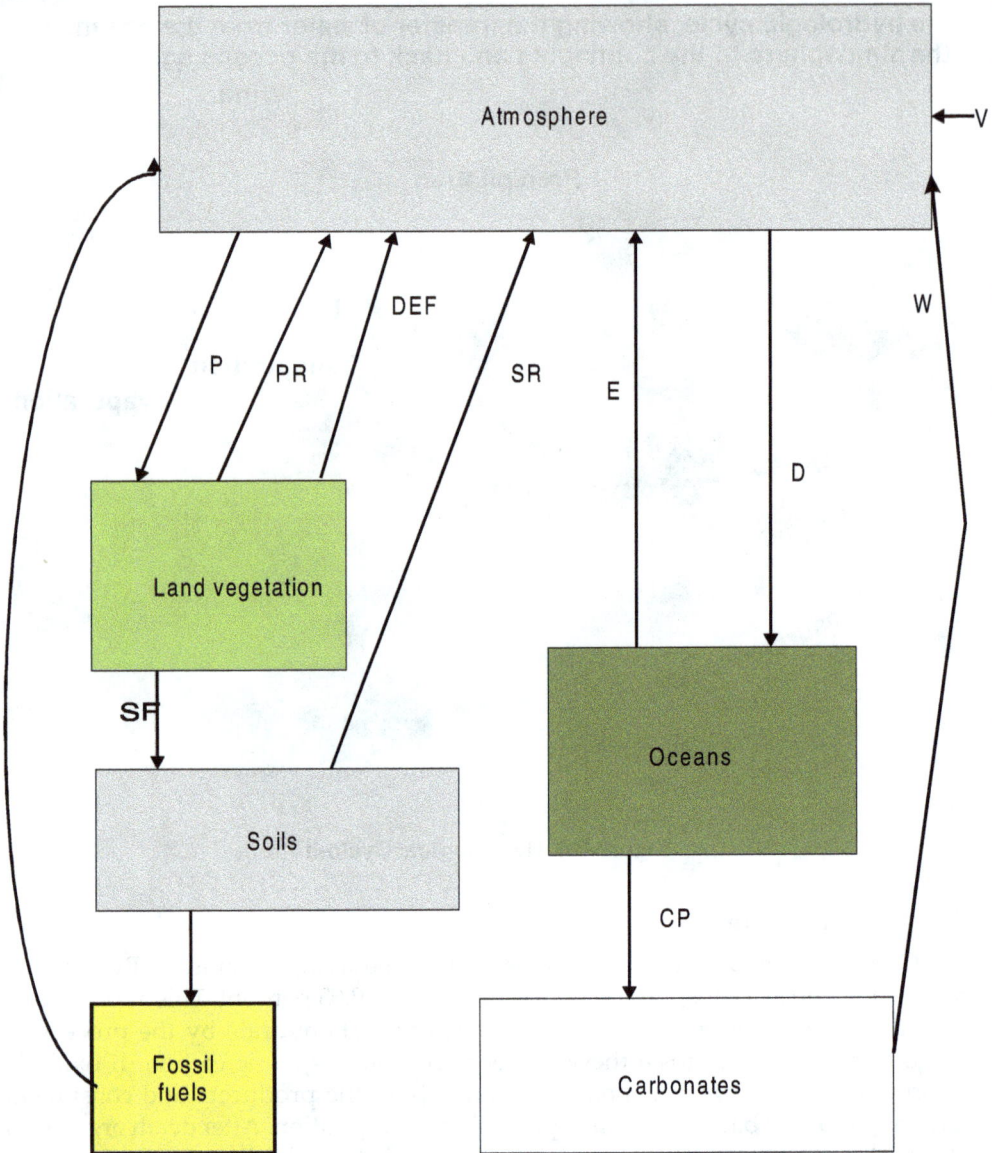

Figure 6.5: Carbon Cycle-I

(NO$_3$) and ammonia (NH$_3$). Conversion of nitrogen into into nitrate can be done both biologically and physically. Ninety per cent of the atmospheric nitrogen is fixed biologically by bacteria (*Azotobacter*). Some symbiotic bacteria (*Rhizobium, Nitrosomonas, Nitrobacter*) live in the root of leguminous plants such as beans, peas, etc. Very little amount of atmospheric nitrogen is fixed physically by cosmic radiations and lightening. Nitrogen fixation results in the formation of ammonia and nitrates soluble in water.

Storage in atmosphere (720 + 3/yr due to burning fossil fuels) *

Volcanoes (0.1/yr)

Burning fossil
fuels (5 – 6/yr)

Land photosynthesis
and respiration
(120/yr)

Oceanic photosynthesis
and respiration (107/yr)

Soil
storage
(1500) *

Coal

Storage in land
plants (560) *

Weathering and
erosion (0.6/yr)

Oil

?

Fossil fuel storage (4000) *

Storage in
shallow and
ocean waters
39,000 *

* Storage units in billions of metric tons
  of carbon

Indicates direction of carbon transfer in billions
  of metric tons/yr

?

Storage in marine sediments
and sedimentary rocks
(100,000,000)

(*a*)

Carbon stored
in the atmos-
phere

Carbon stored in
the land biota,
rocks, soil, and
fossil fuels

Carbon stored in
the ocean biota,
water, and
sediment

(*b*)

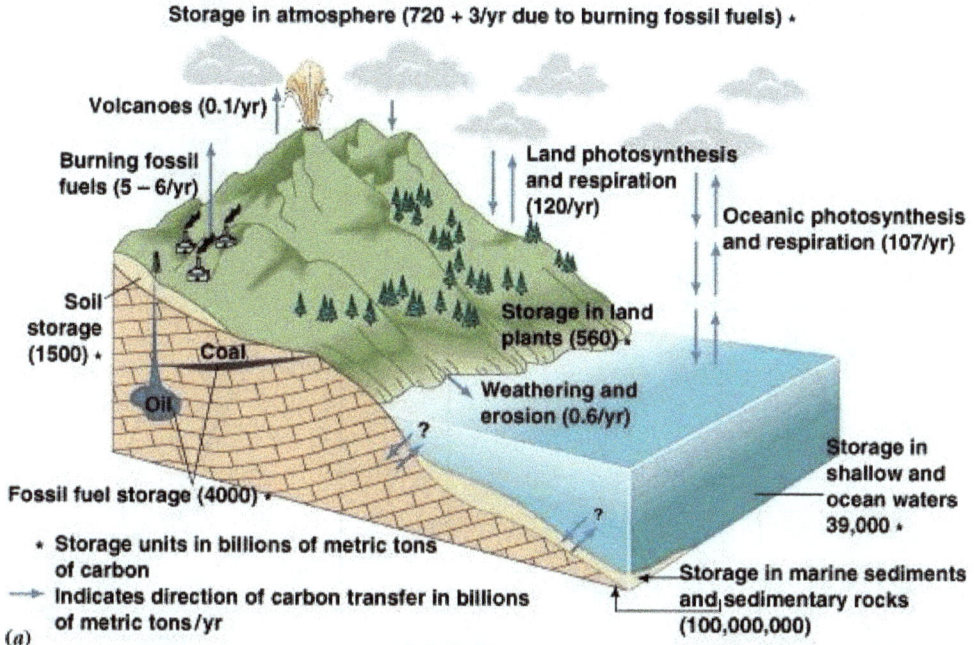

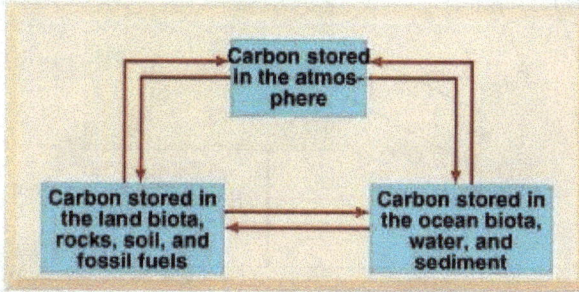

(*a*) Generalized global carbon cycle. (*b*) Parts of the carbon cycle simplified to illustrate the cyclic nature of the
movement of carbon. (*SOURCE:* Modified after G. Lambert, 1987, *La Recherche:* 18, pp. 782-783.)

Figure 6.6: Carbon Cycle-II

Ammonification

The proteins are building blocks of all living organisms. The protein compounds
of the dead organisms are broken into ammonia by decomposers (bacteria, fungi and
blue green algae). The ammonia is released into atmosphere or retained in the soil,
and ultimately absorbed by plants. First of all the, molecular nitrogen breaks into two
atoms (2N), then it combines with three molecules of hydrogen to produce two
molecules of ammonia.

$$N_{2-} \longrightarrow 2N$$

$$2N + 3H_2 \longrightarrow 2NH_3$$

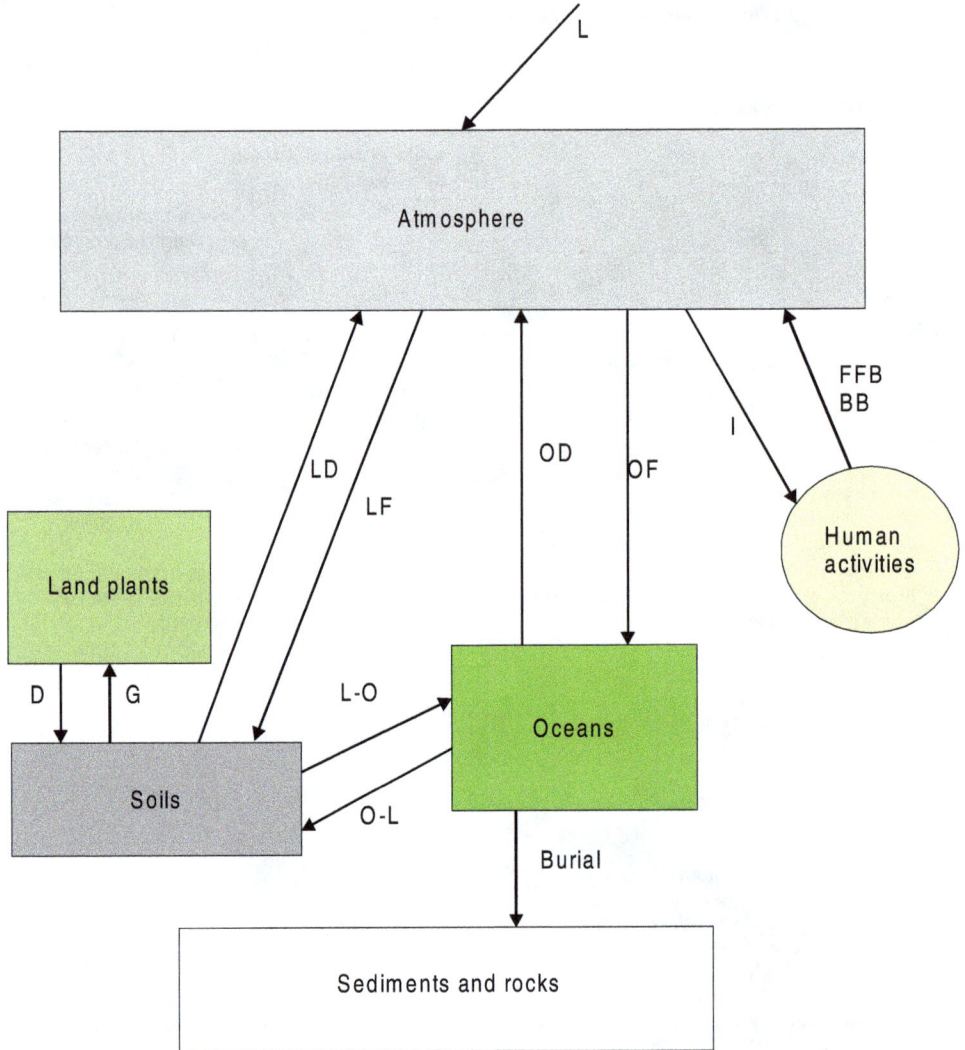

Figure 6.7: Nitrogen Cycle

Nitrification

The conversion of ammonia into nitrates through the activity of nitrifying microbes is called nitrification. Nitrifying microbes are *Nitrosomonas, Nitrococcis*.

Denitrification

Some nitrates of soil due to activity of denitrifying microbes may also be converted to free nitrogen gas. The gaseous form of nitrogen returned back to the atmosphere.

(D) Oxygen Cycle

Oxygen is most essential element for both plants and animals. In the atmosphere, the oxygen content is about 21 per cent. It also remains dissolved in water. It is taken by both animals and plants during respiration and is returned to the environment in the form of carbon di oxide. Oxygen is also produced as by-product of plants during the process of photosynthesis. This oxygen liberated during photosynthesis is used by the organism in respiration.

(E) Sulphur Cycle

Sulphur remains present in all sphere *i.e.*, air, water and soil. Organisms obtain sulphur in the form of inorganic sulphate (SO_4) dissolved in water. Autotrophic plants absorb dissolved sulphate and utilize it in protein synthesis and constitute the protoplasm. From the autotrophs, sulphur enters into the bodies of consumers at various trophic levels. Bodies of dead animals and plants are decomposed by the decomposer bacteria and fungi and the sulphur is released in the form of hydrogen sulphide (H_2S). Some of the H_2S is converted into free sulphur by a bacteria called *Beggiatoa.*

(E) Phosphorus Cycle

Phosphorous is indispensable element of life as it is a basic constituent of nucleic acid, phospholipids, etc. It also affects the metabolic process of energy transfer. Rock and natural phosphate deposits are the chief source of phosphate deposit. The phosphorous is liberated from these deposit by eroding, weathering and mining. Phosphorous is made available to the living organisms as dissolved inorganic phosphate ions. These phosphates are absorbed by producers for their nutrition. From producer (plants), the phosphate is transferred to animals (consumers). After the death of plants and animals, their body decay and phosphates are released to soil. Much of the phosphorus reaches to ocean into sedimentation. Some of the phosphorous deposited in the sea is again brought into the phosphorous cycle by upwelling. Phosphorous is also released in said nature by animal excretion in large quantity.

Seasonal Adaptations

Insects adapt to many adverse environmental conditions such as irregular and unpredictable extremes in temperature, moisture, food, crowding and other factors. Insect response to stresses include a quiescent (torpid) or moving out of the area

Predictable adverse conditions due to seasonal cycle are:

☆ Freezing temperatures in temperate climate

☆ Rain in the tropics

☆ Droughts in deserts

Insects adapt by changing their seasonal cycle. These adaptations includes

☆ Dormancy

☆ Changes in body form or colour

☆ Migration

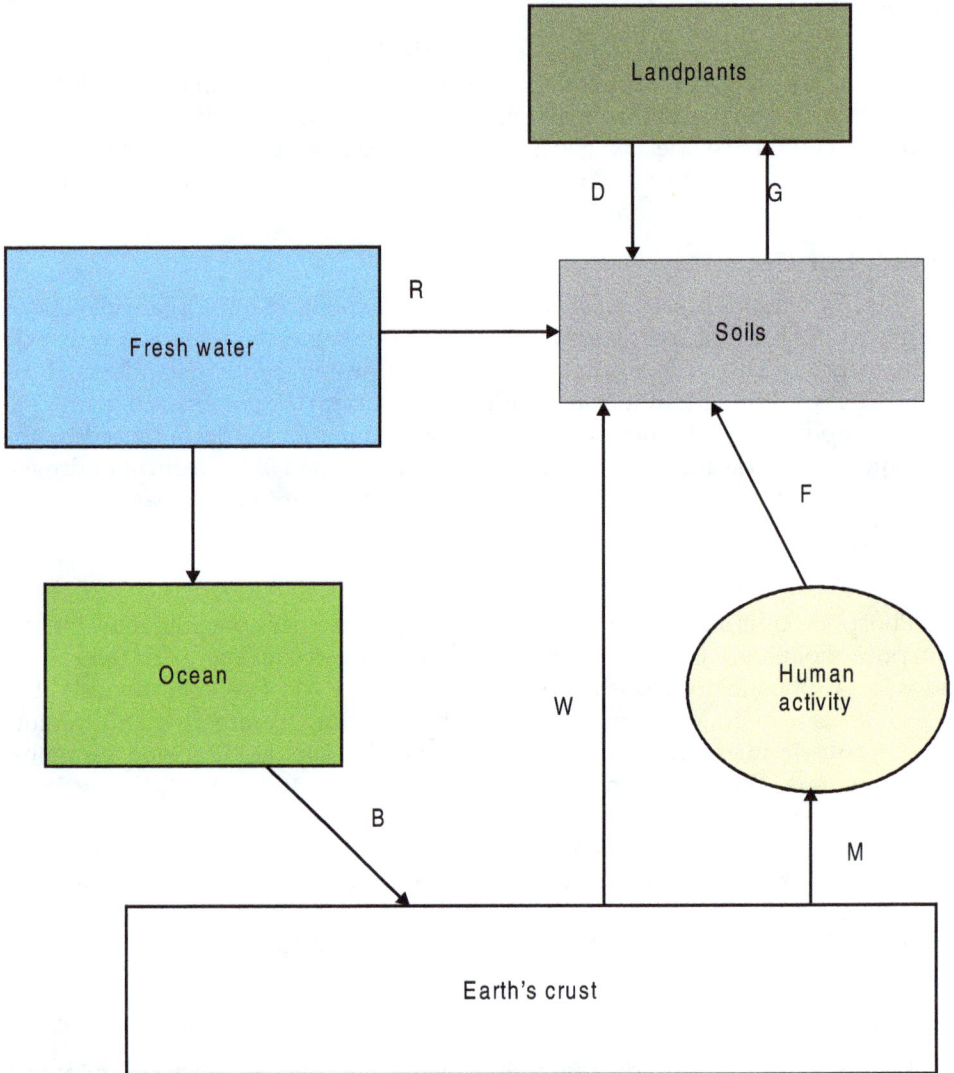

Figure 6.8: Phosphorus Cycle

Dormancy

Dormancy is prevalent among insects as a means for escaping regular seasonal adversities. Dormancy can be defined as a seasonally recurring period in the life cycle when growth, development and reproduction are suppressed.

Dormancy in different seasons is termed as:

☆ Summer Aestivation

☆ Fall Autumnal dormancy

☆ Winter Hybernation

☆ Spring Vernal dormancy

Some insects are dormant for more than one season. *e.g.* Delia platura seed corn maggot undergoes aestivo-hibernation–*i.e.* aestivate most of the summer and go directly into winter hibernation. In temperate climates, hibernation is one of the common characteristics of the seasonal life cycle of insects. Cold hardiness in insects is achieved by physiological processes such as:

1. Super Cooling

Super cooling is a resistance to freezing by lowering the temperature at which freezing of the body fluid begins.

2. Freezing Tolerance

It is a survival despite freezing of body fluid.

Dormancy differs from quiescence in that during dormancy, growth and reproduction may be suppressed even during periods when conditions are temporarily favorable for these conditions. During quiescence, growth and reproduction resume immediately upon return of favorable conditions.

Diapause

☆ Dormancy as well as seasonal migration and morphological change are controlled by a physiological condition called diapause.

☆ Diapause is usually characterized by low metabolism little or no development and increased resistance to environmental extremes and altered (often greatly reduced) behavioural activity.

☆ Diapause is a physiological state of arrested metabolism, growth and development that occurs at one stage in the life cycle.

☆ Diapause is a physiological resting state, wherein there is an obligate, hormonally mediated slowing of physiological development and growth in response to an environmental cue.

The mechanism found to underlie and orchestrate diapause is hormone activity with some of the following involved.

1. Larval and pupal diapause, lack of brain hormone and consequently ecdysone

2. Adult diapause, lack of brain hormone and juvenile hormone

3. Egg diapause, release of a neurosecretory hormone from the suboesophageal ganglia of the parent. Diapause is induced by specific environmental cues called token stimuli. They are given this term because these environmental factors in themselves are neither favourable nor unfavourable but original forthcoming changes in the environment. The most important token stimulus for insects in day length or photoperiod, temperature, moisture and biotic factors also act as token stimuli either alone or in combination.

The onset of diapause occurs well before most environmental conditions change, whereas dormancy is a concurrent response or reaction to changing conditions.

Once diapause is induced, it continues even though favourable environmental conditions prevail.

Usually diapause occurs at a specific stage of the insect's life cycle, with thus stage being genetically fixed.

Examples

Egg diapause–Grasshoppers, corn root worm

Larval diapause–European corn borer, many bark beetles (PBW- larval diapause)

Spruce bud worm Choristoneura fumiferana–diapause in second larval instar.

Pupal diapause–*Hyalophora cecropia* diapause as pupae (RHC–pupal diapause), *Platysamia, Samia, Diparopsis*

Adult diapause–Boll weevil (*Anthonomus grandis*). The stage sensitive to token stimuli may be the diapausing stage (crickets), certain period in the parental generation (silk worms) or even periods in the grand parent generation (some aphids). The diapause phenomenon enhances survival in unfavourable conditions. In univoltine species, every generation has a diapausing stage. In multivoltine species, the phenomenon occurs in a single stage of one of the generations, given the appropriate token stimulus.

Ecotypes

Species with wide geographical ranges often develop locally adapted populations,called ecotypes having different limits of tolerance to temperature,light or other factors compensation along a gradient of conditions may involve genetic races (with or without morphogical manifestations) or near acclimatization reciprocal transplants often reveal whether ecotypes are actually genetic race.

7

Principals of Pest Management

Pest

A pest is any organism, animal, plant or micro-organism that causes damage or annoyance to human beings, their animals, crops or possessions. The word pest is derived from the Greek word *'pestis'* which means to annoy. The major pests of agricultural importance can be broadly divided into the following groups:

Insect and Mite Pests

Insect and mites cause heavy damage to crops. Out of over one million species of insects, only about 200 species can be described as serious pests. A large majority of insects are beneficial to mankind.

Plant Pathogens

Fungi and various micro-organisms such as bacteria and viruses cause diseases in plants.

Weeds

These are plants that either compete with crop plants thus affecting yield and quality, or may interfere with the use of land and water resources.

Vertebrate Pests

These are mainly rodents, birds and some other mammals that cause damage to crops and stored products.

There are three categories of pests:

1. *Regular pests*–They are generally found in abundance during a crop season, *e.g.* aphids, thrips and jassids.
2. *Sporadic pests*–Which assume pest status occasionally in certain years and include locust, grasshoppers, hairy caterpillars, cutworms and cotton semilooper.
3. *Potential pests*–Which normally cause negligible damage but may become highly destructive resulting from some disturbance in the environment and consequent increase their number, *e.g.* armyworm infesting wheat.

Why a species attain the pest status?

1. Entry into a previously uncolonized habitat.
2. Changes in the characteristics of certain species that did not previously compete directly with man.
3. Increased sensitivity of man to an already existing species which may be owing to a change in the habitat of the pest or the activities of man himself.
4. Increase in abundance of a species whose interactions with man were previously negligible because of low numbers.

Type of Losses

The type of loss by an insect pest is influenced by the mechanics, location, velocity and intensity of damage. Mechanics relates to the means or the tools that are being used to cause damage. The feeding and oviposition habits of insects determine this type of loss. Location means the parts on which the insects are active. Velocity or the activity describes the amount of destruction caused by an insect in a unit time (at one or more than one location). Intensity relates to the higher or lower level of quantities or qualitative damage, depending upon the insect numbers operating for a period of time at a known rate of activity.

Mechanics of Damage

The insects pests have the following types of mouth parts:

Chewing Type

Strong mandibles and maxillae help in tearing off and masticating the food. This type is found in caterpillar, grubs and adults of beetles, grasshoppers and locust.

Rasping and Sucking Type

The epidermis of tissues is lacerated and the exuding sap is sucked as in the case of thrips.

Piercing and Sucking Type

A highly specialized form consisting of a tubular structure which encloses needle-like structures. It is characteristic of leaf-hoppers, whitefly, aphids, scales.

Location of Damage

The insects cause damage to different locations (different parts) *viz.*, subterranean parts, stems of herbaceous plants, twigs and buds, leaves, flowers and fruits, etc.

Types of Losses

The losses due to insect pests can be categorized as:

Direct Losses

These relate to decrease in productivity (quantitative) or intrinsic value/ acceptability of the product (qualitative).

Indirect Losses

These are primarily of economic interest as for example decreased purchasing power of the agriculturists and those depending on agriculture owing to reduced production.

Actual Losses

These include the total value of losses, both direct and indirect, the cost of control measures along with the amount spent on development of protection schedule.

Recognized and Hidden Losses

These are subjective terms showing as to whether the factors determining the loss are known or not.

Avoidable and Unavoidable Losses

These are also subjective terms related to the belief as to whether a certain pest can be controlled or not.

Economics of Pest Control

The relationship between density of the pest population and the profitability of control measures is expressed through threshold values. These are economic injury and economic threshold levels.

Economic Injury Level

It is the lowest level of pest population that will cause economic damage, which in turn, is the amount of injury that will justify the cost of artificial control measures.

Economic Threshold Level

It is a population level, lower than the economic injury level, at which pest control measures are to be undertaken for preventing an increasing pest population from reaching the economic injury level.

General Equilibrium Position

It is the average density of any insect population over a period of time which is not affected by the extrinsic temporary interventions. The density of the species tends to fluctuate around this mean level under the influence of environmental conditions.

Principles of Insect-pest Control

Once, it is established that an insect is causing economic losses, it becomes necessary to control it.

The first principle underlying the control of an insect is its correct identification. When it is correctly identified, we can refer to the available information on the biology and the habits of the insect and determine its most vulnerable stage, the appropriate time and the most suitable method or methods to control it.

The knowledge and understanding of the ecological factors, both biotic and abiotic, affecting the population of the insect pest is necessary for planning the proper strategy for controlling it. A lot of attention is paid to this aspect now and elaborate procedures and 'models' have been evolved to pinpoint the 'key factors', dominantly affecting the development and multiplication of a particular pest.

The choice of the proper method or methods of control becomes easier, when the above mentioned information becomes available in respect of a particular pest. Some of the important methods for control of insect pests are discussed below.

The Selection of Proper Methods for Controlling Different Pests

The guiding principles for selecting an appropriate method or methods of control should be that the method chosen must be economical, free from creating any other problem, immediately or in future, should not harm the natural enemies of the pest and should be easy to operate and be readily available to an ordinary cultivator.

The choice of the proper method (or methods) of control becomes easier when the biology and the habits of the pests are known and its most vulnerable stage has been determined. However so far the general tendency among those who are responsible for carrying out the control operations is to employ an easy method, giving quick results and disregarding other unfavourable consequences which may follow the completion of the operations.

Keeping in view the guiding principles in the selection of pest-control methods, it is generally realized that any single approach to the problem of insect control is not feasible. Each method has its own advantages as well as its disadvantages and also has limitations under a particular set of conditions. The current thinking all over the world is to suitably synthesize the use of as many methods as possible to control the insect pests effectively, economically and without any adverse after-effect. This approach is now popularly known as 'Integrated Pest Control'.

Despite a number of serious limitations from which insecticides suffer, it is felt that the use of these chemicals in the control of insect pests will continue for a long time to come and in the strategy for 'Integrated Pest Control', this method will occupy an important place as a curative methods for reducing losses because to pests.

However, the emphasis is to be laid on judicious and proper use of insecticides, so that this method may not impair the efficacy of other methods and create other problems. Keeping this point in view, some more information on insecticides and their proper use is given in the next chapters.

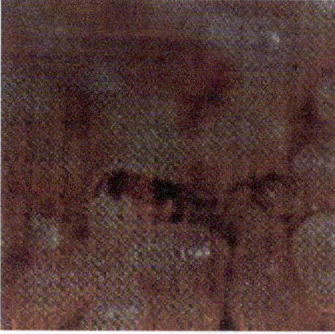

Figure 7.1: Trichogramma Egg Parasitoid

Figure 7.2: Chrysoperla Adult

Figure 7.3: *Bt* Infected Larvae

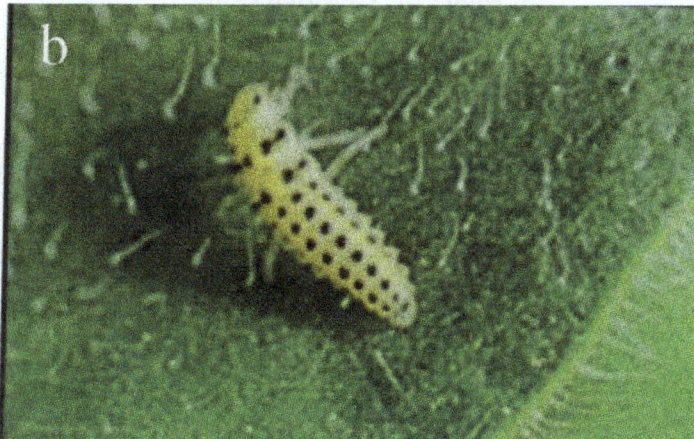

Figure 7.4: Lady Bird Larvae

Figure 7.5: HaNPV Infected Larvae

Figure 7.6: Lady Bird Beetle

Figure 7.7: Praying Mantis

Figure 7.8: Lady Bird Beetle

Figure 7.9: Pheromone Trap-1

Natural Control

All methods that destroy or check the infestation and which do not depend upon man for their success or continuance are collectively known as natural control. The natural control includes:

1. Climatic factors
2. Physical factors
3. Natural enemies
4. Insect diseases

Climatic Factors

Climatic factors include temperature, humidity, rainfall, sunshine, wind velocity and atmospheric pressure which affect directly or indirectly every phase of the insect life. A warm moist environment is favourable for the development of most insects. Extremes of temperature are destructive to insect life. Cold and wet weather, especially when sharply fluctuating temperature changes occur destroy many insects. Optimum temperature is 26°C and aestivation begins at 38°C, while 48°C is fatal.

The sunshine also plays an important role as most of the insects are nocturnal. Besides, there are some insects which fly in bright sunshine only. Insects dispersal to a great extent depends upon wind velocity as they generally fly with air current.

Physical Factors

Physical factors *viz.*, mountain ranges and large bodies of water provide effective barriers to insect spread. The spread of insects, which could not fly, are affected by samller bodies of water *e.g.*, lakes or rivers. Such insects are dependent upon man or other animals for the passage of such barriers. The physical and chemical nature of soil determines the presence or absence of soil dwelling insects, *e.g.*, white grub, which prefers sandy soil for its development and multiplication. Deserts and dense forests also acts as an obstacle for the dispersal of insects from one place to another.

Natural Enemies

This consists of insect predators, insect parasites, other natural enemies *viz.*, reptiles, birds and mammals, which are prevalent in abundance in nature to keep insect population in check. When the population of an insect increases, the natural enemies also begin to increase as they have a larger food supply. The increase continues until most of the hosts are destroyed. Birds, squirrels, moles, newts and snakes provide the most effective measure of control.

Some insects feed upon other insects and keep their population under check. They are predators and parasites.

Insect Diseases

In nature, insects are also attacked by various microbes (bacteria, virus, protozoa, etc.) and these microbs are responsible for various disease in insects. These diseases frequently wipe out insect population. Warm and wet climate is very favourable to disease appearance in the form of epidemics.

Bacteria

Bacillus thuringiensis- causes disease in caterpillar, *B. popillae* causes milky disease in Japanese beetle.

Virus

NPV causes disease in lepidopteron caterpillars.

Protozoan

Nosema spp. causes disease in many insects.

Applied Control

Applied control includes those methods, under the control of man, which are necessary to use when harmful insects have been held in check by natural agencies. The following factors are pre-requisites for the success of insect control:

1. Exact taxonomy and morphology of the insect.
2. Life cycle, natural and seasonal history of the insect.

7.1. Cultural Control

It includes regular farm operations performed to reduce insect population or prevent their injury. For an effective cultural control, it is essential that the life history and habits of an insect should be clearly known. Some important methods under this are:

1. Crop rotation
2. Clean cultivation
3. Variation in time or method of planting and harvesting
4. Tillage
5. Use of resistant varieties
6. Mixed cropping
7. Pruning and thinning
8. Fertilizer and water management
9. Trap crops

1. Crop Rotation

It is important in checking multiplication of insect pests. The same crop, if sown every year on the same field, provides a favorable atmosphere for multiplication of insect pests associated with it. Method of growing crop in rotation not only give agricultural benefits but also check the multiplication of common insect pests which would otherwise continue breeding in the same area throughout the year. Crop rotation gives better results against those insects which are restricted feeders (monophagous), slow breeders, having longer duration in feeding and less migratory power. Sometimes crop rotations are the only means of controlling certain insect. Crop rotation is effective against insects having long life cycle. For effective control by means of crop rotation it depends on the fact that insect concerned cannot migrate to other fields at the time when the change in the crop is made. The effectiveness in crop rotation is governed by three considerations: (*i*) the pest to be controlled must have a limited host range. (*ii*) should be in a non-migratory stage (*iii*) should have a comparatively long life.

2. Clean Cultivation

Regular removal and destruction of all weeds and crop residues is a very important factor in reducing insect population since many crop pests feed on certain weeds when their actual cultivated food plant is not available. Such pests are rice Gundhy bugs, fruit sucking moths, etc. decaying fruits and vegetables have various injurious insects which should be removed or disposed in a way that the insects should be destroyed. The destruction of crop residues is often of great importance as they provide favorable residues for termite pests. Timely hoeing and weeding is therefore recommended.

3. Variation in Time or Method of Planting and Harvesting

There are certain periods when most insects lay maximum eggs if their host plants are available at a stage of crop when it is susceptible and favourable for insect. By adjusting or changing the time of plantation one can avoid egg laying period of a particular insect and get young plants well established before the attack comes thereby giving shorter period of susceptibility during which the insect will attack or even get a crop matured before certain insects become abundant.

4. Tillage

There are certain insects which pass winter in larval or pupal stages in the ground. To make the cultivation most effective, it is very essential that it should be timed with the life history of the insect concerned. Insect population is greatly affected by soil texture. Composition of soil, soil organism, temperature and moisture, since these factors are closely related with the growth of the plant which serves as host of the insects. Certain species of insects which remain as a part of their development in the soil can be easily controlled if the soil is cultivated while they are in their pupal stage, *e.g.* wire worms. The depth of ploughing the soil should also be sometimes varied according to the habit of the insects. Some of the soil infesting insects like white grubs, ploughing during certain periods of the year will lead to the destruction of large number of larvae which are also exposed and serve as food of birds leading to the lowering down the pest population whereas ploughing at other times will be of minor importance. Harvesting has also got its own importance in reducing the insect population, *e.g.* sugarcane harvested up to February and crushed also leads in lowering borer population.

5. Use of Resistant Varieties

Use of crop varieties resistant to pest is considered as a suitable solution particularly in low input oriented cropping system. Resistance to attack if insect pests may vary from complete immunity to less susceptibility where damage caused is much greater in an average case. Degree of resistance or susceptibility among different varieties also depends upon the characteristic which are responsible for the attraction of insects. Physiological condition of the host is also responsible for the degree of resistance against insect and this nature of resistance is a temporary one. Three basic mechanism of resistance to insect pest attack have been identified. These are *non-preference* where host plant possess such characters that do not favor insect pests, *antibiosis* that prevents damage by exerting adverse influence on growth and development of the pest, *tolerance* that denotes replacement of host inspite of normal incidence of the pest.

6. Mixed Cropping

It is also one of the method by which the increase in insect infestation is reduced. When certain crops are sown or mixed, the insect associated with one crop get hinderance in reaching its hosts, since plants of other crops are present which is not the host of the insect. *e.g.* wheat + gram.

7. Pruning and Thinning

Pruning of dead branches, scraping of unhealthy stems, patching up wounds with clay and tar and removal of banks will keep some insects which hibernate or lark in such places as one of their many stages which if not checked may cause damage to the healthy portion in due course, *e.g.* bark eating caterpillars, shoot, bark and stem borers. Plants becoming intense are subject to attack by insects and if the pruning of certain branches are done, the plants suffer less from insect attack, *e.g.* pruning in early stages of crop helps in keeping down the population of cotton aphids, *Aphis gossypii*, pruning the citrus plant in December- January effectively check the incidence of citrus leaf miner, *Phyllocnistis citrella*.

8. Fertilizer and Water Management

Plants which are healthy and vigorous can resist better against insect injury consequently they are less liable to be attacked by insect whereas those plants which are weak due to the deficiency of proper nourishment or disease are liable to be attacked by the insect pests. Therefore, proper cultivation and use of fertilizer and manures provide nourishment for the development of healthy plants which aids in insect control. Thus, judicious use of fertilizers offer possibilities of stimulating plant growth. Application of excess nitrogenous fertilizers to soils tends to increase the susceptibility of sorghum crop to the attack of insect pests due to stimulated plant growth. Water applied to crops either can be beneficial or detrimental to insect pests. Using too much irrigation can cause plants to be especially succulent and attractive to insect pests, and may extend the time to maturity, increasing the duration of vulnerability. Excess irrigation in closely spread rice field is responsible for creation of favourable conditions for BPH, *N. lugens* outbreak.

9. Trap Crops

A trap crop is a small planting often occupying only a few rows, planted earlier than the main planting for the purpose of diverting insects away from the main crop. The trap crop can either be harvested early or cut and used as fodder before the generation of the insect pest gets completed. Growing mustard in cabbage to trap diamond back moth, *Plutella xylostella* is advantageous.

7.2. Mechanical Control

In this method, insects are controlled directly by manual operations or by mechanical devices. The disadvantage of this method is that they are generally costly, destroy the pest when adequate damage has already been done and mostly not suitable for carrying out the control operation on large scale. It includes the following methods:

1. Hand Picking

The destruction of insects on their stages by hand is important practical method and can be employed effectively where labour is cheap.

2. Hand Nets

Hand nets are useful against the flying insects. They may be caught very easily during their migration when they are in large numbers, *e.g.* grasshoppers, jassids, pyrilla, etc.

3. Barriers

Barriers are useful in preventing movement of insect which migrate by moving from one field to another, the following are the common practices which is generally followed:

Tree Banding

Adhesive or sticky bands around the tree trunks are used against those insects which infest trees by climbing the trunks *e.g.* mango mealy bug.

Fencing

It is possible to prevent invasion of a crop by migrating insects through the use of walls around the field so that may not go to another field.

Trenching

The crawling herds of army worms may be stopped by constructing deep furrows, around fields against their marching ways. Trench barriers are also used to stop the locust hoppers.

4. Traps

Many traps have been made to catch insects, some of them are:

Light Traps

Light traps are generally used against those insects which are phototrophic *i.e.* attracted towards light. For this purpose, a strong light may be placed or hanged in the field above a container having kerosinised water. A large numbers of insects attracted to light get killed in the kerosinised water.

Adhesive Traps

Many adhesive traps are used to kill insects and for this purpose, the pieces of bags or cloths are dipped in gum solution or in any adhesive material having insecticide. These pieces are hanged on the trees in the vicinity of fields, the flying insects will come across and attached with the trap and finally die. Some times an attractant is also used in these adhesive traps to attract the insects.

7.3. Physical Control

Physical methods are important in eliminating or preventing the attacks of insect pests. It involves manipulations or changes in temperature, humidity or use of radiant energy to destroy insects. Extremes of both temperature and humidity are effective barrier in insect pest population build up.

Heat

High temperature is frequently used for destroying insect infestations in soil and bulbs, in small lots of seeds and in furniture, beddings and clothing. No insect can survive when exposed to 140-150°F. Majority of insects including insects belonging to stored products are killed by three hours continuous exposure to 120-130°F.

Cold

No damage from insects occur at temperature below 40°F. Most insect pests of stored products are killed in relatively short time even at temperatures considerably above 0°. Fluctuations in temperatures from low to high and back to cold are most effective in destroying insects.

Light

Insects are attracted to different colours of light. It has been observed that many insects are much more strongly attracted by the blue portion of the spectrum and by ultra-violet then by red and yellow portions of the spectrum.

Use of Atomic Energy

Radio-active isotopes have met with success in control and eradication of screw worm, *Callitroga hominicorax* a very serious pest of cattle in *Curacap* Island. The prerequisite of this methods is: (*i*) area to be treated should be localized, (*ii*) habit of female to mate only once, (*iii*) economical breeding of pest in the laboratory, (*iv*) stage of the pest to be released should not be the damaging stage of the pest concerned.

7.4. Legal Control

Legal control is control of insects by controlling human activities. Legislative enactments are made by Government when it is found that some insects or other imported pests are likely to be transferred from place to place helps in getting distributed in areas where they do not exist before. Quarantine laws and regulations have been enacted (*i*) to prevent the introduction of foreign pests, (*ii*) to prevent or retard spread within a country, and (*iii*) to ensure the chemicals used for controlling insects are neither adulterated or misbranded.

Prohibition to prevent the introduction or spread of any dangerous insect or plant disease is called"Quarantine" and according to plant quarantine act of 1912, the government has powered to prohibit the importation or shipment of any class of plants or plant products from any foreign country or locality and from any state or portion of a state or territory in the country. In India, plant quarantine rules and regulations were made under the provision of the Destructive pest, plant disease and noxious weeds act No. 11 of 1914 under the guidance of Directorate of Plant Protection, Storage and Quarantine, Government of India. Before the enforcement of legal control following three important insect entered in India.

1. *Icerya purchasi*- pest of citrus came from Australia.
2. *Aspidiotus perniciosus*- pest of apple and pear came from China.
3. *Eriaosoma lanigerum*- pest of apple and pear came from European countries.

The insect legislation may be broadly grouped into 5 classes:

1. Legislation to prevent the introduction of new pests from foreign countries.
2. Legislation to prevent the spread of established pests within the country or within the states.
3. Legislation to enforce the application of control measures which have been found effective in preventing damage by established pests.
4. Legislation to prevent the adulteration and misbranding of insecticides and to determine the impermissible residue tolerance limits in food stuff.
5. Legislation to regulate the activities of pest control operations and the application of hazardous insecticides.

Quarantine Laws

The quarantine laws are the regulatory measures dealing with the introduction and spread of injurious insects and other pests. In India, these responsibilities rest with central Government under the provision of Destructive Pest and Insect Act of 1914 with the following rules:

1. Prohibiting or restricting the import of plant and plant materials, insects and fungi in India.
2. Prohibiting or restricting the movement of insects or diseases and their hosts from one state to another state of India.

Insecticide Laws

The central Government and state have the laws regulating the sale and usage of insecticides. The insecticide act was enforced in 1968 to regulate the import, manufacture, sale, transportation, distribution and use of insecticides so as to prevent the risk to human beings or animals.

7.5. Autocidal Control

Some control tactics are designed to suppress a pest population by altering its genetic makeup and/or reducing its reproductive potential. As a group, these tactics are frequently known as genetic controls because they affect the accuracy or efficiency with which a pest species passes its genetic material (DNA) from one generation to the next. Genetic control usually works in one of the two ways: either by causing (inducing) reproductive sterility, or by incorporating new and potentially deleterious genes (or alleles) into the genetic makeup of a pest population. In fact, some members of a pest species are transformed into biological time bombs that eventually destroy other members of their own species. Because of the self-destructive nature of these tactics, they are sometimes called autocidal control. Insects can be sterilized by exposing them to certain chemical agents (chemosterilants) or to non-lethal levels of ionizing radiation (X-rays or gamma rays). Chemosterilants are really a form of chemical control. They usually work by blocking the onset of sexual maturity, by inhibiting the production of eggs and/or sperm, or by damaging the chromosomes. These compounds are covered more fully in the section on semiochemical insecticides. Exposure to radiation also damages chromosomes (usually by breakage or mutation). Since cells with damaged chromosomes cannot divide correctly, they do not form normal gametes or produce viable offspring. Although the susceptibility of each insect species is different, a proper dose of radiation administered at an appropriate stage of development (usually to pupae) can often induce sterility without causing other deleterious side effects. Sterile individuals, reared in large numbers and released into the environment, can mate with "normal" individuals but they produce no viable offspring. The more sterile individuals released, the fewer "normal" matings are likely to occur.

The effects of radiation on insect development was first studied in 1916, but this knowledge was not put to practical use until the 1930's when E. F. Knipling suggested that it might be possible to suppress pest populations by flooding the environment with large numbers of males that had been rendered infertile by irradiation. By maintaining a constant population of sterile males that was large in comparison to the number of virgin females, Knipling calculated that the number of "normal" matings would decrease each generation until the population was forced into extinction. The first large-scale test of Knipling's sterile male technique was conducted in 1954 on the island of Curaçao against the screwworm, *Cochliomyia hominovorax*, a Dipteran pest of livestock. After just 13 weeks (four to five generations of the fly) and release of nearly a million sterile males, the screwworm population was completely eradicated from this tiny Caribbean island.

The dramatic success of Knipling's first project led to larger screwworm eradication programs in the southeastern and southwestern United States and to intense interest in the sterile male technique for eradication of other pest species, including Mediterranean fruit flies (*Ceratitis capitata*), boll weevils (*Anthonomus grandis*), horn flies (*Haematobia irritans*), tsetse flies (*Glossina* spp.), pink bollworms (*Pectinophora gossypiella*), and codling moths (*Cydia pomonella*).

Overall, the sterile male technique has had a spotty record of achievement. Despite isolated successes, no other eradication program has ever measured up to the outstanding performance of the screwworm project. From a theoretical standpoint, the use of sterile males is highly attractive because it is non-polluting, species specific, and becomes more effective as the target population decreases. But in practice, the technique simply doesn't work unless a pest population meets the following criteria:

Easy to Mass-Produce

A sterile insect release program cannot even be considered unless large numbers of insects are produced at a relatively low cost.

Females Mate Only Once

This characteristic is uncommon among insects, but it is important for success of the sterile male technique because multiple matings only increase the probability that a female will receive sperm from a fertile male.

Males can be Sterilized without Loss of Competitive Vigor

Sterile males must be able to compete successfully for mates with normal (fertile) males. Sterile males must be physically and behaviorally identical to normal males.

Low Initial Population

The pest population must be small enough so the initial release of sterile males will outnumber the population of normal males. In some cases, alternative control tactics may be used to reduce initial populations to practical levels.

Restricted Geographic Range

Sterile males must disperse to all parts of the range. Small, localized populations are easier to control than large dispersed populations.

Within the last few years, geneticists have begun to devise new genetic control strategies that work by altering DNA or by adding new genes or alleles into the genetic makeup of a pest population. Most of these new tactics are still untested in field applications, but they do offer intriguing possibilities for the future. One such approach involves breeding members of a pest species that have been genetically altered to make them more susceptible to cold winter (or hot summer) temperatures. This type of genetic trait is known as a conditional lethal mutation; it causes mortality only when triggered by extreme environmental conditions. Genetically altered individuals are mass-reared, released into the environment, and allowed to breed with other members of their species. Hopefully, the mutation will spread throughout much of the pest population before the "carriers" are killed by cold (or hot) temperatures.

Another promising approach involves cytoplasmic incompatibility. Although insects lack an immune system, some species do have different "strains" that are not reproductively compatible: the egg cytoplasm of one strain may contain substances that block or inhibit sperm from another strain. By mass-rearing and releasing large numbers of these "incompatible" insects, it may someday be possible to eradicate a pest population and replace it with a more "benign" strain.

Other Autocidal Insect Control Techniques

Some species of insects have not proved to be good candidates for the sterile insect technique, so other methods of using the insect to control itself (autocidal control) have been investigated. The following approaches have been investigated.

Inherited Sterility

This effect has also been referred to as: delayed sterility, F_1 sterility, partial sterility, etc. This phenomenon is of use in organisms, such as Lepidoptera and Homoptera that contain polycentric chromosomes. It involves the transmission of aberrant chromosomes (usually in the form of translocations) from the released population to the native population.

This can be accomplished in two ways: (*i*) by treating a laboratory-reared population of insects with a dose of radiation which is sufficient to break chromosomes, but low enough to produce little direct sterility, or (*ii*) by culturing stocks of the target species which contains homozygous translocations (this is rather difficult and tedious to accomplish). When the aberrant chromosomes are passed on to the native populations, the individuals carrying the aberrations in the heterozygous state will show sterility. The amount of sterility will depend on the number and size of the translocations carried by each individual. Some advantages of this method include: (*i*) enhanced reproductive competitiveness of the partially sterile released individuals compared to the fully sterilized individuals used in SIRM; (*ii*) the F1 individuals are produced in the native population, thus freeing rearing facilities for the production of higher numbers of the primary release organism.

7.6. Host Plant Resistance

R.H. Paineter defined *resistance of the plants to insect attack as the relative amount of heritable qualities possessed by the plant which influence the ultimate degree of damage done by the insect. Pest resistance is any inherited characterstic of a host that lessens the effect of attack.*

Mechanism of Resistance

Most widely accepted classification of mode of resistance is that proposed by R.H.Painter. These modes or mechanisms include nonpreference, antibiosis, and tolerance.

Nonpreference

Nonpreference refers to plant characterstics that lead insects away from a particular host. Recently, the term antixenosis, has been suggested as a more appropriate term for non-preference. Non-preference in a cultivar may have either a allelochemical or morphological basis. Allochemicals are chemicals used in interspecific communication.

Allelochemic Nonpreference

Nonpreference of the host is because of the presence or absence of allelochemicals. Spotted cucumber beetle, *Diabrotica undecimpunctata howardi* and cucurbitacins (tetrayclic terpenes).

Morphological Nonpreference

Nonpreference results from plant structural characterstics that interfere with normal behaviour of insects.Corn earworm, *Heliothis zea*, hairiness in cotton and resistance to jassids.

Limitations/Advantages of Non-preference

Dependent on presence of other preferred host. May breakdown if alternate preferred hosts/genotypes are absent. Non-preference especially if based on morphological characters, may be long-lasting, less chances of resistance breakdown because of lack of selection pressure.

Antibiosis

Antibiosis refers to all adverse effects on the insect lifehistory which result when a resistant host plant variety or species is used for food.

Some effects of antibiosis mode of resistance are:

☆ Reduced survival.

☆ Reduced growth rate, Insects take longer time to complete the life cycle.

☆ Reduced weight and size

☆ Shortened adult lifespan.

☆ Reduces fecundity.

☆ Morphological abnormalities.

Allelochemicals and primary metabolites are generally associated with antibiosis, *e.g.* Rice cultivar aspargine and brown planthopper, *Nilaparvata lugens*.

Tolerance

Tolerance includes all plant responses resulting in the ability to withstand insect infestations and yield satisfactorily inspite of injury levels that would debilitate nonresistant plants. Unlike nonpreference and antibiosis, only plant response is involved in tolerance. Some of the components of tolerance mechanism are:

☆ General vigor.

☆ Compensatory growth in infested plants.

☆ Quick wound healing.

☆ Chances in photosynthate partitioning.

☆ Mechanical support in tissues and organs.

Disadvantage

Insect populations are allowed to build up, this may affect other crops in the same area. Plant tolerance is also strongly affected by environmental extremes.

Apparent Resistance

Apparent resistance also referred as ecological resistance or pseudoresistance. Apparent resistance is not heritable and is dependent heavily on environment. The characterstics of this resistance are temporary and cultivars involved are potentially susceptible. There are three types of apparent resistance- host evasion, induced resistance and host escape.

Genetic Nature of Resistance

Breeding programme aimed at developing resistant varieties require a knowledge of genetic nature of the resistance. Such knowledge provides a quantative basis for designs to recombine genes and select for proper characters.

Epidemiological Types of Resistance

This classification takes into consideration the effectiveness and stability of resistant varieties that are determined by plant genes which confer resistance and the insect genes that allow the resistance to be overcome.

Vertical and Horizontal Resistance

Vertical Resistance–refers to cultivars with resistance limited to one or few genotypes of insects.

Horizontal resistance–describes cultivars that express resistance against a broad range of genotypes.

Resistance Classes Based on Mode of Inheritance

Classes of plant resistance can also be distinguished based on mode by which resistant is inherited *i.e. monogenic, oligogenic* and *polygenic* resistance.

Factors Affecting the Expression of Resistance

Although resistance is governed primarily by genetic, physical ad biotic factors in the environment which also influence its expression.

Physical Factors

Weather, soil, plant architecture, and cultural practices can influence plant's physical environment. These factors can affect plant resistance by influencing elements such as temperature, light intensity, soil fertility, soil pH.

Temperature

Abnormally high or low temperatures for a period of time may cause loss of resistance. Spotted alfalfa aphid on alfalfa- varietal performance is affected by temperature. Loss of resistance to the Hessian fly has been found in some wheats at temperatures above 18°C.

Light Intensity

Shade-induced loss of resistance has been found in several insects including stem sawfly, *Cephus cinctus*, in wheat, green peach aphid, *Myzus pesicae* resistance in sugar beets, and colorado potato beetle, *Leptinotarsa decemlineata*.

Soil Fertility

Changes in soil-nutrient levels may influence expression of resistance in some plants. Potassium fertilizers generally have negative effects on insect populations. Application of potassium fertilizers to rice increases resistance to yellow stem borer and brown plant hopper and a number of other lepidopterans.

Soil pH

Soil pH can alter nutrient availability and thereby affect varietal resistance. Fecundity, longevity and population build-up of whitebacked planthopper, *Sogatella furcifera* were significantly higher on rice plant grown under salinity stress.

Biotic Factors

Many biological factors related to host plants and insects can affect the expression of resistance.

Plant Factor

Physiological responses in plants vary with age and these can lead to changes in the expression of cultivar resistance. Plant density may also affect the expression of resistance. Damage by sorghum midge *Contarinia sorghicola* was higher in plots with lower plant densities.

7.7. Biological Methods

Practically, every crops pest has its natural enemies in the form of parasites, predators and disease causing organisms. The biological control involves a large scale multiplication of and liberation of such agents, or creating conditions under which the naturally occuring agents can act effectively. This type of control cannot be undertaken by individual farmers and has necessarily to be carried out by specialised agencies.

Some very outstanding successes have been achieved by using this method, but the method suffered a set-back owing to the large-scale and indiscriminate use of insecticides. The approach at present is to evolve methods by which the biological and chemical methods can be intergrated, so that the harmful effects of insecticides do not interfere with the activities of the natural enemies.

The introduction, encouragement and artificial increase of predators and parasites, other animals and diseases to control injurious insects is called biological control. It is related to natural control in which man plays an important role.

Biological control may be defined as destruction or suppression of undesirable insects, other animals or plants by the introduction, encouragement or artificial increase of their natural enemies. Thus, man is responsible in doing something to make the work of natural enemies of insects more effectively then it is called as biological control. These natural enemies include insectivorous birds, reptiles, frogs, etc. among higher animals and spiders, centipedes, crabs, etc. and insects are among the lower animals and diseases caused by fungus, bacteria, viruses, nematodes and protozoa.

The insect enemies of insects are the key in the biological control. Predators, such as lady beetles and lacewings, are mainly free-living species that consume a large number of prey during their lifetime. Parasitoids are species whose immature stage develops on or within a single insect host, ultimately killing the host. Many species of wasps and some flies are parasitoids. Pathogens are disease-causing organisms including bacteria, fungi, and viruses. They kill or debilitate their host and are relatively specific to certain insect groups. Each of these natural enemy groups is discussed in much greater detail in following sections.

The behaviors and life cycles of natural enemies can be relatively simple or extraordinarily complex, and not all natural enemies of insects are beneficial to crop production. For example, hyperparasitoids are parasitoids of other parasitoids.

A high reproductive rate is important so that populations of the natural enemy can rapidly increase when hosts are available. The natural enemy must be effective at searching for its host and it should be searching for only one or a few host species. Spiders, for example, feed on many different hosts including other natural enemies. It is also very important that the natural enemy occur at the same time as its host. For example, if the natural enemy is an egg parasitoid, it must be present when host eggs are available. No natural enemy has all these attributes, but those with several characteristics will be more important in helping maintain pest populations.

There are three broad and somewhat overlapping types of biological control: conservation, classical biological control (introduction of natural enemies to a new locale), and augmentation.

Conservation

The conservation of natural enemies is probably the most important and readily available biological control practice available to growers. Natural enemies occur in all production systems, from the backyard garden to the commercial field. They are adapted to the local environment and to the target pest, and their conservation is generally simple and cost-effective. With relatively little effort, the activity of these natural enemies can be observed. Lacewings, lady beetles, hover fly larvae, and parasitized aphid mummies are almost always present in aphid colonies. Fungus-infected adult flies are often common following periods of high humidity. These natural controls are important and need to be conserved and considered when making pest management decisions. In many instances, the importance of natural enemies has not been adequately studied or does not become apparent until insecticide use is stopped or reduced. Often the best we can do is to recognize that these factors are present and minimize negative impacts on them. If an insecticide is needed, every effort should be made to use a selective material in a selective manner.

Classical Biological Control

In many instances, the complex of natural enemies associated with an insect pest may be inadequate. This is especially evident when an insect pest is accidentally introduced into a new geographic area without its associated natural enemies. These introduced pests are referred to as exotics. To obtain the needed natural enemies, we turn to classical biological control. This is the practice of importing, and releasing for establishment, natural enemies to control an introduced (exotic) pest, although it is also practiced against native insect pests. The first step in the process is to determine the origin of the introduced pest and then collect appropriate natural enemies (from that location or similar locations) associated with the pest or closely related species. The natural enemy is then passed through a rigorous quarantine process, to ensure that no unwanted organisms (such as hyperparasitoids) are introduced, then reared, ideally in large numbers, and released. Follow-up studies are conducted to determine if the natural enemy successfully established at the site of release, and to assess the long-term benefit of its presence.

There are many examples of successful classical biological control programs. One of the earliest successes was with the cottony cushion scale, a pest that was devastating the California citrus industry in the late 1800s. A predatory insect, the vedalia beetle, and a parasitoid fly were introduced from Australia. Within a few years the cottony cushion scale was completely controlled by these introduced natural enemies.

Classical biological control is long lasting and inexpensive. Other than the initial costs of collection, importation, and rearing, little expense is incurred. When a natural enemy is successfully established it rarely requires additional input and it continues to kill the pest with no direct help from humans and at no cost. Unfortunately, classical biological control does not always work. It is usually most effective against exotic pests and less so against native insect pests. The reasons for failure are often not known, but may include the release of too few individuals, poor adaptation of the natural enemy to environmental conditions at the release location, and lack of synchrony between the life cycle of the natural enemy and host pest.

Augmentation

This third type of biological control involves the supplemental release of natural enemies. Relatively few natural enemies may be released at a critical time of the season (inoculative release) or literally millions may be released (inundative release). Additionally, the cropping system may be modified to favor or augment the natural enemies. This latter practice is frequently referred to as habitat manipulation.

An example of inoculative release occurs in greenhouse production of several crops. Periodic releases of the parasitoid, *Encarsia formosa*, are used to control greenhouse whitefly, and the predaceous mite, *Phytoseiulus persimilis*, is used for control of the two-spotted spider mite.

Lady beetles, lacewings, or parasitoids such as *Trichogramma* are frequently released in large numbers (inundative release). Recommended release rates for *Trichogramma* in vegetable or field crops range from 5,000 to 200,000 per acre per week depending on level of pest infestation. Similarly, entomopathogenic nematodes are released at rates of millions and even billions per acre for control of certain soil-dwelling insect pests.

Habitat or environmental manipulation is another form of augmentation. This tactic involves altering the cropping system to augment or enhance the effectiveness of a natural enemy. Many adult parasitoids and predators benefit from sources of nectar and the protection provided by refuges such as hedgerows, cover crops, and weedy borders.

Mixed plantings and the provision of flowering borders can increase the diversity of habitats and provide shelter and alternative food sources. They are easily incorporated into home gardens and even small-scale commercial plantings, but are more difficult to accommodate in large-scale crop production. There may also be some conflict with pest control for the large producer because of the difficulty of targeting the pest species and use of refuges by the pest insects as well as natural enemies.

Examples of habitat manipulation include growing flowering plants (pollen and nectar sources) near crops to attract and maintain populations of natural enemies. For example, hover fly adults can be attracted to umbelliferous plants in bloom.

Scopes of Biological Control

1. Biological control has its permanent effect.
2. Biological control agents are non poisonous, hence they do not have any toxic residual effect on crop plants.
3. Some biological control agents can be produced cheaply and conveniently.
4. Insects never become resistant against biological control agents.
5. Some biological control agents are sprayed along with insecticides so that the insect is certainly killed either by the insecticide or biological control agents.
6. Beneficial insects do not have any adverse affect as in the case of chemical control.

Limitations of Biological Control

1. Biological control involves living organism, hence a qualified person is required for it.
2. Biological control method have narrow spectrum as they chiefly kill insects of a particular species.
3. Biological control agents are slow in action so it requires sufficient time. Besides, they should be available in large numbers.
4. It requires sufficient money and time.
5. Few biological control agents are effective only in particular environmental conditions so one is required to wait for the favorable conditions.
6. Exotic biological control agents require favourable climatic conditions and sometimes they could not survive under the prevailing climatic conditions.

8

Chemical Control

The chemical which either kill the insects or repel them are used in chemical control. Insecticides are included under this group of control and are those substances which are used for protecting plants from harmful insects. The chemicals that kill the pests are known as pesticides which may be classified as under:

Pesticides	Against which Used
Insecticides	Insects
Nematicides	Nematodes
Acaricides	Mites
Rodenticides	Rodents
Molluscicides	Snails and slugs
Fungicides	Fungal diseases
Weedicides	Weeds

After the discovery of insecticide properties of DDT in 1939 by Paul Muller, the chemical control of insects has become most popular. As a matter of fact, the method has become so popular that most of the cultivators and extension workers speak only about insecticides, whenever the question of insect control is raised. The main reason for its popularity is the spectacular and immediate results obtained by use of such chemicals.

Classification of Insecticides

Classification of insecticides according to their chemical nature:

1. Inorganic

Arsenic and fluoride compounds were the mainstay of chemical pest control from the mid 1800's to the mid 1900's. These materials [*e.g.*, lead arsenate, arsenic trioxide, and copper acetoarsenate (Paris green)], are persistent in the environment and highly toxic to all forms of animal life. Most inorganic compounds were phased out after World War II. They were replaced by synthetic organic compounds that were more effective and less hazardous. *e.g.* Sodium fluoride (NaF) and cryolite (sodium fluoroaluminate, Na_3AlF_6) are stomach poisons that still have commercial applications.

2. Organic

2.1. Plant Origin

Plant extracts have been used to kill insects since ancient times. These compounds represent a wide range of chemical structures and activities.

Nicotine
  ☆ *Source*: Tobacco (*Nicotiana* spp.)–Solanaceae. Sold commercially as a fumigant (nicotine) or as a dust (nicotine sulfate).

  ☆ *Activity*: Mimics acetylcholine in the nerve synapse, causing tremors, loss of coordination, and eventually death.

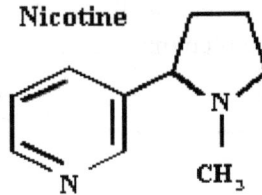

Nicotine

Sabadilla
  ☆ *Source*: *Schoenocaulon officinale*–Liliaceae. Grown commercially in Venezuela.

  ☆ *Activity*: Contains alkaloids (primarily cevadine and veratridine) that act as nerve poisons.

Rotenone
  ☆ *Source*: Tropical legumes–Leguminaceae. *Derris* spp. grown in Malaysia and the East Indies and *Lonchocarpus* spp. grown in South America

  ☆ *Activity*: Rotenone is a metabolic poison. It inhibits electron transport during aerobic respiration by blocking the action of an essential enzyme, $NADH_2$dehydrogenase. Rotenone is extremely toxic to fish.

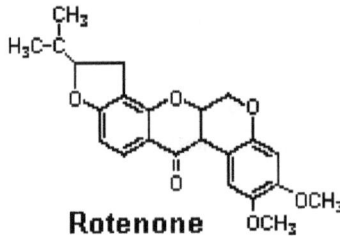

Rotenone

Ryanodine

- ☆ *Source*: *Ryania speciosa*–Flacourtaceae.
- ☆ *Activity*: Ryanodine acts as a muscular poison by blocking the conversion of ADP to ATP in striated muscles.

Pyrethrum

- ☆ *Source*: *Chrysanthemum* spp.–Asteraceae. Most of the world's supply is derived from two species grown commercially in Kenya.
- ☆ *Activity*: Pyrethrums change the permeability of sodium channels in the nerve axon. This typically results in excitation, lack of coordination, and paralysis. Since the effects of pyrethrum poisoning are often reversible, commercial insecticides usually contain a synergist, such as piperonyl butoxide, that blocks the insect's detoxication pathway.

Pyrethrin II

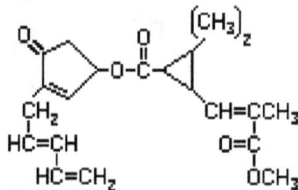

2.2. Synthetic

Organochlorines

These were the first synthetic organic insecticides ever discovered. The group includes DDT and its relatives (*e.g.*, methoxychlor and kelthane), lindane, toxaphene, and cyclodiene insecticides (*e.g.*, aldrin, dieldrin, endrin, mirex, chlordane, heptachlor, and endosulfan). Most organochlorines (they are also known as chlorinated hydrocarbons) are relatively stable compounds that act as nerve poisons. They are several thousand times more soluble in fat (lipid) than in water. This means they tend to accumulate in fatty tissues and concentrate in organisms at the top of a community's food chain (bioaccumulation). Most uses of organochlorines have been banned or discontinued in the United States because of environmental problems associated with their long persistence and bioaccumulation. Lindane is still used as a seed dressing and sold by prescription for control of human lice; endosulfan and kelthane are still used for agricultural pest control.

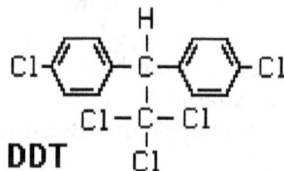

Organophosphates

These compounds were first discovered by German scientists in the late 1930's as a by-product of nerve gas research. They are much less persistent than the organochlorines and do not accumulate in fatty tissues. The group includes many general purpose insecticides with a wide range of mammalian toxicity (*e.g.* malathion, parathion, diazinon, chlorpyrifos, azinphosmethyl, acephate, phorate, and phosmet) as well as compounds that work as fumigants (*e.g.*, DDVP) and as systemics (*e.g.* dimethoate, disulfoton, demeton, and ronnel). All the organophosphates are nerve poisons. They block the active site of an enzyme (acetylcholinesterase) that breaks down and removes a neurotransmitter substance (acetylcholine) from the nerve synapse. The excessive build up of acetylcholine results in symptoms of hyperactivity, including tremors, convulsions, and eventually death. Examples-Parathion, Malathion, Phosphamidon, Dimethoate, Monocrotophos, Disulfoton Phorate, Dichlorvos

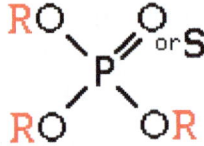

$$RO-\!\!\!\overset{\displaystyle O}{\underset{\displaystyle OR}{\overset{\|}{P}}}\!\!\!\overset{O}{} {}_{or} S$$

Carbamates

First developed in the early 1950's, this large family of insecticides is comparable in many respects to the organophosphates, sharing a common mode of action, biodegradability, low solubility in fat, and a wide range of mammalian toxicity. Carbaryl (Sevin) is probably the best known and most widely used carbamate. It is also least toxic to humans. Other carbamates include carbofuran, propoxur, methomyl, bendiocarb, formetanate, oxamyl, and aldicarb (the last three are plant systemics). One advantage, carbamates have over organophosphates is reversibility of their inhibitory reaction with acetylcholinesterase. Chronic exposure to carbamates, therefore, is less likely to cause illness than chronic exposure to organophosphates. Carbaryl, Isolan, Dimetilan, Carbofuran

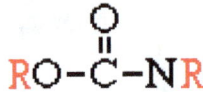

$$RO-\overset{\displaystyle O}{\overset{\|}{C}}-NR$$

Synthetic Pyrethroids

Modeled after the natural product found in chrysanthemums, these insecticides are highly toxic to insects yet relatively safe to humans. Natural pyrethrum is nearly useless outdoors because it breaks down rapidly in sunlight. The first light-stable analogue was synthesized in 1967 (resmethrin), and by 1976, permethrin and fenvalerate were commercially available as agricultural insecticides. All of these compounds act as nerve poisons; they appear to disrupt sodium transport in axons in much the same way as some of the organochlorines.

Nitroguanidines

Imidacloprid is the first member of this new class of chemical insecticides. It has both systemic and contact activity against a wide variety of sucking insects like aphids, leafhoppers, and whiteflies. Imidacloprid's mode of action appears to be similar to that of nicotine: it mimics the action of acetylcholine in the nerve synapse, causing tremors, loss of coordination, and eventual death.

Avermectins

The chemicals in this new class of compounds have been isolated from a soil-dwelling fungus, *Streptomyces avermitilis*. These pesticides (*e.g.*, avermectin, abamectin, and ivermectin), cause paralysis and death by inhibiting transmission of nerve impulses across the neuromuscular synapse. The avermectins have a rather narrow activity spectrum: they affect only insects, phytophagous mites, and certain plant-parasitic nematodes.

Classification of Insecticides According to Mode of Action

Insecticides can be classified according to their mode of action as physical, protoplasmic, chemical and stomach poisons. Insecticides belongs to the category of chemical poison are further divided as metabolic and neuroactive, the former inhibiting certain metabolic processes of the body and the letter interfering with the conduction of nerve impuses either along the axons or at the synapses.

1. Physical Poison

Exclusion of air, abrasion/absorption of cuticular lipids, *e.g.* Heavy mineral oils, inert dusts (silica aerogel).

2. Protoplasmic Poisons

The chemical cause precipitation of protoplasm. *e.g.* Heavy metals, acids.

3. Chemical poisons

3.1. *Metabolic inhibitors*: These are respiratory poisons, inhibitors of MFO (mixed function oxidases), inhibitors of carbohydrate metabolism, inhibitors of growth and reproduction, *e.g.* HCN, Co, H_2S, arsenic and copper compounds, rotenone, sodium fluoroacetate, etc.

3.2. *Neuroactive poison*: Chemicals affects the anticholinesterases, effectors of ionpermibility, etc., *e.g.* organophosphates, carbamates, pyrethroids.

4. Stomach Poison

The chemical disrupts gut epithelium, causes gut paralysis. *e.g.* Toxins of *B.t.*

Classification of Insecticides According to Mode of Entry

Insecticides can be classified according to their mode of entry in the following categories.

1. Contact Insecticides

e.g. malathion, methyl parathion, diazinon, ethion, DDVP, parathion, etc.

2. Systemic Insecticides

These are those chemicals which are absorbed by the plant system and translocated in the whole parts of the plants. The systemic insecticides have the following merits–(*i*) equal coverage is possible, (*ii*) protecting new plant growth (*iii*) having less harmful effects on natural enemies and beneficial organisms iv) increasing the residual effect as less affected by weathering. Some of the important insecticides are: dimethoate, mevinphos, monocrotophos, formothion, phorate, carbaryl, carbofuran, aldicarb, etc.

3. Fumigants

Gaseous poisons used to kill insects are called as fumigants. Their application is generally restricted to plants or products in tight enclosures or to soil. They may be used to control all types of insects since the gas enters the insect body through the spiracles during respiration. The fumigants may be divided into three broad groups:

1. Gases at ordinary temperature, *e.g.* methyl bromide, ethylene oxide.
2. Liquid at ordinary temperature, *e.g.* ethylene dichloride, ethylene dibromide and carbon tetra chloride.
3. Solids at ordinary temperature, *e.g.* sodium cyanide, aluminium phosphide.

Insecticide Formulations

An insecticide can be manufactured in a variety of formulations, each tailored for specific applications. Some of the more common formulations include:

Granules (G) or Pellets (P)

Coarse particles (*e.g.*, clay, ground corn cobs, or walnut shells) can serve as a carrier for certain types of insecticides. The toxicant slowly leaches out of the carrier, minimizing its movement within the ecosystem and maximizing its active life (persistence). Granular formulations are commonly used for controlling soil-dwelling insects and in systemic insecticides (see above) that are applied around the base of plants. Since, granular formulations do not blow or drift, they are considered relatively safe from the standpoint of accidental human exposure.

Dusts (D)

These dry powders are usually formulated with inert particles of ash, chalk, talc, or clay. They are designed to be sprinkled or blown onto target surfaces. The powder sticks to feet, legs, and other body parts of passing insects, and the toxicant is eventually ingested when the insect cleans itself. Insecticides that act as stomach poisons are often formulated as dusts. These formulations are not always suitable for outdoor applications because they have a tendency to drift (blow away with the wind).

Soluble Powders (SP) or Wettable Powders (WP)

Dry formulations that are mixed in water to form homogeneous spray solutions. A soluble powder dissolves completely in water, whereas a wettable powder contains an emulsifier that produces a uniform suspension (colloid). Many foliar insecticides are formulated as wettable (or soluble) powders because they are easy to transport and generally have a long shelf life.

Emulsifiable Concentrates (EC)

These liquid formulations contain the toxicant(s) and an emulsifier dissolved in organic solvent. The concentrate is diluted with a large volume of water to produce the final spray mixture. Some foliar insecticides are formulated as emulsifiable concentrates. Unlike most wettable powders, they do not leave a visible residue on fruits and vegetables. Sensitive plants, however, may be injured by organic solvents in the mixture. Emulsifiable concentrates are also commonly used in sprays for urban and industrial pests.

Aerosols (A)

Insecticides that are formulated together with a solvent may be pre-packaged in pressurized spray cans or sold unpressurized for use in special fogging machines. Spray cans are relatively expensive (per pound of active ingredient) but they are convenient, easy to store, and have a long shelf life. Commercial foggers are typically used indoors (*e.g.*, greenhouses and warehouses) or for control of biting flies in community-wide pest control operations.

Ultralow-Volume Concentrates (ULV)

These highly concentrated formulations (more than 8 pounds of active ingredient per gallon) are designed to be used in specialized spray equipment that atomizes the concentrate droplets. ULV spray equipment is used by most aerial applicators (airplane sprayers) who treat forested lands or large agricultural acreages.

Common Types of Adjuvants in Pesticide Formulations

Solvents	Organic molecules are not always soluble in water. Solvents such as xylene, acetone, or oils may be necessary to produce a homogeneous formulation.
Emulsifiers	Like the dish detergent that lifts grease off a dirty plate, these soap-like compounds break up oily solvents into microscopic "bubbles" and suspend them as colloids in a water solution.
Spreaders	Often called wetting agents, these compounds reduce the surface tension of water, allowing droplets to cover a larger surface area and to penetrate further into small cracks.
Stickers	Also known as adhesives, these compounds are added to an insecticide to improve its adhesion to target surfaces. Spray stickers help prevent wash-off during a rain.
Buffers	Some insecticides are sensitive to pH (acidity or alkylinity) of water. Buffers can be used in a formulation to help stabilize pH of the final spray solution within an acceptable range.
Thickeners	In some cases, it may be desirable to increase the viscosity of an insecticide formulation. Used in a spray solution, these adjuvants help reduce drift by increasing droplet size.
Baits	Insecticides that must be ingested are often formulated with some type of bait or feeding stimulant. Buffalo gourd root powder, for example, is added to some formulations as a feeding stimulant.
Synergists	By themselves, these chemical agents are not toxicants, but when mixed with an insecticide, they significantly enhance its lethal effect. Synergists are typically used with pyrethroid insecticides to block the activity of detoxification enzymes.
Abrasives	Certain substances, such as silica or pumice, can improve the efficacy of a formulation by abrading the protective wax layer from an insect's exoskeleton and improving the dermal penetration of toxicants.

New Approaches of Chemical Control

Semiochemicals

Much of an insect's behavior is mediated by chemicals in its environment. By turning these chemicals to our own advantage, it is often possible to attract pests to traps or baits, or repel them from our homes, our crops, or our domestic animals. Behavioral messages are delivered by a wide array of chemical compounds. As a group, these compounds are known as semiochemicals. In some cases, they may facilitate communication between the members of a single species (*e.g.*, pheromones) or between members of different species (*e.g.*, allelochemicals). Functionally, semiochemicals may have a wide range of activity. They may serve as attractants or repellents, they may stimulate or inhibit feeding, they may provoke flight or inhibit it, or they may simply elicit behavior patterns at inappropriate times.

Attractant pheromones and allelochemicals can be used as lures or baits in a wide variety of insect traps, or they can be mixed together with toxicants to produce an "elixir of death". Protein hydrolysates, for example, serve as feeding attractants for fruit flies (*Rhagoletis* spp.). These chemicals can be applied to sticky traps to improve catch, or combined with an insecticide and sprayed on fruit crops to suppress active infestations. Phenethyl propanoate, eugenol, and geraniol can be mixed in a 3.5:3.5:3 ratio and used as an attractant for Japanese beetles (*Popillia japonica*). These are the active ingredients in the "floral attractant" found in popular bag traps for Japanese beetles. In some cases, chemists have produced synthetic compounds that are even more attractive than naturally occurring chemicals. Trimedlure, a synthetic substitute for alpha-copaene, is produced commercially as an attractant for the Mediterranean fruit fly (*Ceratitis capitata*). Improved food lures and baits are among the most promising new developments for controlling cockroaches (Blattoidea) in homes and businesses. These are the active ingredients in a new generation of "roach motels" where the insects "check in but don't check out."

Sex pheromones are among the most powerful of chemical attractants. Ever since they were first discovered by A. A. Budenandt in 1959 (from silkworm moths, *Bombyx mori*), these chemicals have aroused great interest because of their potential as pest control agents. During the past 30 years, chemists have identified the sex pheromones for over 300 insect species. Many of these compounds are now sold commercially. In some cases, pheromones are packaged (or encapsulated) in slow-release dispensers (rubber septa, hollow fibers, or rope wicks) that are used as lures in traps of various designs. At low densities, these pheromone traps are a valuable monitoring tool, providing information on the density and distribution of pest populations. At high densities, they can be used for mass trapping sexually active adults (usually males) in efforts to reduce population density and lower a pest's reproductive potential.

Slow-release formulations of sex pheromones can also be used for mating disruption. By increasing the concentration of pheromone in an insect's environment, it may be possible to make everything smell like a prospective mate. Males wear themselves out courting inanimate objects or become habituated to the odor and stop responding to it. This approach is variously known as air permeation or the innundation technique, has shown promise for controlling a number of fruit and vegetable pests, including the codling moth (*Cydia pomonella*), the cabbage looper (*Trichop lusiani*), the oriental fruit moth (*Grapholita molesta*), and the peachtree borer (*Synanthedon exitiosa*).

Chemical repellents and feeding deterrents are also useful tools for managing insect behavior. As their name suggests, these compounds cause insects (or other arthropods) to disperse or to discontinue normal feeding behavior. Repellents such as dimethyl phthalate, benzyl benzoate, and N,N-diethyl-m-toluamide (DEET) have been developed to protect humans from biting flies, ticks, and chiggers (immature trombiculid mites). Other compounds, like di-n-butyl succinate or butoxypolypropylene glycol, are used as fly repellents for cattle. Moth balls and flakes (paradicholorobenzene or alpha-naphthalene) are placed in drawers and closets to prevent infestation by a variety of insects that feed on stored products and natural fibers. Woolen cloth can be manufactured with colorless dyes (*e.g.*, Eulans and Mitin FF) that bond permanently to the fabric and make it unpalatable to clothes moths and carpet beetles. Wood preservatives, such as pentachlorophenol, act as feeding deterrents to termites and other wood-dwelling insects.

The neem tree, *Azadirachta indica* (Meliaceae) is a promising new source of feeding repellents that may be developed for use on selected non-crop plants. The leaves, twigs, and seeds of this tree, which is grown commercially in India, contain at least 25 biologically active compounds that act as insect repellents, feeding deterrents, or growth regulators. Azadirachtin, the most abundant of these active ingredients, is now commercially available all over the world.

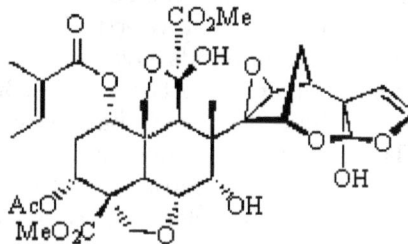

Azadirachtin

Chemosterilants

There are over four hundred chemical substances that are known to cause reproductive sterility in insects. Some of these compounds inhibit ovarian growth and development, while others appear to induce fundamental changes in the chemical structure of nucleic acids (DNA and RNA). These changes (mutations) prevent cell division or obstruct normal embryonic development. Chemosterilants belong to several major chemical groups (see Table 1). These compounds are applied directly to the insect or incorporated into food that serves as a bait.

All chemosterilants are extremely hazardous compounds. Their effects are not restricted to insects; they also cause cancer, birth defects, and other mutations in humans and domestic animals. Clearly, these chemicals cannot be dispersed in the environment like other pesticides. Instead, they must be applied under controlled laboratory conditions, usually to insects that are mass reared and released as part of a sterile release program. Although there is much interest in finding a chemosterilant whose effects are limited to insects, no such compound has yet been found.

Insect Growth Regulators: Chemical Control of Development

The enzymes and hormones that regulate developmental processes within an insect's body can sometimes be exploited as chemical control weapons. These compounds, often known as insect growth regulators (IGRs), can be used to stimulate development at inappropriate times or inhibit it at other times. The major groups of IGR compounds include:

Chitin Inhibitors

These chemicals (*e.g.*, diflubenzuron and teflubenzuron) inhibit the molting process (apolysis) by blocking the activity of chitin synthetase, an enzyme needed by epidermal cells when constructing a new exoskeleton. Because of this mode of action, chitin inhibitors are highly specific to arthropods. They act rather slowly (2-5 days), but eventually disrupt any process that involves construction of new cuticle (*e.g.*, molting, hatching, pupation). They are most effective when used against the immature stages of a pest. Diflubenzuron, currently registered under the trade name Dimilin, is used for controlling gypsy moths, boll weevils, and various other pests.

Molting Hormone Analogues

Ecdysteroids stimulate the molting process by mimicking the action of molting hormone. Applied to the surface of an insect's body or incorporated into its food, these compounds work by initiating premature ecdysis during the immature stages of development. Ecdysteroid-like compounds have been found in some plants where they evidently serve as a defense against insect herbivores. But despite their potential as insect growth regulators, the ecdysteroids have never been developed into commercial products. Their chemical structural is similar to that of human reproductive hormones (estrogen, progesterone, and testosterone), and like many other steroid compounds, they have the potential to cause cancer and birth defects.

Juvenile Hormone Analogues

Juvenile hormone (JH) and related compounds act as insect growth regulators by inhibiting the developmental changes associated with embryogenesis, morphogenesis, and reproduction. During normal development, JH levels are elevated in larvae (or nymphs) and decrease prior to pupation (or adult eclosion). Contact exposure to JH analogues during the egg stage or after the last larval (or nymphal) molt can inhibit development, delay maturation, and eventually result in death. Since, the onset of mortality is usually quite slow (days to weeks), JH analogues have limited utility in agriculture. But several compounds (*e.g.*, hydroprene, kinoprene, and methoprene) have been successfully incorporated into household products for controlling ants, fleas, and other household pests.

Anti-juvenile Hormones

These unique compounds (the precocenes) were first isolated in 1976 from a common houseplant (*Aegeratum houstonianum*). Precocenes are cytotoxic agents. They become activated by enzymes in the insect's corpora allata, selectively destroying these glands, and preventing all subsequent production of juvenile hormone. In immature insects, exposure to anti-JH compounds may result in premature (precocious) development of adult structures or behaviors. In adults, precocenes can cause sterility because the presence of juvenile hormone is necessary for normal production of eggs and sperm. Anti-JH compounds seem to be most effective against Hemipterans. Despite their unique mode of action, these IGRs have never been developed into commercial products because they break down too rapidly in the presence of oxygen.

9

Integrated Pest Management

Integrated Pest Management (IPM) is a system approach and the idea was conceived in 1954 by an American Scientist. In 1962 Rachel Carson in her book *Silent Spring* aroused worldwide concern about the excessive use of pesticides which eventually led to the concept of IPM as an environmentally sound alternative to the sole use of chemicals. Agenda 21 of the United Nations Conference on Environment and Development (UNCED) at Rio de Janeiro in June, 1992 identified IPM in agriculture as one of the requirements for promoting sustainable agriculture and rural development.

In 1967, FAO (Food and Agriculture Organization of United Nations) panel of expert defined IPM as *"a pest management system that, in the context of the associated environment and the population dynamics of the pest species, utilize all suitable techniques and methods in as compatible manner as possible and maintains the pest population at level below those causing economic injury"*. IPM emphasizes the need for simpler and economically safer measures for the pest control to reduce environmental pollution and other problems caused by excessive and indiscriminate use of pesticides.

History of Integrated Pest Management

☆ Michelbacher and Bacon (1952) coined the term "integrated control"

☆ Stern *et al.* (1959) defined integrated control as "applied pest control which combines and integrates biological and chemical control"

☆ Geier (1966) coined the term "pest management"

☆ Council on Environmental Quality (CEQ, 1972) gave the term "Integrated Pest Management"

☆ Food and Agricultural Organization (FAO, 1967) defined IPM as "a pest management system, that, in the context of associated environment and population dynamics of the pest species, utilizes all suitable techniques and methods in as compatible a manner as possible and maintains pest populations at levels below those causing economic injury".

☆ In 1989, IPM Task Force was established and in 1990, IPM Working Group (IPMWG) was constituted to strengthen implementation of IPM at international level.

☆ In 1997, Smith and Adkisson were awarded the World Food Prize for pioneering work on implementation of IPM.

Components of IPM

1. Identify the pest to be managed in crop production system.
2. Define management unit.
3. Develop pest management strategy.
4. Develop reliable monitoring techniques.
5. Establish economic threshold level (ETL).
6. Develop descriptive and predictive models.

Principles and Strategies of Integrated Pest Management

1. Monitoring Insect Pests and Natural Enemies

Pest surveillance and forecasting are essential tools in IPM which help in making management decision.

2. Concepts of Injury Levels

ETL (Economic threshold level) and EIL (Economic injury level) concepts are followed to reduce the use of insecticide and their impact on environment.

3. Integration of Pest Control Tactics

Proper choice of compatible tactics and blending them so that each component complements the other.

The strategy of applying pest management tactics is similar to that of human medicine. *i.e.* Preventive practice and Curative practice.

Goals of IPM

Increase Farm Profitability (Increase Net Profit)

☆ Prevent or avoid crop and pest problems before economic losses occur.

☆ Eliminate crop input expenses by avoiding unnecessary management actions.

☆ Improve the efficiency of management actions by adopting better application practices.

Improve Environmental Quality

☆ Judicious use of pesticides and fertilizers based on identified needs.

☆ Use selective chemicals or application methods whereever possible to reduce risk to non-target organisms.

Improve Public Image of Agriculture

☆ Far-reaching "side benefits" of reducing further regulatory and societal restrictions on the use of pesticides.

Principles of IPM

The management unit is the agroecosystem and any management action may produce unexpected and undesirable effects–this notion forms the basis of the systems or holistic approach to IPM.

☆ *Any pest exists at some tolerable level*–This notion forms the basis of the economic injury level concept.

☆ *Natural control factors regulate pest populations and are maximized* in IPM as the primary means of management; if this strategy fail to maintain pests below economic levels, then pesticides in combination with other tactics are used as a last resort.

☆ *Less than 100 per cent control is desirable to leave a permanent pest residue* for natural enemies and as a refuge for susceptible pests to reduce the chances of resistance development.

Key Components or Steps in the Implementation of IPM

1. Correct Pest Identification

What pests and stages are causing damage. This is a foundation of all decision-making.

2. Understanding of Pest and Crop Dynamics

Must have enough information about the biology of the pest encountered to assess the potential risk that the pest posess and determine the best possible management strategy.

3. Planning Preventive Strategies

As the preferred management strategy in IPM; a careful examination of field history and all aspects of the crop production system should be made to determine if the crop can be grown or treated to prevent pest populations from exceeding economic level.

4. Monitoring

Involves periodic assessment of pests, natural control factors, crop characteristics, and environmental factors to the need for control and the effectiveness of any management action. Different methods and sampling frequencies are used, depending on the type of pest and monitoring objective. Involves direct and indirect means: field scouting to make visual counts or assessment of damage, use of trapping devices (pheromone traps, light traps).

5. Decision Making

Involves an evaluation of the monitoring information to assess the relevant economic benefits versus the risks of pest management actions. Estimates of pest population size are compared to "economic thresholds" or "action thresholds" which serve as references for loss potential at particular crop growth stages or sets of crop conditions.

6. Selection of Optimal Pest Control Tactics

To manage the problem while minimizing economic, health and environmental risks.

7. Implementation

Once the management options are selected, they should be deployed on a timely manner with precision and completeness. Concept to remember for chemical control: Proper timing and placement is often more important than the rate.

8. Evaluation

Always take time to follow-up and evaluate pest control actions to determine if you got your money's worth. Review what went wrong but more importantly what went right.

*Was the choice of control action appropriate?

Use of the Economic Threshold in IPM

Complete control of pests is neither necessary in most cases for maximum yields nor appropriate for IPM. Nearly all crops can tolerate a certain amount of pest damage without appreciable effects on vigor and yield. For most of the key pests, quantitative studies of the amount of damage versus reduction in crop yield have established allowable levels of damage or population density. These measures of tolerable damage or density are referred to as economic injury levels" or "economic thresholds", which are fundamental to the goals of IPM. Without an estimate of the pest density that can be tolerated, there can be no reasonable safeguard against either over treatment with pesticides or unacceptable crop damage.

Economic Injury Level (EIL)

The lowest pest population level that will cause economic damage or the critical population density where the loss caused by the pest equals in monetary value to the cost of management.

Economic Threshold or Action Threshold (AT)

The point at which management actions should be taken to prevent an increasing pest population from exceeding the economic injury level. The ET always represents a pest density or level of pest damage lower than the EIL.

General Equilibrium Position (GEP)

The average population density of a pest over a long period of time, unaffected by interventions of pest management. This level fluctuates about a mean level as a result of biotic and abiotic regulating factors.

The position of the EIL in relation to the GEP defines the status of a pest.

The economic injury level concept is flexible and may vary from area to area, crop variety to crop variety, and even between to adjacent fields, depending on crop growth stage and specific agronomic practices.

$$EIL = C/VIDK$$

where,

    C: Management costs

    V: Market value of crop

    I: Injury per pest

    D: Crop damage per unit injury

    K: Reduction in injury or percent control

Most actual thresholds used in IPM today are more complicated and dynamic than a simple fixed level. Many are presented as decision making guidelines or rules which give variable levels, depending on a number of factors in addition to the pest population density or damage. Action thresholds can be expressed as the number of pest stages in the crop, pest damage, or a relative measure of pest activity by trapping or other indirect sampling methods.

Monitoring in IPM

IPM can not be implemented effectively without accurate estimates of pest and natural enemy population densities, or without reliable assessments of crop damage and its effects on yield.

The amount and frequency of monitoring required for decision making depends upon the crop and its pests. Almost invariably, uniformity of pest infestations does not occur, so it is essential to take a representative sample that overcomes the lack of uniformity. Also, it is important to make a representative survey of a field in the least amount of time.

Field scouting is the primary means of obtaining information to make management decisions. Upon entering any field, there are certain general procedures that must always be followed:

Frequency of Scouting Visit

Generally, each field should be scouted at least once a week. Although, the incidence of pest problems may not require that all fields be scouted each week, some fields may require checking more than once per week when infestations are borderline and may be approaching economic levels. It is important not to waste time making detailed counts when pest problems are not present. Efficient use of time will come with experience.

Scouting Pattern

Samples must be taken from representative areas of the field. There are many sampling patterns that may be used when scouting a field. Generally, one should move about 50 feet beyond the end rows before making counts. Border rows should be avoided unless there are special reasons for surveying these areas. For example, certain pests invade fields move from outside areas and thus may be expected in these areas first.

There are three basic patterns for pest infestations in a field and sampling should be arranged accordingly.

Pattern 1

Pests expected to be uniformly spread over the field. When scouting for a pest with this distribution, the sample sites are chosen so as to be evenly distributed over the field, excluding obvious influencing factors such as field edges. In a square field this might mean one sample in each corner and one in the center. Pests fitting into this pattern are the alfalfa weevil, potato leafhopper, European corn borer, corn rootworm adults, corn leaf aphids, corn earworm, most foliar diseases, etc.

Pattern 2

Pests expected to be concentrated in particular areas of a field. Examples: many annual and perennial weeds, early season infestations of Mexican bean beetles, black cutworm, white grubs, *Phytophthora* root rot or other root diseases may be distributed in high or low spots, or other distinguishable features of the field. If pests are detected in one spot and not in others, sub-samples should be made in that region to determine the extent and severity of the problem more accurately.

Pattern 3

Pests expected to appear at field edges first. Examples are spider mites, common stalk borer, armyworm, grass sawfly, and grasshoppers. Sample for these pests by walking fields borders or waterways.

It may be necessary to combine 2 or more patterns on one sampling date. For example, scouting corn for armyworms and stalk borers can be accomplished by walking the field borders followed by sampling throughout the field for weeds, European corn borer, and other pests.

Sampling Unit

Once the pattern for sampling has been established, the method of selecting a sub-sample or sampling unit then becomes important. A sampling unit consists of a specific number of plants, a specific number of feet of row, or a specific area in square feet. Regardless of the sampling unit used, the first plant or site to be examined is chosen at random, *i.e.*, wherever the trowel falls when thrown, etc. The number and size of the sampling units are dependent upon the pests being sampled, their distribution patterns, and other factors.

If the sampling unit involves a specific number of plants at each sub-sample site, two methods to select those plants are used. Consecutive plants are examined when the pest will not be disturbed by your action to adjacent plants. For example, with European corn borer, cutworms, stalk rot or seedling diseases, a series of consecutive plants are examined. Random plants are examined when mobile insects are being surveyed. In this case, the next plant to be examined will not be adjacent to you but will be some distance away. For example, random sampling would be used with corn rootworm adults.

Scouting and Field Size

Most recommended scouting procedures are designed to provide an accurate assessment of pest activity in 40 acres of a crop field. Many fields will be much larger or smaller than 40 acres. Small fields should be scouted using the full sampling plan. Larger fields should be divided into smaller units of approximately 40 acres for scouting. For example, an 80 acre field can be broken into two 40-acre sections for scouting purposes and a complete sample should be taken in each section.

Fields should be divided into smaller units if parts of the field are under different management systems. For example: 20 acres of corn and 20 acres of soybeans last year planted as one 40-acre corn field this year should be scouted as two 20-acre sections because weed, insect, and disease problems may differ, depending on the preceding crop.

Types of Pest Monitoring Methods

Absolute Methods

Estimates of pest population density are expressed as a level per unit of crop area or as a percentage of the sampling units affected. Examples are direct visual counts per plant or per foot of row or per unit of area.

Advantages

Broad range of applicability, less influenced by spatial patterns and changes in pest behavior and sampling efficiency, easier to predict potential crop damage.

Disadvantages

More time-consuming

Relative Methods

Estimates of pest population activity per unit of effort or time but not expressed with units of the crop area. Examples include visual searches, sweep net sampling, beating or shake cloth estimates, blacklight traps, pheromone traps, visual sticky traps, and bait traps.

Advantages

Yield more data given the same effort, less time-consuming, easier to implement.

Disadvantages

Efficiency is affected by pest behavior, diurnal activity, weather conditions, the crop habitat being sampled, and variations in the way the methods are deployed; requires more information to relate relative estimates to potential crop damage.

Population Indices

Estimates of crop damage or the frequency of pest infestations which indirectly reflect the size of the pest population. Examples are percentage of plants infested or diseased, percentage of defoliation, percentage of damaged fruits, visual ratings of root or foliage injury, etc.

Advantages

Less time-consuming and easy to implement, more directly related to crop yield losses.

Disadvantages

Can not be used alone to make control decisions, may not allow enough time to take management actions.

Preventive Methods of IPM Include the Following

Many IPM practices are used before a pest problem develops to prevent or stall the buildup of pests.

Cultural Controls

Cultural controls are those that disrupt the environment of the pest. Ploughing, crop rotation, removal of infected plant material, sanitation of greenhouse equipment, and effective manure management are all cultural practices that are employed to deprive pests of a comfortable habitat. The management of urban and industrial pests has improved when sanitation programs have been improved, pest harborages eliminated, garbage pickup frequency increased, or when lights are installed that do not attract insects.

Structural Modifications

By preventing support timbers from soil contact, damage from several different wood destroying pests can be avoided. Wood absorbs moisture and is more susceptible to attack by carpenter ants and termites when in direct contact with the soil.

Construction Site Sanitation

Removing tree stumps and lumber scraps from construction sites, which are prime food sources for subterranean termites, can prevent problems in the future.

Biological Controls

Using natural enemies (biological control agents) to keep pests in check can be put into place before pest problems increase. Examples of biological control agents are beneficial mites that feed on mite pests in orchards, the milky spore disease that kills harmful soil grubs, and En*carsia formosa*, a wasp that parasitizes the greenhouse whitefly. Many biological control agents are commercially available.

Physical Barriers

Physical barriers such as netting over small fruits and screening in greenhouses can prevent crop loss. Physical barriers are important in termite, house fly, and rodent control.

Use of Pheromones

Use of pheromones (natural insect scents) has become widely used in pest management. Sometimes a manufactured "copy" of the pheromone that a female insect emits to attract males can be used to confuse males and prevent mating. This technique is used in curbing damage from the grape berry moth.

Pest-Resistant Varieties

Pest-resistant varieties are those that are less susceptible than other varieties to certain insects and diseases. Use of resistant varieties often means that growers do not need to apply as many pesticides as with susceptible varieties. Potato growers control the golden nematode by planting resistant cultivars. Apple growers can save up to eight fungicide applications a year by growing Liberty and Freedom cultivars, which resist diseases. Farmers growing alfalfa and wheat keep several pests at bay by planting resistant varieties.

Curative Methods of IPM Include the Following

1. Physical and mechanical methods
2. Inundative method releasing biocontrol agents
3. Chemical insecticides, IGR

   ☆ Preventive methods can be used, irrespective of the level of pest incidence. It can be followed as a routine, even if the pest is at a low level.

   ☆ Curative methods have to be followed only when the pest attains economic threshold level (ETL).

The Following are Some Examples of Successful IPM Programmes Worldwide

1. In Philippines in 1993, IPM farmers obtained 4.7 to 62 per cent higher rice yield and reduced pesticide use by 15 per cent compared to non-IPM.
2. In India in 1995, IPM farmers obtained 6.2 to 42.1 per cent increased rice yield, and reduced pesticide use by 50 per cent compared to non-IPM farmers.
3. In India on cotton crop, adoption of IPM technology resulted in 73.7 and 12.4 per cent reduction in the number of insecticide sprays against sucking pests and bollworms. Inspite of reduction in pesticide sprays, 21-27 per cent increase in seed cotton yield was obtained in IPM areas compared to non-IPM. Natural enemy population also increased 3 folds.
4. In Thailand in 1993, adoption of IPM technology resulted in 145 per cent increase in net profit in IPM fields over non-IPM fields in cruciferous vegetables.
5. IPM is useful and economical in high value, plantation crops like Coconut, Coffee, Tea, Cashewnut and Arecanut.

Constrains in IPM

1. Institutional Constraint

IPM requires interdisciplinary approach to solve pest problem. Lack of coordination among different institution is a constraint. Research programme based on farmer's need–is lacking.

2. Informational Constraint

Lack of information on IPM among farmers and extension worker. Lack of training on IPM.

3. Sociological Constraint

Some farmers feel it is risky to adopt IPM compared to use of pesticides alone. Our farmers are habituated of using more pesticides.

4. Economic Constraint

Lack of funds for training farmers and extension workers on the use of IPM.

5. Political Constraint

☆ Vested interest associated with pesticide trade

☆ Pesticide subsidy by Government. These are the constraints for the implementation of IPM.

Options/Strategies in IPM Implementation

Acceleration of IPM implementation requires the following:

1. Farmer's Participation

Farmers must be encouraged to participate in IPM and give their views.

2. Government Support

Government can remove subsidies on pesticides and allot more fund for IPM implementation.

3.. Legislative Measures

Suitable legislation (law) may be passed for adopting IPM by all farmers (IPM will be successful only if adopted on community basis).

4.. Improved Institutional Infrastructure

National level institution for implementation of IPM is a must. Data base on role of biotic and abiotic factors on pest population, crop yield are required.

5. Improved Awareness

Awareness should be created at all levels on IPM *i.e.* Policy makers, farmers, consumers and general public. NGOs (Non Governmental Organisation) should be made aware of the advantages of IPM.

(10)
Biotechnological Approaches to Pest Management

Biotechnology can be broadly defined to include all practical uses of living organisms. As such, biotechnology has been practiced since the beginning of recorded history through endeavors such as fermentation of microorganisms for production of beer, selective breeding of crops, beekeeping for the production of honey, and maintenance of silkworms for the production of silk. Laboratory techniques developed within the last 20 years that enable transfer of genes from one organism to another have resulted in tremendous scientific and commercial interest and investment in biotechnology. The word biotechnology is now commonly used to refer to manipulation of organisms at the molecular level.

Biotechnology in Pest Management

Use of molecular biology techniques for the management of insect pests. The following are some strategies.

1. Wide hybridization

This technique involves transfer of genes from one species to other by conventional breeding. The genes for resistance are transferred from a different species, *e.g.* WBPH resistant gene has been transferred to *Oryza sativa* from *O. officinalis*.

2. Somaclonal Variability

The variation observed in tissue culture derived progeny, *e.g.* Somaclonal variants of sorghum resistant to *Spodoptera litura* has been evolved.

3. Transgenic Plants

Transgenic plants possess one or more additional genes. This is achieved by cloning additional genes into the plant genome by genetic engineering techniques. The added genes impart resistance to pests.

Transgenic plants have been produced by addition of one or more following genes.

(a) *Bt* endotoxin from *Bacillus thuringiensis*

(b) Protease inhibitors

(c) α-Amylase inhibitors

(d) Lectins

(e) Enzymes

(a) *Bt* endotoxin Gene

The gram positive bacteria *Bacillus thuringiensis* produces a crystal toxin called δ (delta) endotoxin. The δ endotoxin is a stomach poison and kills the lepidopteran insects, if consumed. The gene (DNA fragment) responsible for producing δ endotoxin is isolated from Bt and cloned into plants like cotton, potato, maize, etc. to produce Transgenic plant, etc.

Transgenic Bt plants and target insect pests

Crop	Pest
Cotton	Bollworms
Maize	European corn borer
Rice	Leaf folder, stem borer
Tobacco, Tomato	Cut worms
Potato, Egg plant	Colarado potato beetle

(b) Protease Inhibitors (PI)

Protease inhibitors (PI) gene Insects have proteases in their gut which are enzymes helping in digestion of protein. Protease inhibitors are substances which inhibit the proteases and affect digestion in insects. The protease inhibitor gene are isolated from one plant and cloned into another to produce transgenic plants, *e.g.* Cowpea trypsin inhibitor (CpTI) is a PI isolated from cowpea and cloned into tobacco. This transgenic tobacco is resistant to *Heliothis virescens*.

(c) α-Amylase Inhibitor Gene

Amylase is a digestive enzyme present in insects for digestion of carbohydrate. α-Amylase inhibitor, affect digestion in insects.Transgenic tobacco and tomato expressing α-amylase inhibitor have been produced which are resistant to Lepidopteran pests.

(d) Lectins Genes

Lectins are proteins that bind to carbohydrates. When insect feed on lectins, it binds to chitin in peritrophic membrane of midgut and prevents uptake of nutrients, *e.g.* Transgenic tobacco containing pea lectin gene is resistant to *H. virescens*

(*e*) Enzyme Genes

Chitinase enzyme gene, and cholesterol oxidase gene have been cloned into plants and these show insecticidal properties. pyramiding genes Engineering transgenic crops with more than one gene to get multi-mechanistic resistance is called pyramiding of genes, *e.g.* The CpTi gene and pea lectin gene were cloned to produce a transgenic tobacco. Transgenic potato which express lectin and bean chitinase have been produced.

4. Transgenic Arthropod Natural Enemies

Recombinant DNA methods may be applied to produce improved strains of natural enemies such as predatory arthropods and parasitoids, but techniques are in the early stages of development. For example, the western predatory mite, *Metaseiulus occidentalis* (Acari) is among the group of mites that are mass reared for the control of spider mites. However, pesticides applied for control of other pest species often wipe out the predatory mites. Engineering beneficial insects such as the western predatory mite with insecticide resistance genes would in theory provide protection from chemical sprays applied for control of insect pest species.

5. Engineered Insect Pathogens for Pest Control

Insect pathogenic bacteria, viruses, fungi and nematodes have been used for the management of insect pests in various niche markets. However, each agent suffers from at least one major limitation such as susceptibility to environmental stress, temperature extremes, desiccation or solar radiation. Most work has been done on the genetic enhancement of bacteria and viruses in part because of the relative ease of genetic manipulation of these organisms. Genetic engineering to enhance the insecticidal properties of entomopathogenic nematodes and fungi is in infancy. Genetic engineering has been used to enhance the insecticidal efficacy of various strains of *Bt* by increasing virulence, extending host range and increasing field stability and by introducing alternative toxins to a facilitate resistance management. Techniques have been developed for production by genetic means of new strains of *Bt* with new combinations of toxin genes.

Advantages of Biotechnology in IPM

1. Slow development of resistance against transgenic *Bt*, protease inhibitors, lectins.

2. All plant parts express toxin and so no need for insecticide spray.

3. No need for continuous monitoring.

4. No environmental pollution, safe to natural enemies, non-target organism.

11

Insect Vectors of Plant Diseases

It is a well known fact that certain insects act as vectors of plant diseases. As early as in 1912, scientists have proved that fire blight of fruit plants is transmitted by insects. Plant diseases transmitted by insects may be grouped as follow:

1. Fungal diseases: 15-20 per cent
2. Bacterial diseases: 8-10 per cent
3. Viral diseases: 60-80 per cent
4. Other diseases

In nature insects depend on plants for the food to complete their life cycle and they are known to transmit different diseases in plants of economic importance. While obtaining food from plants, insects make tunnels, wounds, etc. which serve as a media for the growth of microbes (fungi, bacteria and virus) which are responsible for diseases in plants. Insects transmit various diseases in plants by the following ways:

1. Insects as direct vectors of plant diseases.
2. Insects as indirect vectors of plant diseases.
3. Insects as intermediate host of plant diseases.
4. Insects as catalyst of plant diseases.
5. Insects as vectors of plant diseases.

1. Insects as Direct Vectors of Plant Diseases

Several insects carry within their body micro-organisms of plant disease and they visit from one plant to another in search of food. Bacterial and fungal spores easily adhere in different parts of insects body and as such transmit diseases from one plant to another. Aphids, honeybees and butterflies are good examples of this class of insect vectors.

2. Insects as Indirect Vectors of Plant Diseases

In indirect approach, insects damage the plants by making holes, wounds, tunnels, etc. on which micro-organisms easily attack and transmit various plant diseases. If the plants are healthy and not damaged by insects, micro-organisms will not penetrate the bark or epidermis of the plant and as such, they transmit various diseases in the plants indirectly, *e.g.* early blight of potato by flea beetle of potato and moulds in maize after the attack of corn ear worm.

3. Insects as Intermediate Host of Plant Diseases

Insects act as media of food for several micro-organisms and as such insects are not there, the life cycle of these micro-organisms will not be completed. Among such micro-organisms, bacteria are the main which grow in adverse conditions in insect body and in favourable conditions they are known to transmit diseases in plants. Cucumber beetle is a good example. On the body of these beetles, micro-organisms live and cause cucurwilt disease in plants of cucurbitaceous family in favourable conditions.

4. Insects as Catalyst of Plant Diseases

Certain insects act as catalyst of plant diseases, *e.g.* black rust of wheat stem caused by *Puccinia graminis tritici*. Insects play important role in transmitting this disease.

5. Insects as Vectors of Plant Diseases

Most of the viral diseases of the plants are transmitted by the insects. The moment insects move on the healthy plants after taking their food from infected plants, they also carry within themselves the virus to the healthy plants. Some viruses complete their life cycle inside the body of specific insects and act as a disease carrier. Some of the viral diseases are transmitted through contact by insects after taking their food. Such viruses are transmitted through having piercing and sucking type of mouth parts, *i.e.* leaf curl disease of tomato and chilli is transmitted by white fly. About 300 plant diseases transmitted by viruses are making all over the world which are mostly spread by insects. Hemipteran insects play an important role in their spread. Among *hemipterans*, aphids are most prevalent in spreading about 90 per cent viral disease followed by hoppers and whitefly.

Common Insect Vectors of Plant Diseases

A. Aphid Vectors

The biology, feeding behaviour and worldwide distribution of aphids make them ideally suited for transmitting plant viruses. Over 200 species of aphids have been reported as vectors of plant viruses. The mouthparts are the direct means by which aphids pierce the plant tissues, extract sap and in the process acquire and inoculate plant viruses. The mandibular and maxillary stylets have series of projections which serve to anchor the stylets into the plant tissue during penetration. Secretion of the salivary sheath continues as the stylets penetrate intercelularly in the phloem.

Two classification systems are in use to describe virus transmission by aphids. In the first one, viruses are grouped into non-persistent (NP) and persistent (P), on the basis of retention of infectivity of the vector following acquisition.

The second widely accepted system of grouping plant viruses is based on the route of virus transport classifying them as stylet borne, circulative and propagative viruses. Aphid transmissible viruses affect a large number of cultivated crops, fruit trees, ornamentals and medicinal plants besides infecting numerous weeds.

Among the graminaceous crops, rice and wheat are free from aphid-borne viruses but barley yellow dwarf virus (BYDV) has been recorded from India. The potato viruses (PVY and potato leaf roll virus) are major constraints to the cultivation of solanaceous crops particularly potatoes, Chilli mosaic, a common disease is caused by chilli mosaic virus, PVY and CMV. Papaya mosaics and Citrus tristeza are common diseases of fruit plants. While many plant viruses are transmissible through several aphid vectors, banana bunchy top and *foorkey* of large cardamom are so far known to be transmitted by single vector species.

The transmission of mosaic streak of large cardamom by *Brachycaudus helichrysi* (Kaltenbach); broad bean mosaic by *Macrosiphoniella sanborni* (Gillette); papaya mosaic by *Uroleucon sonchi* (L); radish mosaic by *Rhopalosiphum maidis* (Fitch), *Schizaphis graminum* (Rondani), *Toxoptera odinae* and *T. citricidus* (Kirkaldy); and the transmission of citrus tristeza by *Uroleucon (Uromelan) jaceae* (L.) are some classic examples of this kind.

B. Hopper Vectors

Two aspects in particular, have raised much interest in hopper borne plant viruses; (a) most of them multiply both in plants and their vectors and (b) plant reoviruses and rhabdoviruses closely resemble animal viruses of the same taxonomic groups. The number of hopper borne viruses is about 60 and that of their vector species is 55-60. With few exceptions, hopper borne viruses are not transmissible through sap inoculations.

The great majority of hopper borne viruses infect rice and maize. In Japan, Rice dwarf virus (RDV) over winters in diapausing leafhoppers and is transmitted to early rice nurseries and the yield loss varies from 20-80 per cent.

A number of hopper borne viruses have been reported infecting wheat, barley, oats, etc. BCTV is the virus of most consistent economic importance that infects dicotyledons.

Survival of hopper-borne viruses falls into two categories. In first viruses are maintained in susceptible crops and hopper where it vectors throughout the year without any obligatory passages through intermediate host plants. In the second category, viruses survive in wild host plants for limited period of the year and are introduced annually into crops.

Plant Diseases Spread by Insects

Name of Disease	Host Plant	Insect Vector	Mechanism of Transmission	Causal Organism
		VIRAL DISEASE		
Tobacco mosaic	Tobacco	Aphid (Aphis gossypii)	Virus rear inside body and by food	Filterable viruses
Cruciferae mosaic	Mustard, cole crops, radish, turnip	Aphid (Bravicorn brassicae)	Virus rear inside body and by food	Filterable viruses
Potato leaf roll and crinkle disease	Potato	Aphid (A. rahmnii, Myzus persicae)	Rearing inside body and by food	Filterable virus
Sugarcane mosaic	Sugarcane, maize, sorghum	Aphid (A. maidis)	Rearing inside body and by food	Filterable viruses
Cucumber mosaic	Cucurbits and tobacco	Aphids and beetles	Rearing inside body and by food	Filterable viruses
Bean mosaic	Bean	Aphid (Myzus persicae)	Rearing inside body and by food	Filterable viruses
Streak disease	Maize, millet, sorghum	Leaf hoppers (Sendulling spp.)	Rearing inside body and by food	Filterable viruses
Stunt or dwarf disease	Mango	Leaf hopper (Nephotettix spp.)	Rearing inside body and by food	Filterable viruses
Curly blossom of mango	Mango inflorescence	Leaf hopper (Amritodus atkinsoni)	Rearing inside body and by food	Filterable viruses
Spindle tuber	Potato	Flea beetle, locust, Colorado potato beetle and bugs	Rearing inside body and by food	Filterable viruses
Spotted or yellow wilt	Tomato and pineapple	Thrips	Making wound by feeding	Filterable viruses
Potato mosaic	Potato	Aphid (Myzus persicae)	Rearing inside body	Filterable viruses
Yellow vein mosaic of okra	Okra	White fly (Bemisia tabaci)	Rearing inside body	Filterable viruses
Tomato leaf curl	Tomato	White fly (Bemisia tabaci)	Rearing inside body	Filterable viruses
Yellow mosaic of bean	Bean	White fly (Bemisia tabaci)	Rearing inside body	Filterable viruses

Contd...

Contd...

Name of Disease	Host Plant	Insect Vector	Mechanism of Transmission	Causal Organism
		BACTERIAL DISEASES		
Tobacco leaf curl rot	Tobacco, cauliflower	Aphid and cabbage looper	By mouth parts	*Bacillus campestres*
Bacterial soft rot	Cauliflower and potato	Cabbage maggot and seed corn maggot	Rearing inside body or at the time of egg laying	*B. carotovorus*
Cucurbit wilt	Bottle guard, musk and water melons	Spotted cucumber beetle	Rearing inside digestive system, while taking food and excreting	*B. tracheidhilus*
Bacterial wilt	Maize, millet	Flea beetle, root worm and seed corn maggot	Rearing inside digestive system, while taking food or by remaining wounds	*Aplanobactor stewarti*
Fire blight	Bean, pear	Honey bee, wasp, aphid, flies, leaf hoppers	Wound made while taking food or by direct entrance	*B. amylovorus*
		FUNGAL DISEASES		
Root rot	Sugarcane	Mucid and sarcophagids	By digestive system or body parts. Flies give spores on leaves by eating then, in the form of excreta	*Ithyphallus cardlloides*
Ergot	Wheat, barley, grass	Honey bees and flies	By digestive system or body parts. Flies give spores on leaves by eating then, in the form of excreta	*Claviceps purpurea*
Early blight	Potato and tomato	Beetle (*Apitrix cucumcris*)	Fungi are deposited beneath the epidermis of leaves while eating	*Alternaria solani*
Perennial canker	Apple	Wooly aphid (*Eriosoma lanigerum*)	Fungi are deposited beneath the epidermis of leaves while sucking fruit juice	*Gleoporium perennans*
Black rot	Apple	Codling moth (*Carpocapsa pomonella*)	On the wounds of flowers while feeding	*Physalopora obtusa*
		MLO DISEASE		
Phyllody	Sesamum	Hoppers (*Orosius aibicinctus*)	While sucking the sap of leaves	*MLO*

A few hopper borne plant viruses, *viz*. Chlorotic striate mosaic (CSMV), Fiji disease (FDV), Pangola stunt (PaSV) can reproduce vegetatively but none is a seed borne. Thus, for most of them, the only efficient means of dispersal is by hopper vectors.

Some viruses like rice ragged stunt virus (RRSV), cereal chlorotic mottle (CCMV), colocassia bobone disease (CBDV) and wheat dwarf (WDV) are transmitted by only one hopper species, while others *viz*. rice dwarf (RDV), rice gall dwarf (RGDV) and maize streak virus (MSV) are transmitted by several species.

C. Whitefly Vectors

Whitefly transmitted diseases mainly occur in tropical countries but are also found in sub-tropical and temperate countries. Whitefly transmitted diseases have become important on legume crops, tomato, cotton and chilli in various parts of the world. Three species of whiteflies *Bemisia tabaci*, *Trialcurodes vaporariorum* and *T. abutilonia* are known to transmit plant viruses. There are about 1156 known whitefly species worldwide. *Bemisia tabaci* is responsible for transmitting a large number of viruses.

Management Tactics of Insect Vectors

1. Cultural Control

Effective control of tobacco leaf curl disease was obtained by keeping the fields fallow for three weeks after destroying the remains of tobacco crop. This prevented the whitefly vectors from acquiring virus from infected roots of the previous crop. Yellow vein mosaic of Okra can be managed by rouging of diseased plants in early stages and removal of alternate host plants. The use of straw mulching in cucumber fields to prevent the spread of whitefly-borne bottle gourd mosaic virus was found effective. The susceptibility of tobacco plants to leaf curl virus decreases with increase in phosphorous content.

2. Use of Chemotherapeutants

Use of growth regulator (Gibberellic acid) on tobacco was useful in reducing the severe curling and puckering of tobacco leaves. Further, root dipping of tomato seedlings in gibberellic acid and 2-thiouracil at 50 ppm was useful in reducing the severity of leaf curl disease of tomato.

3. Use of Resistant Varieties

Cotton cultivar Lambert is resistant to leaf curl virus. Pearl harbour variety of tomato is resistant to tomato leaf curl virus. The chilli varieties, *viz*. Puri red and Puri orange were found resistant to chilli leaf curl. A wild species of Okra *Abelmoschus manihot* was immune to yellow vein mosaic virus, while IC-1542, selection 1-1 and selection 2-2 exhibited some degree of tolerance. Sesamum varieties *viz*. T 13-3/2, 65-1/1-1 and 67-B-1/2-1 were found moderately resistant to phyllody.

4. Biological Control of Vector

In Israel, a mite *Amblyseius rubini* was recorded as effective predator of *Bemisia tabaci* Gen., *Encarsia formosa*, a chalcid parasite of whitefly, also showed promise.

5. Use of Oils

Mixture of oil (Mineral oil 2 per cent) + endosulfan (0.025 per cent) + Oxy-demeton methyl (0.05 per cent) was found successful in controlling the whitefly *B. tabaci* in pulse crops.

6. Use of Insecticides

Rouging of infected plants in combination with weekly spray of pyrocolloid (1 part in 800 parts of water) and derriphyton (1 per cent) was effective in suppressing okra yellow mosaic virus. Similarly, the spread of virus was also restricted by controlling the whitefly *B. tabaci* in okra by systemic insecticides and incidence of yellow mosaic virus was reduced. Foliar applications of oxy-demeton methyl or dimethoate (4 to 6 per cent) reduce the vector population effectively. The application of systemic insecticides, *viz.* dimethoate as spraying and phorate as soil application are useful in controlling the vector born diseases in many crops.

12

Insect Resistance to Insecticides

When certain insecticides are used to kill insects continuously and intensively, they lead to development of resistant strains/races of insects. The ability of insects to develop resistance to insecticide action was first observed with San-jose scale against lime sulphur in 1914. After that, several insect pests were found showing resistance to number of insecticides. At present more than 504 species of insect have developed resistance against insecticides, worldwide. The wide spread usage of organic insecticides has resulted in the development of resistance in mosquitoes, houseflies and cockroaches. Resistance is seen normally in field operation as a progressive inability of a given compound at a fixed application rate to achieve control. Resistance is expressed in relative degrees such as 10 folds, 20 folds, 50 folds. It refers to amount of insecticide required to kill 50 per cent resistant population and is 10, 20 and 50 times more than the quantity required to kill 50 per cent of the normal population.

Resistance may be defined as "the added ability to withstand insecticide acquired by breeding of those individuals which survive exposures to that particular toxicant insufficient to wipe out the whole colony."

The official definition of resistance is "development of an ability in a strain of insects to tolerate doses of toxicant which would prove lethal to the majorities of individuals in a normal population of the same species".

FAO defines resistance as a decreased response of a population of animal/plant species to a pesticide or control agent as a result of their application.

Development of Pesticide Resistance in Insect and Mite Pests

Year	No. of Resistant Pests
1953	07
1954	23
1960	137
1967	224
1975	264
1975	364
1986	447
1993	504

Origin and Development of Resistance

Generally two explanations are given for the occurrence of the resistance in insects against insecticides.

1. Pre-adaptation

It explains the genetic differences already present in the population and the insecticides act as selective agent favouring the resistant genotype.

2. Post-adaptation

In which the resistance is physiological and does not depends on genetic constitution.

The pre-adaptation explanation has been accepted widely. The origin and development of resistance is biphasic. In the first phase, those variants which carry pre-adaptive genes are selected and in the second phase the enhancement of resistance taken place in those variants by the presence of pre-existing detoxifying enzymes which breakdown the toxicant much faster. The rate of development of resistance in a population depends upon the following factors:

1. The frequency of resistant genes present in population.
2. The intensity of the selection pressure of the toxicant.
3. The rate of breeding of the species.

Pesticides do not produce resistance; they merely select resistant individuals and confer resistance to their progeny in the genes so succeeding generation of insects become resistant to the pesticides. In majority of cases, the pesticides probably do not induce mutations, which confer resistance, though this may be true for warfarin-resistant rats.

When an insect population (A) is treated with an insecticide, it simply eliminates the susceptible (S) individuals (strains) of the population, leaving mostly the R individuals, (B) with further treatments, only R individuals are left out, (C) which increase in number and since they are not killed by the same insecticides, we say that they have become resistant to that insecticide.

Different Terms

Behavioural Resistance

Resistance occurs when an insect strain develop the ability to withstand lethal dose, rather than developing a physiological difference to the same.

Extra Vigour

It is the added ability to withstand a toxicant which appears to be gained from improved nutrition, extra weight or any other factor usually associated with, may be called extra vigour.

Vigour Tolerance

A strain developed by breeding only from those individuals which survive exposure to such diverse stress as extremes of temperature, lack of moisture, abnormal food or an injurious chemicals will have an altered ability to withstand many kinds of stress including exposure to chemicals. This change from the normal tolerance is called vigour tolerance.

Cross Resistance

The phenomenon where by a strain is subjected to selective pressure with one compound becoming less sensitive to the other compound to which it has not been exposed before is called cross resistance.

Multiple Resistance

Multiple resistance is the co-existance of different defense mechanism in the same strain. The multiple resistance develops due to simultaneous or consecutive use of several insecticides under field conditions.

Monogenic Resistance

When a singal gene is involved in the development of resistance, the resistance is called monogenic. In this case the level of resistance may be very high, *e.g.* 200 folds resistance to organophosphates in spider mite.

Polygenic Resistance

In polygenic resistance, several genes are involved for the development of resistance.

Mechanism of Resistance

1. Behaviouristic and ecological consideration
2. Genetic consideration
3. Biochemical and physiological consideration

1. Behaviouristic and Ecological Consideration

It is the ability of a species to avoid doses of toxic substances, which would otherwise, be lethal. Some insects are able to develop resistance due to change of their behaviour and ability to prolong the exposure of the insecticides, *e.g.* mosquitoes, avoid surface treated with DDT, 3ʳᵈ instar larvae of *Helicoverpa* avoid malathion ULV droplets of a suitable size on cotton leaves, the population of housefly has developed a characteristic of not coming readily to malation in sugar bait.

Behaviorist resistance is the ability of insect through protective habits or behavior to avoid lethal contact with a toxicant. This type of resistance is found in codling moth with due to its acquired habit, rejects the first bite prior to penetrate in the fruit. Thus, it avoids ingestion of toxicant sprayed over the fruit. Behaviorist resistance is also found in mosquitoes, DDT causes irritability to anophele mosquito with the result that mosquitoes avoid the treated surface. Nocturnal migration of houseflies from sprayed houses to near by unsprayed vegetation has been reported.

2. Genetic Consideration

It has been established that insecticides are not mutagenic agent. It is therefore, clear that changes that occurs in a population when exposed to insecticides must be due to mutation occurring for the reasons other then insecticides exposure. The mutation in a population is probably due to the spread of genetic character present in a small part of the original population. Genetic resistance is chromosomally determined. In houseflies, the resistance to DDT is due to recessive gene called Kdr (knockdown resistance). The gene is located on chromosome 3 and is responsible for causing low nerve sensitivity to DDT. Another gene imparting resistance in housefly to DDT is DDTase gene which is located on chromosome 5 and is different from Kdr. It is associated with the production of DDT-dehydrochlorinase (DDTase), which is responsible for the oxidative detoxification of DDT. Synergist like DMC blocks the activity of DDTase. The resistance due to DDTase gene is, therefore, abolished by the use of DMC with DDT. On the other hand, Kdr strains are not affected by the application of DDT with synergists. This gene also confers resistance to methoxychlor.

In *Drosophila melanogastor* a singal dominant recessive factor present on chromosome 2 is responsible for DDT and BHC resistance. It is likely that several genes on each chromosome are responsible for resistance (Polygenic).

3. Biochemical and Physiological Consideration

Since the toxicant undergo various biochemical reaction before bringing about its action, insects develop certain phenomenan, which ultimately leads to decreased response against a pesticide.

(a) Impermeability of Pesticides through Cuticle

Some insects are able to withstand the poison because of greater impermeability of the cuticle than the rest, which depends largely on the thickness of cuticle. Heavily sclerotized insects are relatively resistant to contact insecticides. There is correlation between cuticle thickness and resistance to pyrithrins. The rate of penetration also affects the degree of resistance. The insecticide is also stored in lipid layer of cuticle and hence more quantity of insecticide is needed to kill the insect.

(b) Detoxification of Chemicals/Conversion of Insecticides to Non Toxic Metabolites

The most prevalent type of resistance in insects by converting the insecticides to metabolites which are nontoxic (detoxification). The main cause for resistance to

DDT in resistance strain is the conversion of DDT to DDE by the enzyme DDTase which catalyses the dehydro chlorination of DDT in the presence of glatuathione. The high concentration of DDTase is found in the fat body and brain tissues. The occurrence of this enzyme in fat body is very significant since fat body in insects functions like a mammalian liver and, therefore, is rich in detoxifying enzyme. Inspite of ample evidences in support of relationship between DDTase and resistance in housefly, there are three main points of disagreement:

1. In the tissues of surviving flies, large quantity of charged DDT is found which is sufficient enough to kill the susceptible flies.
2. Metabolism of DDT is slow to cause resistance.
3. Cross-resistance in resistant flies against these insecticides, cannot be dehydro-chlorinated.

The possible explanation for these points is:

1. The presence of uncharged DDT does not mean that the toxicant is in active form but because it may be rendered inactive by binding or storing in the non sensitive tissue or may be distributed throughout the body in concentration below threshhold of poisoning.
2. The rate of DDT metabolism varies from strain to strain.
3. Prolan-resistance flies metabolise prolan to neutral derivatives but they also dehydro chlorinate DDT by a different mechanism. In the same way DDTase may detoxify other insecticides via a different mechanism.

Demethylation of methyl parathion and fenitrothion is an important detoxification mechanism in resistant houseflies. Diazinon and diazoxon in this insect are converted into diethyl phosphorothioic acid and diethlphosphoric acid, respectively by hydrolytic reaction. Malathion is first oxidized to malaxon with an equal rate in resistance and susceptible housefly but malaoxon is degraded more rapidly by the action of phosphatase in resistant fly. In *culex tarsalis* the resistant larvae convert malathion to its nontoxic metabolites by carboxyesterase at much higher rate than susceptible larvae. In the bollworm larvae, the resistance to dimethoate is attributed to the conversion into non-toxic derivatives by the action of amydase and the rapid excretion of metabolic products. In some insects before the toxicant reaches to site of action, it gets metabolized by certain enzyme and rendered as non-toxic compounds, *e.g.* DDT gets detoxified to DDTE by an enzyme DDT-hydrochlorinase (DDTase) in houseflies. Chlorinated compounds other then DDT are converted to non-toxic or less toxic compounds by the same enzyme. Likewise, malathion is detoxified to malathion mono and diacids in mammals by the action of carboxyesterase. All the degradation reaction leads to detoxification.

(c) Storage and Dispersal of Toxicant

It is well established that many insecticides particularly chlorinated hydrocarbons and some of their metabolites are deposited in the adipose tissues of insects. The lipid content may be responsible for picking up the insecticides and depositing them in various tissues of the resistant strains. This results in the reduction of critical concentration at the site of action. This mechanism, however, can afford only slight protection to insects and cannot be considered as a major cause of resistance.

These lipoids seem to play an important role in the development of resistance. Resistant insects have a good capacity to store insecticides in various fat and lipoid layer of body, hence the insecticide is stored and does not reach to site of action.

(*d*) Excretion of Toxicant

The rate of excretion of toxicants and their active metabolites plays an important role in developing resistance in insects. It has been observed that when toxicant is applied on resistant and susceptible strains, the rate of penetration of active metabolite is same in both the cases but in susceptible strains, the excretion is probably impeded because of biochemical lesion causing its knockdown. On the other hand, excretion of the metabolite is very fast in resistant strains.

(*e*) Dietary Factors

Dietary factors have a marked effect on insect resistance to insecticides. The two-spotted mite is more susceptible to malathion when fed on bean plant containing high level of phosphorus. The larvae of tobacco budworm show resistance against DDT when their diet contains high level of ascorbic acid and lipids. The effect of dietary factors on the resistance to certain insecticides, thus might be associated with the increase of DDTase or cytochrome P-450 or MFO activity in various species.

(*f*) Higher Concentration of Sensitive Mechanism

The insecticide affects a higher concentration of sensitive mechanism and hence more insecticide are required to cause death or severe injury. If concentration of chemical is increased then more quantity of insecticide are required to kill the insect.

(*g*) Low Concentration of Sensitive Mechanism or Decreased Sensitivity of Sensitive Mechanism

If sensitive mechanism is altered slightly and become less sensitive to the toxicant than resistance is developed (system is less affected by toxicant). The nerve of DDT and BHC resistance strains of cockroaches were less sensitive to poisoning effect. Thus, it is more or less protective manner by which the sensitive mechanism is altered slightly to less sensitive mechanism.

(h) *Failure of Activation System*

All the latent inhibitor requires to be activiated. If the insect develops such a system that there is no activation of toxicant than insect is said to be resistant.

(*i*) Certain by Pass System of by Passing of an Sensitive Mechanism and Inhibitor Mechanism

If toxicant affects one system and the other would rapidly or slowly compensate the effect of toxicant. In certain flies strain where DDT cause partial inhibition of cytochrome oxidase and some flies have higher capacity to compensate the inhibition of remaining cytochrome oxidase.

Tactics for Prevention and Management of Resistance

It is a well known fact that continuous wide spread use of insecticides inevitably leads to selection of resistant population and once the resistance has set in, it is irreversible in an ecologically closed population. The present and widely used chemical based pest control approach if left to continue without intervention of comprehensive resistance management, efforts are feared to limit the usefulness of many present and future pesticides very soon. The resistance management is therefore a rational solution and an essential part of integrated pest management (IPM). This implies the optimum long term use of all pest control tactics in compatible manner. Some resistance management tactics are as follows:

1. Variation in Dose and Low Frequency of Application

Resistance may be delayed or minimized by individuals or alleles using low rates of a given pesticides so as not to select against heterozygotes where resistance is recessive. Fewer or less frequent application that reduce the selection pressure over time should reduce the rate and probability of resistance development.

2. Local Rather than Area Wide Application

Control of a pest with a particular pesticide in a single field or site rather than over a large area can leave refuges in surrounding areas, thus diluting the frequencies of resistance.

3. Treatment Based on Economic Threshold

This tactic delays pesticide application until the economic threshold is reached and may allow a certain level of crop damage to occur. This is a means of reducing the selection pressure for resistance.

4. Use of Less Persistent Insecticides

Insecticides with short residual lives tend to slow the development of resistance due to reduced exposure but success may depend on the nature of insect and insecticides.

5. Life Stages of Pest

This tactic is based upon using a pesticide against the life stage of the target pest that is not so likely to develop resistance, *e.g.* in some lepidopterans, the adults and early larval stages are apparently less able to metabolite insecticide than older ones. Therefore, insecticides may be targeted to adults or very early stages thereby reducing the selection pressure as well as the development of resistance.

6. Mixture

A mixture of insecticides with an independent action have been suggested as a counter measure for resistance, *e.g.* BHC-DDT mixture for malaria mosquitoes. BHC to kill DDT resistant species and DDT to kill BHC resistant species.

7. Alternation, Rotations a Sequences of Pesticide Application

This means that the same insecticide should not be used for a long time particularly when resistance has been detected. After a time lapse one may return to the first insecticide.

8. Pesticide Formulation Technology

Additional research is needed to substantiate this tactic. It can reduce the dose or rate of pesticide application. Synergists, adjuvant penetrates and material that improve bait attractancy can be incorporated into pesticide.

9. Synergists

It usually locks the action of the detoxifying enzymes allowing the insecticide to function normally. In reality, it does not enhance the killing power or potency of an insecticide, *e.g.* piperonyl cyclonene when mixed with DDT prevents its detoxification to DDE in the resistant strain housefly.

10. Use of Insect Pheromones and Insect Hormone

By using sex pheromones insects are driven to the poison baits. The regular growth and reproduction can be interfered using insect hormone at wrong times, *i.e.* when they are not needed.

11. New Toxophores with Alternate Site of Action

The discovery and development of new pesticides have often been viewed as a major approach to management of resistance to earlier insecticides. However, priorities should be given to the further pesticides for their new or alternate site of action that have lower risk of resistance development.

12. Exploiting Unstable Resistance

Pesticide resistance often carries with it, especially during its original development some deficiencies in fitness, vigour, behaviour or reproductive potential. Unstable resistance can be exploited by using other insecticides or control programs to control resistance insect preferentially or selectivity until resistance diminishes.

13. Protection and Use of Natural Enemies with Pesticide Tolerance

Introduction of the predators especially those with natural or induced tolerance to the specific pesticide has become a tactic of much interest in resistance management. Genetic engineering offers even more potential in this area. Endograma, a *Trichogramma* resistant to endosulfan has been developed by PDBC, Bangalore.

14. Reintroduction of Susceptible Pest

This method shifts the population away from a critical frequency of resistance. The reintroduction of the susceptible must be in such a way that the number of the susceptible should be numerous enough to swamp the endemic resistant population.

Agricultural and Horticultural Pests and their Management

Insect-Pests of Cereals

Rice

Brown Planthopper, *Nilaparvata lugens* **(Hemiptera: Delphacidae)**

Marks of Identification

Adult hopper is 4.5-5.0 mm long and has a yellowish brown to dark brown body. The wings are sub hyaline with a dull yellowish tint. It has two characteristic wing morphs: macropterous (long winged) and brachypterous (short winged). (Wing morphism is influenced by various factors *viz.*, crowding during the nymphal stage and reduction in the quality and quantity of food, short day length and low temperature which favour macroptery).

Nature of Damage

Both the nymphs and adults remain at the ground level and suck the plant sap. It is a typical vascular feeder primarily sucking phloem sap leading to hopper burn. At early infestation, circular yellow patches appear which soon turn brownish due to drying up of the plants. The patches of infestation then may spread out and cover the entire field. The grain setting is also affected to a great extent. During sustained feeding, it excretes a large amount of honeydew. It also acts as vector of the virus diseases like grassy stunt, wilted stunt and ragged stunt. (Transmission of persistent ragged stunt and grassy stunt virus require more time. Sheath blight and stem rot incidence was high in BPH infested plants). Symptoms will not be visible from outside in the early stages, but if we enter the field and tap the plants large number of this insect can be seen. They are visible only when the damage has been severe, the plants present a burnt up appearance, hopper burn, in circular patches.

Life Cycle

Adult female lays eggs in groups towards the lower side of the plant. It makes an incision and inserts the eggs inside the leaf sheath. The incision can be marked as a brownish patch from out side the stem. Each female lays about 100-200 eggs during its life span. Eggs hatch after 7-10 days of oviposition and the 1st instar nymph emerges out. The insect has five nymphal instars during its developmental period. Nymphal development takes about 2-14 days to become adult. Two forms of adult stages in both male and female insects have been found *i.e.* macropterous (winged) and brachypterous (wingless) forms. The adult female BPH starts egg laying after 2-3 days of emergence. When the population of insect in a area becomes more, macropterous forms migrate and infest new crop areas.

Management

1. Adopt planting with formation of alleys of 25 cm at intervals of 2 mts to provide good aeration and sunlight.
2. Avoid dense planting, planting 33 hills in kharif and 44 hills in rabi per sq.m. may be followed.
3. Excess application of N fertilizers may be avoided.
4. In vegetative phase of the crop growth, periodical drying and wetting may be followed for short period to create disturbance in micro climatic conditions favorable to pest development.
5. Grow resistant varieties *viz.*, Chaitanya, Krishnaveni, Chandan, Triguna, Deepthi, Nandi, Vijeta,Pratitha, Vajram, etc.
6. Egg parasitoid like *Angrus* spp. and nymphal and adult parasitoid like *Pseudogonatonus* spp., were observed to exercise control to the extent of 10 to 40 per cent. Mirid bugs, *Cyrtorlinus livipennis*, is one of the most important predators of BPH in rice ecosystem.
7. Application of Carbofuran 3 G @ 25 kg/ha or spraying of ethofenprox 10 EC at 2 ml or chlorpyriphos 20 EC at 2 ml per liter of water were found effective.

Green Leafhopper, *Nephotettix virescens* (Hemiptera: Cicadellidae)

Marks of Identification

Adults are 3-5 mm long, bright green with variable black markings, wedge shaped with a characteristic diagonal movement. Male insect has a black spot in middle of the forewings that is absent in females. The insect is active during July to September. The nymphs are soft bodied, yellow white in colour. Gradually, the colour changes to green.

Nature of Damage

Both nymphs and adults suck the plant sap from leaf and leaf sheath (It is a phloem feeder). Amino acid content is high in phloem sap than xylem. The xylem and phloem vessels are plugged with their stylet sheath causing disruption in the transport of food substances in the vessels). Mild infestation reduces vigour of the plant and the number of reproductive tillers. Heavy infestation causes withering and complete drying of the crop. Plants are predisposed to fungal and bacterial infection through feeding and ovipositional punctures. Nymphs and adults exude sticky, whitish honeydew, which attracts sooty mould (that reduces the photosynthetic rate). It also transmits plant diseases such as dwarf, transitory yellowing, yellow dwarf and rice tungro virus (Tungro is transmitted during short feeding period). Affected plants become pale yellow in colour and get stunted in growth. If the plants are tapped, large number of leafhoppers may be seen jumping to water.

Life Cycle

The female lays eggs on the inner surface of the leaf-sheath in groups of 3-18. The eggs hatch in 3-5 days and the nymphal stage is completed in 12-21 days. The adult live for 7-22 days. There are 6 generations in a year. The insect overwinters in the adult stage.

Management

1. Adopt planting with formation of alleys of 25 cm at intervals of 2 mts to provide good aeration and sunlight.
2. Avoid dense planting, planting 33 hills in kharif and 44 hills in rabi per sq. m. may be followed.
3. Excess application of N fertilizers may be avoided.
4. In vegetative phase of the crop growth, periodical drying and wetting may be followed for short period to create disturbance in micro climatic conditions favorable to pest development.
5. Grow resistant varieties *viz.*, Chaitanya, Krishnaveni, Chandan, Triguna, Deepthi, Nandi, Vijeta,Pratitha, Vajram, etc.
6. Egg parasitoids like *Angrus* spp and nymphal and adult parasitoids like *Pseudogonatonus* spp., were observed to exercise control to the extent of 10 to 40 per cent. Mirid bugs, *Cyrtorlinus livipennis*, is one of the most important predators of BPH in rice ecosystem.
7. Application of Carbofuran 3 G @ 25 kg/ha or spraying of ethofenprox 10 EC at 2 ml or chlorpyriphos 20 EC at 2 ml per liter of water were found effective.

Whitebacked Planthopper, *Sogatella furcifera* (Hemiptera: Delphacidae)

Marks of Identification

The adult hopper is 3.5-4.0 mm long. The forewings are uniformly hyaline with dark veins. There is a prominent white band between the junctures of the wings. Macropterous males and females and brachypterous females are commonly found in the field.

Plate 13.1: BPH–Adults

Plate 13.2: Rice–BPH

Plate 13.3: Rice Gall Midge

Plate 13.4: BPH–Hopper Burn

Plate 13.5: Rice Borer

Plate 13.6: Rice Gall Midge Affected Crop

Nature of Damage

WBPH is more abundant during the early stage of the growth of rice crop, especially in nurseries (It attacks less than four-month old plants in fields with standing water and shows a marked increase with the age of the crop. Rice is more sensitive to attack at the tillering phase than at the boot and heading stages). Damage is caused through feeding and oviposition. Gravid females cause ovipositional punctures in leaf sheaths. Both nymphs and adults suck phloem sap causing reduced vigour, stunting, yellowing of leaves and delayed tillering and grain formation. (Rice crop fails to produce complete grains [seedless glumes] and this condition is known as red disease in Malaysia.) Feeding puncture and lacerations caused by ovipositor predispose the plants to pathogenic organisms and honeydew excretion encourages the growth of sooty mould. It is not a vector of any viral disease.Heavy infestation cause outer leaves of a hill to show burn symptoms. Damage in the form of hopperburn appears uniformly in a rice field, whereas it appears as circular patches in the case of BPH.

Life Cycle

Pre mating period is of 2 days, pre-oviposition and oviposition periods are of 2 and 45 days, respectively. Eggs are laid in clusters inside the leaf sheath after injuring the surface with the ovipositor. A female lays about 85 egg masses and about 546 eggs. The incubation period is 7-8 days. Nymphal period is 15 days. There are 5 nymphal instars. Male longevity is 40 days, whereas female survives up to 50 days. Total life cycle is 25 days (egg to egg).

Management

1. Timely planting, and judicious use of nitrogenous fertilizers.
2. Grow resistant varieties like Tulasi, Tripti, Salivahan, Haryana Basmathi-1 and IR-62.
3. The following spiders and insects are potential predator of the WBPH. Wolf spider (*Lycosa chaperi*), jumping spider (*Plexippus paykulli*), orb weaver or garden spider (*Araneus sinhagrdensis*), crab spider (*Thomisus cherapunjeus*), four jawed spider (*Tetragnatha mandibulata*), lynx spider (*Oxyopes pandae*), ground beetle (*Casnoidea indica*), ladybird beetle (*Coccinella arcuata*), etc. encourage their activities in the field.
4. Spray chlorpyriphos or monocrotophos @ 0.5 per cent a/i. per ha or carbofuran @ 1.0 kg a. i. per ha to control the pest, effectively.

Gundhi Bug, *Leptocorisa acuta* (Hemiptera: Coreidae)

Marks of Identification

Adults are greenish yellow, long and slender, above ½ inch in length with a characteristic buggy odour.

Nature of Damage

Both adults and nymphs do the damage. The nymphs start feeding 3 to 4 hours after hatching. They feed on the leaf sap near the tip/on milky sap in developing spikelets at milky stage. Sucking of the milky sap causes ill-filled/partial filled and chaffy grains. Serious infestation can reduce the yield by 50 per cent. The straw gives off-flavour that is unattractive to cattle. Leaves turn yellow and later on indicate rusting from tip downwards. Appearance of numerous brownish spots at the feeding sites/shrivelling of grains. In the case of heavy infestation, the whole earhead may become devoid of mature grains. Its presence in the field is made out by its strong smell.

Life Cycle

The pest breed all the year on main crop and grasses. The female lay 24-30 round yellow eggs in row on the leaves. The eggs hatch in about 6-7 days and the nymphs grow to maturity in 6 stages within 2-3 weeks. The adult bug live for 33-35 days. Many generations are completed in a year.

Management

1. Removal of weeds in the vicinity of paddy crop as the pest breeds on a variety of grasses prior to its migration to rice crops.
2. Collection of bugs by hand netting is useful. Sweeping of rice plants with winnows smeared with sticky material like castor.
3. Spraying of monocrotophos 1.0 ml or endosulfan 2.1 ml/lit of water once at flowering and another at grain hardening stage or dusting of endosulfan 4 per cent @ 25 kg/ha in evening hours. Repeat this after 10 days, if needed.

Paddy Stem Borer, *Scirpophaga (=Tryporyza) incertulas* (Lepidoptera: Pyralidae)

Marks of Identification

They exhibit remarkable sexual dimorphism. The female moth is bright yellowish brown with a black spot at the centre of the forewing and a tuft of yellow hairs at the anal region. The male is small in size and brownish.

Nature of Damage

The insect may start attacking the plants in the nursery especially long duration varieties. The incidence is mild in the season June to September, but later on gets intensified from October to January–February. The caterpillar enters the stem and feeds on the growing shoot. As a result the central shoot dries up and produces the characteristic dead heart. The tillers may get affected at different stages. When they are affected at the time of flowering, the earheads become chaffy and are known as white ear. A number of stem borer moths seen dead and floating on the water in the fields. In the vegetative stage, dead hearts are seen in the affected tillers and in the reproductive stage, white ear may be seen. The full-grown caterpillar measures about 20 mm, white or yellowish white in colour with a conspicuous prothoracic shield.

Life Cycle

The female lays eggs on the under side of the leaves. A single female lay about 120-150 eggs. The eggs are covered with yellowish brown hairs of the female tuft. They hatch in 6-7 days. The larva grow in 6 stages and is full-fed in 16-27 days. It then constructs an emergence hole which is always located above the water level and pupates inside the attacked plant. Pupal period lasts 9-12 days. The life cycle is completed in 31-46 days. There are 2-5 generations in a year in different agro climatic zones.

Management

1. Paddy stalks should be harvested close to the ground.
2. The field should be ploughed in the summer months and the stubbles should be collected and burnt.
3. Sowing and planting time should be adjusted to avoid peak borer infestation.
4. Crop rotation with pulses, vegetables or groundnut after kharif rice should be followed to prevent continuous build up of borer population.
5. Heavy pasturing of stubble in field and ploughing in summer which expose the roots of stubbles also reduce borer population.
6. Since the eggs are laid near the tip of the leaf blade, clipping the seedlings before transplanting reduces the carry-over of eggs from seed bed to the transplanted field. Harvesting at ground level or ploughing after harvesting remove majority of larvae and pupae.
7. The use of sex pheromone traps for mass trapping of male moths.
8. *Telonomus* spp., *Tetrastichus* spp and *Trichogramma* spp are identified as dominant complex stem borer egg parasitoids, they could be utilized.

9. Sasyasree (RNR 446), Ratna and Kaveri were observed tolerant to this pest in many parts of India.

10. Spray chlorpyriphos 2 ml/lit and repeat the same at 10-15 days interval. Apply cartap hydrochloride 4 G @ 20 kg/ha or carbofuran 3G @ 25 kg/ha.

Paddy Gall Midge, *Orseolia oryzae* (Diptera: Cecidomyiidae)

Marks of Identification

The adult fly is yellowish brown and mosquito like. The male is ash grey in colour. Adults feed on dewdrops.

Nature of Damage

The maggot bores into the growing point of the tiller and causes abnormal growth of the leaf sheath, which becomes whitish tubular and ends bluntly. It may be pale green, pink or purplish. Further, growth of tiller is arrested. This is called onion shoot, silver shoot or anaikomban. The feeding by the maggot and the larval secretion, which contains an active substance called cecidogen, is responsible for cell proliferation of the meristematic cells and gall formation. It is a pest in irrigated and wet season crop. Tillers in 35 to 53 days old crops are preferred. The central shoot instead of producing leaf produces a long tubular structure. When the gall elongates as an external symptom of damage, the insect are in pupal stage and ready for emergence.

Life Cycle

Adults are nocturnal, phototropic and live for 2-5 days. Mating usually takes place soon after emergence and oviposition initiates a few hours latter. Female lays 100-200 eggs, either singly or in groups of 3-4 on the ligules or in their vicinity on the leaf blade, or on the leaf sheath. Incubation time is 3-4 days. The newly hatched maggots of both sexes are about 1 mm long and can live in water upto 3 days without any adverse effect. They crawl down the leaf sheath to the growing points of the tillers and reach the interior of the bud, where they lacerate the tissues and feed until pupate. The feeding stimulates tillers to grow into tubular gall that resembles an onion leaf. The average larval period is 15-20 days. The maggot moults thrice at 3, 5 and 7 days for 1st, 2nd and 3rd instars, respectively. Pupation occurs inside the gall near the base. Pupal period is 5-8 days. Life cycle completes in 19-21 days in rainy season but in winter months, it takes 32- 39 days. There are 3-5 overlapping generations on the same crop and 5-8 in a year.

Management

1. Advancing the planting date needs to be adopted to avoid the peak activity of the pest to protect the crop. Late and closer planting predispose the crop to gall midge infestation.

2. Off-season activity of gall midge in rice stubbles or alternate weeds like *Echinocloa crus-galli* and *Leersia hexandra* can be checked by removal of these plants.

3. Use resistant varieties *viz.,* Shakti, Samalei, Kakatiya, Surekha, Phalguna and Vikram and Rajendradhan.

4. Seed treatment with chlorpyriphos or isofenphos (0.05 per cent) for 3 hours or seed mixing with chlorpyriphos (0.75 kg a. i. per 100 kg seeds) or imidacloprid (0.05 kg a. i. per 100 kg seeds) provides protection for 30 days in the nursery.

5. Seedling root dip for 12 hours before planting in chlorpyriphos, isofenphos or chlorfenvinphos at 0.02 to 0.04 per cent concentration has provided most effective and economical control.

6. In case of late infestation, based on the economic threshold level, effective granular insecticides like phorate, carbofuran or quinalphos can be broadcasted in one or two rounds.

Rice Hispa, *Dicladispa armigera* (Coleoptera: Chrysomelidae)

Marks of Identification

The adult beetle is somewhat square shaped about 1/6 to 1/8″ in length and width. Dark blue or blackish in colour with spines all over the body.

Nature of Damage

The grub mines into the leaf blade and feed on the green tissue between the veins. Adults also feed in the green tissue; they scrape the green matter of the tender leaves. Generally the plants are affected in the young stage. The mining of the grubs will be clearly seen on the leaves. White parallel line will be clear on the leaves.

Life Cycle

The female starts egg laying in the nurseries. The eggs are embedded in the leaf tissue towards the tip. On hatching, the grubs feed as leaf-miners, between the upper and the lower epidermis. The attacked leaves die. When the grubs are full-fed they pupate inside and finally emerge as black beetles. There are 2-6 generations in a year.

Management

1. The pest is suppressed if the infested leaf tips are clipped off and destroyed, while transplanting.

2. Spray chlorpyriphos 20 EC or fenitrothion 50 EC at 2 ml/1 litre of water.

Wheat

Wheat Aphid, *Macrosiphum miscanthi* (Hemiptera: Aphididae)

Marks of Identification

The adults are green, inert, louse like and first appear on the young leaves. The nymphs and the females look alike, except that the latter are larger. The winged forms appear only in early summer.

Nature of Damage

It attacks wheat, barley and oats. The nymphs and adults suck the sap from plants, particularly from their ears and decrease yield of crop.

Life Cycle

The female gives birth to young ones and are capable of reproducing without mating. During the active breeding season, there are no males and the rate of reproduction is very high.

Management

1. Excessive use of nitrogenous fertilizers should be avoided.
2. Wheat varieties *viz.*, HD-2329, Raj-3077, HD-2285, UP-2121, WH-542, HD-4645 are resistant to aphid.
3. The grubs and adults of ladybird beetle *Coccinella septumpunctata* feed voraciously on the nymphs and adults of aphid. Similarly, the maggots of syrphid fly feed very actively on aphid. Normally no control method for wheat aphid is required when these bio control agents are active.
4. Spray imidacloprid 17.8 per cent at 3 ml per 10 liter of water.

Armyworm, *Mythimna separata* (Lepidoptera: Noctuidae)

Marks of Identification

The adult moths are pale brown in colour. The newly emerged larvae are dull white and later turn green.

Nature of Damage

The early stage larvae feed on tender leaves in central whorl of the plant. As they grow, they feed on older leaves of plants and skeletonize them totally. In case of severe attack, whole leaves, including the mid-rib are consumed and the field looks as if grazed by cattle.

Life Cycle

The female lays eggs singly in rows or in clusters on dry or fresh plants, or on the soil. They hatch in 4-11 days depending upon weather. The larval period lasts for 13-14 days in spring and 88-100 days in winter. The insect spins a cocoon and pupates usually in soil. The pupal period lasts for 9-13 days. The adult lives for 1-9 days.

Management

1. The pest can be suppressed by collecting and destroying the caterpillars.
2. Wheat varieties *viz.*, HD-1982, UP-2009, Sonilika, Kalyan Sona, UP-1008, UP-319, UP-215, UP-301, UP-368, WL-711 and C-306 have been found resistant to armyworm.
3. *Apantelis flavipes* is a larval parasitoid of the armyworm.
4. Spray 500 ml of dichlorovos 100 EC or one liter of quinalphos 25 EC in 425 liter of water to suppress this pest.

Ghujhia Weevil, *Tanymecus indicus* (Coleoptera: Curculionidae)

Marks of Identification

The weevils are earthen grey and measure about 6.8 mm in length and 2.4 mm in width. Their forewings are oblong and hindwings are more or less triangular. They can not fly.

Nature of Damage

The damage is caused by the adult weevils only where in they cut the germinating seedlings at the ground levels. The damage is particularly serious during October-November.

Life Cycle

The female lays 6-76 eggs in 5-11 installments in soil under clods in the ground. The eggs hatch in 6-7 weeks and young grubs enter the soil. The grubs feed on soil humus. They are full grown in 10-18 days and pupate in soil in earthen chamber. The pupal period lasts for 7-9 weeks. The pest complete only one generation in a year.

Management

1. Plough the field in summer to expose and kill the pupae.
2. Drenching of the soil should be done with chlorpyriphos 20 EC @ 4.5 liter per ha.
3. Mix thoroughly malathion 5 per cent dust at 25 kg per ha with 12-25 cm deep layer of soil.

Maize, Sorghum and Pearl Millet

Sorghum Shoot Fly, *Atherigona varia soccata* (Diptera: Anthomyiidae)

Marks of Identification

The adult is a small dark fly. Female fly has whitish grey head and thorax, while the abdomen is yellowish with paired brown patches. Male is darker in colour.

Nature of Damage

The maggots bore into the shoot of young plants, a week after germination to about one month and as a result the central shoot dries up. If the plants are attacked at the initial stages, the mother plant may produce profuse side tillers, but the tillers also may be attacked. The infestation often goes as high as 60 per cent. The high yielding hybrid varieties are severely attacked. In South India, crop is damaged during October to December as also in summer. Dead hearts or drying of central shoots or production of profuse side tillers in main plants is a characteristic damage of the pest.

Life Cycle

The fly lays white minute (0.72 x 0.16 mm) rice grain shaped eggs sculptured with horizontal longitudinal ridges singly or in small batches on the stem above the ground or just below ground level and also on the under surface of cotyledonary or first leaf of young. The egg, larval and pupal stage of this pest lasts for 1-3, 15-18 and 8-9 days, respectively. Total life cycle is completed in 17-33 days. The adult flies survive for 1-2 weeks.

Management

1. Use higher seed rate to minimize the damage.

2. Later thinning out of infested and extra plants is also useful in areas which have low infestation.

3. Destruction of wild sorghum in the vicinity of field reduce the natural breeding place of the pest.

4. Parasitoids *viz.*, *Tetrastichus nyemitawus*, and *Neochryso-charis* spp. (Eulophidae) regulate pest population effectively under the field conditions.

5. A number of granular insecticides like carbofuran, fensulfothion, disulphoton, phorate, aldicarb, trichlorphon and carbaryl in doses ranging as high as 1.0–2.5 kg a.i. per ha as soil furrow application at sowing time are effective.

Maize Stem Borer, *Chilo partellus* (Lepidoptera: Pyralidae)

Marks of Identification

Moth is medium sized and straw coloured. Male has pale brown forewings provided with dark brown scales forming a dark area along the coastal margin. Hindwings are light straw in colour. Female possesses forewing of a lighter colour and nearly white hind wings.

Nature of Damage

The caterpillar bores into the stem and feeds on the central shoot. There may be more than one caterpillar in a single plant. In early stages, the caterpillars make circular holes on unfolded leaves and later central shoot dries up producing dead heart. Later it acts as an internode borer and is found till the time of harvest. Young cobs may also be attacked. Yield is affected much and the quality of the fodder is also reduced. The damage caused to the crop by this pest is estimated to range between 70–80 per cent. Presence of circular holes on the unfolded leaves and dead hearts in the early stages are the main symptoms. The bored holes may be visible in contrast to the dead heart caused by the stem borer. When grown up plants are attacked, the symptoms are not quite visible.

Life Cycle

The average fecundity per female is 87 to 110 eggs. Egg are laid in masses of 20-25 on the underside of the leaves. The egg, larva, pupal stage lasts for 4-5, 18-20, 6-8 days, respectively, and life cycle is completed in 29-33 days in case of non-hibernating stage during summer, while corresponding figures are 8-9, 28-93, 9-24, 83-210 days, respectively. There are 6 larval instars. The full grown caterpillar of the last generation hibernate in stubble, stalks, etc. and remain there till the next season. There are probably 5 generations in a year.

Management

1. The manipulation of the date of sowing, higher plant population and use of sorghum as a trap crop can help in achieving the targeted yield of maize.

2. Plough up the field soon after harvesting the maize, *Jowar* and *Bajra* crop. Collect and burn the stubbles. Use the stalks by the end of February and chop the remaining stalks for subsequent use. The cores of maize cobs may

harbor the borer larvae. These should also be destroyed by burning by the end of February. For seed, keep only healthy cobs free from borer.

3. Clipping of lower leaves of maize (up to fourth) on which most of eggs of *C. partellus* are laid, reduces damage to some extent.

4. Use recommended seed rate.

5. Remove and destroy the plants showing severe borer injury while hoeing the crop.

6. Growing maize in association with legumes significantly reduce pest incidence in maize. Also intercropping maize with soybean give considerable reduction in pest attack.

7. Spray the crop 2-3 weeks after sowing or as soon as borer injury to the leaves is noticed with fenvalerate 20 EC, cypermethrin10 EC, or deltamethrin 2.8 EC.

Pink Stem Borer, *Sesamia inferens* (Lepidoptera: Noctuidae)

Marks of Identification

Adults are stout, straw coloured and are nocturnal in habit. The fully developed caterpillar is cylindrical, pinkish dorsally and whitish ventrally. Larvae can migrate from plant to plant.

Nature of Damage

The young larvae after hatching, congregate inside the leaf whorls and feed on folded central leaves causing typical 'pin hole' symptoms. Severe feeding results in killing of the central shoot and consequent dead heart formation. Usually the second instar larvae migrate to neighboring plants by coming out from the whorls and suspending themselves from the plants by silken threads, these are then easily blown off by wind to other plants. These larvae penetrate in the stem and cause tunneling resulting in stunting, infested plants become weak and bear very small earheads. The weakened stems, especially of tall local varieties, break easily during heavy rains or with high velocity winds.

Life Cycle

The female lays beaded like eggs in 2 to 3 longitudinal rows within the lower most leaf sheaths preferably young plants of maize. The incubation period is 6 to 7 days. The larval period varies from 24 to 36 days. The pupal period is about 10 days. The total life cycle varies from 40 to 80 days. The larval period is prolonged in severe winter. The adults live for 4-6 days

Management

1. Removal and destruction of dead hearted plant is recommended.

2. Sex pheromone traps with (Z)-11-hexadecenyl acetate and (Z)-11-Hexadecen-1-0 L in a ratio of 4:1 has been proved effective method for forecasting the occurrence of *S.inferens*.

3. Maize varieties 'Syn A 226' Syn E 13, 'Puerto Rico Gr 1 x TAD', Puerto Rico Gr 1 x E 13; Ant. 40; Kitole III; H 2; JML 22; Syn B 19; Thai DMR; Warangal local; TAD x Syn 13; DHM 103; EH 4025; Mex 17; Hybrid Odesski 50; Temp. Yellow dent; Ganga 5; Antigua Gr.1 etc. are fairly resistant.

4. Pink stem borer is parasitized by *Telenomus* sp. and *Trichogramma australicum* in egg stage, *Apanteles flavipes*, *Bracon chinensis* in larval stage and *Tetrastichus ayyari* and *Xanthopimpla* sp. in pupal stage.

5. Spray the crop 2-3 weeks after sowing or as soon as borer injury to the leaves is noticed with fenvalerate 20 EC, cypermethrin 10 EC, or deltamethrin 2.8 EC at 0.05, 0.003 and 0.025 per cent concentrations, respectively.

Sorghum Earhead Bug, *Calocoris angustatus* (Hemiptera: Miridae)

Marks of Identification

The adult is a slender green elongate bug about 1 cm long and active flier. The newly hatched nymphs have light orange red abdomen, which changes to green in the advanced instars.

Nature of Damage

The adults and nymphs live inside the earhead and suck the milky fluid from the tender ripening grains. Due to the feeding, the grains get shriveled and chaffy and thus unfit for sowing and for consumption. No damage is caused to fully ripened grains. A reduction of 15–30 per cent in the yield is estimated due to its attack. Usually high yielding varieties with compact earheads are subjected to more infestation than the loose earheads. No external symptom are visible. The earheads should be tapped either on the palm or a piece of cardboard. A number of brownish or greenish nymphs and adults can be seen. On the developing grains small brownish spots are visible. In severe infestation, the grains get shriveled without maturing and the earheads appear uneven.

Life Cycle

The adults appear on sorghum crop as soon as the ears emerge from the leaf sheaths. The bugs lay eggs under the glumes or in between anthers of florets by inserting its ovipositor. The female lays 150-200 cigar shape eggs. The eggs hatch in 5-7 days and the nymphs start feeding on developing grains in the milk stage. The nymphs pass through 5 instars and develop into adults in about 3 weeks. The insect completes its life cycle in about one month and there are many generations in a year.

Management

1. Avoid staggered sowing in area which favour multiplication of bug.
2. Setup light trap till midnight to attract and kill adults.
3. Apply malathion 5 per cent or phosalone 4 per cent dust (at 25 kg per ha) 3 and 18 days after panicle emergence.

Insect-Pests of Pulse Crops
Red gram, Black gram, Green gram, Lablab, Cowpea and Chickpea

Gram Pod Borer, *Helicoverpa armigera* (Lepidoptera: Noctuidae)

Marks of Identification

It is a medium-sized light brown coloured moth. On the fore wings, there is a speck that forms a V-shaped mark. Hind wings are dull grey coloured with a black border on the distal end. Female moth is bigger than male and presence of tuft of hairs on the tip of the abdomen is the characteristic identification of the pest.

Nature of Damage

Young larva feeds on tender leaves, buds, flowers, and subsequently it bores into the pods and feeds on the seeds with its head and part of the body only thrust inside, the rest remaining outside. A single larva may destroy 30-40 pods before maturity. In the early stages, plants seen defoliated. Boredholes seen on the pods and affected pods have no seeds.

Life Cycle

Egg is spherical in shape with a flattened base, giving dome shaped appearance, and surface is sculptured in the form of longitudinal ribs. Yellowish-white, glistening and change to dark brown, before hatching. They hatch in 2-6 days. Newly hatched caterpillar is sluggish and whitish-green in colour. Full-grown larva is 3.5-4.0 cm in length with pale-green body colour. However, the colour varies according to the food intake. Dorsal surface bears dark broken stripes. Head is reddish-brown. Larva is highly cannibalistic and readily eats one another. The larvae is full fed in 13-19 days. It pupates in soil in earthen cell. Pupa is obtect type. Freshly formed pupa is greenish yellow in colour and darkened prior to emergence of moths. The pupal period lasts for 8-15 days. There may be as many as 8 generations in a year.

Management

1. Timely sowing *i.e.,* upto mid October or growing early maturing varieties which complete podding by first week of March.
2. Mixed or inter cropping with non- preferred host plants *viz.* barley, wheat, mustard and linseed.
3. Apply *Helicoverpa* NPV @ 250-500 LE per ha alone or half dose of endosulfan 35 EC (1.25 liter per ha).
4. Install pheromone traps for monitoring the pest population (10 per ha).
5. Spray endosulfan 35 EC (0.07 per cent) or fenvalerate (0.05 per cent) or cypermethrin (0.003 per cent) for effective control.

Blue Butterfly, *Lampides boeticus* (Lepidoptera: Lycaenidae)

Marks of Identification

It is medium sized butterfly. The colour of the wings is violet metallic blue to dusky blue. The tail of hind wings is black and tipped white. The female is slightly bigger than the male. In males, the abdomen is slender and tapering, while in female it is long and broader at the tip.

Nature of Damage

The larva bores into the buds, flowers and green pods just within couple of hours after hatching and feeds inside the developing grains.

Life Cycle

Eggs are laid on the buds, flowers, green pods and on shoot and leaves. Greenish white in colour, round in shape with a slight depression at the top. Newly hatched larva is yellowish green in colour with black head and a dark-brown patch on the prothorax and cylindrical body with scattered hair. Full-grown larva is yellowish green to yellowish red sometimes light purple in colour, ventral surface is light green. Whole larva is covered with small setae and marked with irregular black markings. It looks like a slug. Brownish mid-dorsal and yellowish lateral lines are well marked. Pupa are green in colour, later on it darkens and wings are also visible. Five weeks are required to complete one generation in field conditions.

Management

1. Spraying 5 per cent neem seed kernel suspension gives good results against this pest.
2. Spray endosulfan (0.07 per cent) to manage this pest in case of severe infestation.

Plume Moth, *Exelastis atomosa* (Lepidoptera: Pterophoridae)

Marks of Identification

It is lightly built and light brown in colour, wings deeply fissured, the forewings longitudinally cleft into two plumes and hindwings into three plumes. Forewings are extremely elongate. Legs are long and slender. Abdomen is dark-brown in colour.

Nature of Damage

Pods are scrapped in the early stages, later boredholes seen on the pods and seeds are eaten away. Young larvae bore into unopened flower buds for consuming the developing anthers. Grown up larvae first scrap the surface of the pods and then bore into pods. The larvae never enter the pod completely.

Life Cycle

The female moths lays 17-19 eggs singly on tender parts of the plants. The eggs hatch in 2-5 days and the young larvae feed on pods and become full grown in 10-26 days. Pupation takes place outside the pod on its surface or in the entrance hole itself. The pupal period lasts for 3-12 days. The life cycle is completed in 17-42 days.

Management

1. Spray endosulfan 35 EC (0.07 per cent) or fenvalerate (0.05 per cent) or cypermethrin (0.003 per cent).

Plate 13.7: Helicoverpa Egg

Plate 13.8: Helicoverpa Full Grown Larvae

Plate 13.9: *Helicoverpa armigera*–Adults

Plate 13.10: Pea Leaf Miner

Plate 13.11: *Helicoverpa armigera*

Plate 13.12: Pigeonpea Pod Borer

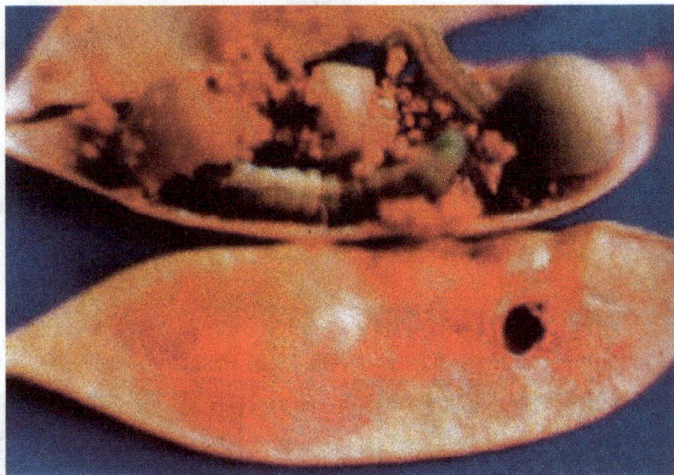

Plate 13.13: Pea Fruit Borer

Plate 13.14: *Maruca* **sp.**

Spotted Pod Borer, *Maruca testulalis* (Lepidoptera: Pyralidae)

Marks of Identification

Dark brown with a white cross band in the middle of the forewings and the hind wings are white with a dark border.

Nature of Damage

Presence of semi-solid excreta at the junction of the boredhole. Young shoot with dried tip, large scale dropping of flowers. Larva present inside the webbing of leaves, flowers and young pods, faecal material accumulates outside the boredhole. It feeds on the seeds by boring into the pods.

Life Cycle

Eggs are elongate oval in shape and light yellow in colour and are laid singly on or near flower buds of host plants. Young caterpillar feeds on reproductive parts of flowers and move from one flower to another. Later they web inflorescences with adjacent leaves and developing pods and feed within by boring into the flowers and pods. Full grown caterpillars are light brown in colour with irregular brownish black dorsal, lateral and ventral spots. Incubation period is 2-3 days. Larval stage lasts for 8-14 days and pupal period lasts for 6-9 days.

Management

1. Spray endosulfan 35 EC (0.09 per cent) or fenvalerate (0.05 per cent) or cypermethrin (0.003 per cent).

Spiny Pod Borer, *Etiella zinckenella* (Lepidoptera: Phycitidae)

Marks of Identification

Greyish brown moth, distinct pale-white band along the costal margin of the forewings, hind wings are semi-transparent with a dark marginal line. Prothorax is orange in colour.

Nature of Damage

Entrance hole in the green pod disappears and leaves little evidence that the pod is infested. In pods, the larva devours many seeds. The pod always contains a mass of frass and held together by a loosely spun web. Young larva bores into floral parts, making rough and irregular incision.

Life Cycle

A female can lays 50-200 eggs. The eggs are laid singly or in clusters on any part of the plant including pods. Incubation period lasts for 5-6 days. The newly hatched larvae are greenish first feed on flowers and then bore into the pods to feed the seeds. The larvae is full-fed in 10-13 days. Pupation occurs at a depth of 2-4 cm under the soil. Pupal period lasts for 9-20 days. Overwintering pupae emerge in February or March. Adults live for 5-7 days.

Management

1. Spraying 5 per cent neem seed kernel suspension give good results against this pest.
2. Spray endosulfan (0.07 per cent) to manage this pest in case of severe infestation.

Tur Podfly, *Melanagromyza obtusa* (Diptera: Agromyzidae)

Marks of Identification

It is slightly bigger than the male. Its wings are also slightly broader. Colour of a newly emerged adult is dull-white and smoky patches at places, but gradually it acquires the normal black colour with slight bright greenish tinge. Abdomen is glossy black, but in some cases it is slight bronzy, while in other cases, it has a greenish-blue background.

Nature of Damage

Shriveled pods and seeds. Damaged seeds become unfit for consumption and also do not germinate. However, the attack of the fly remains unnoticed by the farmers due to the concealed mode of life of this insect within the pods. Young maggot attaches itself on the immature seed inside the pod. In the beginning it feeds on the surface and thereafter mines into seeds and makes galleries just under the seeds epidermis causing a ring like track (One seed is enough for the development of a maggot. It never leaves the pod in which it enters once and completes its maggot stage).

Life Cycle

Freshly laid eggs are white, broad and round at its posterior end which is embedded in the tissues of the pod and narrowed anteriorly into a somewhat elongated egg sheath. Eggs hatch in 3-5 days. Freshly hatched maggot is white with dark-brown mouth hook. Full-grown maggot is cylindrical in shape and is narrower at the head end, which bears black mouthparts. It is creamy white but acquires a yellowish tinge just before pupation. Maggots become full-fed in 8-12 days. Pupation takes place in hard chitinous puparium, which is found sticking to the side of pod or in the groove eaten into the grain by the larva. It is cylindrical with broadly rounded ends. Fresh pupa is yellowish white, but becomes darker subsequently. Pupal period lasts 7-10 days. The life cycle is completed in 11-27 days and several generations are produced in a year.

Management

1. Spray of endosulfan (0.07 per cent) at 15-days intervals during flowering and pod-formation gives good control.

Stem Fly, *Ophiomyia phaseoli* (Diptera: Agromyzidae)

Marks of Identification

Light brown when freshly emerged, but fully developed adult is metallic-bluish or greenish-black in colour with light brown eyes. Wings are transparent. Female is slightly bigger than the male.

Nature of Damage

Drooping of the tender leaves and yellowing is the serious damage to young plants. The sites where maggot and pupae are present become swollen and start rotting. Older plants show stunting but are not usually killed. Maggot is the damaging stage. It mines sub-epidermally through the leaves. Plants are most seriously affected at the seedling stage, where stem is tunneled.

Life Cycle

The female lays eggs in holes made on the upper surface of young leaves, especially near the petiol end of the leaf. On hatching, the maggot forms a short linear leaf mine and further on, it tunnels underneath the epidermis of the leaf until it reaches one of the veins which leads it to the midrib and then to the leaf stalk and the stem. Pupation takes place inside the stem. The total life cycle takes 2-3 weeks. As many as 7 generations have been completed in a year.

Management

1. Spray monocrotophos (0.04 per cent) during the flowering stage.
2. Soil application of 10 kg of phorate 10G is effective upto 40 days of sowing.

Blister Beetle, *Mylabris pustulata* (Coleoptera: Meloidae)

Marks of Identification

Medium sized, 12.5-25.0 mm long. Conspicuous in appearance and are moderately robustly built. Beetles are bright metallic blue, green, black and yellow or brown in colour.

Nature of Damage

The adult beetles feed on flowers, leaves and tender panicles, thus preventing grain formation.

Life Cycle

Eggs are laid on the ground or in the soil. First stage larva is 'triungulins' (long-legged) and actively searches for the host. They moult to become eruciform or caraboid (Hyper metamorphosis). Pupates in the soil.

Management

1. Manual picking and destruction of adult blister beetles is often the only available practical control measure.
2. In case of severe infestation spray monocrotophos (0.04 per cent) provides good result.

Bean Aphid, *Aphis craccivora* (Hemiptera: Aphididae)

Marks of Identification

Apterous females are shiny, dark brown or black. Alate forms are greenish black with transparent wings.

Nature of Damage

Colonies of nymphs and adults found on leaves, terminal shoots and pods and suck the plant sap. They also act as vector of stunt disease in chickpea, rosette of groundnut. Serious pest when the rainfall is low.

Life Cycle

The common mode of reproduction is through vivipary and parthenogenesis, though reproduction by ovipary has been also recorded. A single apterous, parthenogenetic female produced 29 nymphs. The nymphs generally undergo 4 moults before reaching the adult stage. The duration of each instar is usually one day, though in some cases it was even 3 days. Within a day after it becomes an adult, the apterous female starts producing its brood. A female reproduced upto a maximum of 12 days.

Management

1. Encourage the activities of the predators, *viz.*, *C. septunpunctata*, *M. sexmaculata*, *B. suturalis* and *X. scutellarae*.

2. Spray methyl-o-demeton (0.05 per cent).

Pod Bug, *Clavigralla gibbosa* (Hemiptera: Coreidae)

Marks of Identification

The adult bugs are greenish-brown in colour, 2 cm in length, with spines on either side of the middle of the prothorax. Female bug is bigger and has a round and swollen abdomen in comparison with a narrow and pointed abdomen of the male.

Nature of Damage

Nymphs and adults cause substantial damage to pods and also to stem, leaves and flower buds. Attacked pods show pale-yellow patches. When the attack is heavy, the pods shrivel up. The grains in the attacked pods remain shriveled and extremely small. Both the nymphs and adults cause damage by sucking cell sap. The pest assumes serious proportions on the pods before the maturity of the crop.

Life Cycle

The eggs are preferentially laid on pods. They hatch in 7-8 days. On hatching, the nymphs collect at a point on any part of the plant to feed gregariously including pods. The nymphal development completes in 17-20 days. Adult bug live for 6-7 days. The pest is in the months of October-May.

Management

1. The eggs of this pest are parasitized by Gryon (= Hydronotus) antestiae (Hymenoptera: Scelionidae).

2. Spray the crop with monocrotophos (0.04 per cent).

Insect-Pests of Oilseed
Mustard

Mustard Aphid, *Lipaphis erysimi* (Hemiptera: Aphididae)

Marks of Identification

Minute, soft bodied, light green and pear shaped insects having a pair of short tubes called cornicles or honey tubes or siphons on the postero-dorsal region of the abdomen.

Nature of Damage

Both nymphs and adults damage the plants by sucking sap from leaves, buds and pods. Due to very high pest population, the vitality of plants is greatly reduced, leaves become pale and curled up, flowers fall to form pods and developing pods fail to form seeds so that the yield may reduce upto 25-40 per cent.

Life Cycle

The main season of activity of the aphid coincides with the general growing period of cruciferous crops, extending from October to March. During this period, it completes on an average 11 generations. The pre-reproduction and reproduction period is 8-18 and 5-7 days, respectively. The total life cycle completes in 23-67 days.

Management

1. Early sowing of mustard is preferred.
2. Under field conditions *D. rapae* is very effective parasitoid of this aphid. Predators *viz.*, *C. septumpunctata*, *M. sexmaculatus*, *B. suturalis* and *X. scutellarae* check the population of this pest effectively in field.
3. Mustard cultivars *viz.*, T-59, B-85 glossy, RW white glossy and RL-18, RLM-198 are resistant to this pest.
4. Spray neem seed kernel extract (5 per cent) or neem oil (0.5 and 1.0 per cent).
5. Spray dimethoate (0.03 per cent) or endosulfan (0.05 per cent) or methyl-o-demeton (0.025 per cent).

Mustard Sawfly, *Athalia lugens proxima* (Hymenoptera: Tenthredinidae)

Marks of Identification

Small orange yellowish insects with black markings on the body and smoky wings with black veins. Larvae are dark-green with black dorsal strips and wrinkled body.

Nature of Damage

Rapeseed and mustard seedlings are seriously affected because the grub either riddle the entire leaves or cause numerous shot holes by excessive feeding. This pest causes 12-13 per cent loss in the yield.

Life Cycle

About a week after the first mating, the female starts laying eggs, which are inserted singly in the tissues of leaves around the periphery, within incisions with the help of its ovipositor. The incubation period varies from 3-6 days. The larval period lasts from 7-19 days. One female lays about 40 eggs. There are six larval instars. The full grown grubs enter the soil and form elongated oval cocoons. The larvae prefer sandy soil with 15 per cent moisture for pupation. Pupal period lasts 7-10 days. The adults live upto a maximum of 20 days. The pest completes 3-4 generations in a year.

Management

1. Saw fly infestation could be minimized by sowing mustard crop late (5th November).
2. Pest incidence could be minimized by intercropping mustard with gram.
3. The activities of the grub could be reduced by irrigating the crop after 15 days of sowing.
4. The nymphs and the adults of *Cantheconida furcellata* attack the grubs. Larval pupal parasitoid, *Peritissus cingulator* (Hymenoptera: Ichneumoidae) parasitize the larvae of the pest. The bacteraim, *Serratia marcescens* cause mortality of the grubs.

5. Spray neem seed kernel extract (5 per cent) or Mahua (*Madhuca indica*) (5 per cent) suppress the pest population effectively.

6. Spray insecticides *viz.*, fenvalerate (0.005 per cent) or monocrotophos (0.04 per cent) for effective control of this pest.

Painted Bug, *Bagrada cruciferarum* (Hemiptera: Pentatomidae)

Marks of Identification

The full-grown nymphs are about 4 mm long and 2.66 mm broad. They are black with a number of brown markings. The adult bugs are 3.71 mm long and 3.33 mm broad. They are sub-ovate, black and have a number of orange or brownish spots.

Nature of Damage

Both nymphs and adults of this bug suck sap, especially from cruciferous plants and devitalize them, when attacked young plants wilt and die. On larger plants, bugs cluster over the leaves and pods and suck the sap. These bugs also feed on mustard seeds.

Life Cycle

A female on an average lays 32-100 eggs in her life span of 2-4 weeks. The egg stage is 3-20 days. The nymphal period is 14-36 days and the nymphs pass through 5–6 instars. There are 9 generations in a year.

Management

1. The bug is parasitized by a tachinid, *Alophora* sp. while the eggs are parasitized by Scelionids, *Gryon* sp., *Trissoleus samueli* and *Typhodytes* sp.

2. Spray of *Heeng* (*Asafoetida*) @ 20 g per liter or neem based pesticides effectiviely controls this pest.

3. Dust endosulfan (4 per cent) or methyl parathion (2 per cent) @ 20 kg per ha at the time of seedling stage.

Groundnut

Aphid, *Aphis craccivora* (Hemiptera: Aphididae)

Marks of Identification

Apterous females are shiny, dark brown or black. Alate forms are greenish black with transparent wings.

Nature of Damage

The adults and nymphs feed on the growing tips before the crop starts flowering. Later they migrate to the floral shoots, seriously hampering pod formation. It is estimated that this pest alone reduces the yield by about 40 per cent. It is also a vector of *rosette*, a viral disease of groundnut

Life Cycle

The common mode of reproduction is through vivipary and parthenogenesis, though reproduction by ovipary has been also recorded. A single apterous, parthenogenetic female produce 29 nymphs. The nymphs generally undergo 4 moults before reaching the adult stage. The duration of each instar is usually one day, though in some cases it is even 3 days. Within a day after it becomes an adult, the apterous female starts producing its brood. A female reproduced upto a maximum of 12 days.

Management

1. Groundnut intercropped with cereals (*e.g.* millets or maize) reduce the incidence of this pest.
2. Encourage the activities of the predators, *viz.*, *C. septumpunctata*, *M. sexmaculatus*, *B. suturalis* and *X. scutellarae*.
3. Spray demeton-o-methyl (0.05 per cent).

White Grub, *Holotrichia* spp. (Coleoptera: Scarabaeidae)

Marks of Identification

The full grown grubs are about 35 mm long and white, having brown head and prominent thoracic legs. The adult beetles are dull brown and measure about 18 mm in length and 7 mm in width.

Nature of Damage

The grub eat away the nodules, the fine rootlets and may also girdle the main root, ultimately killing the plant. The beetles also defoliate other plants *viz.*, neem, banyan, etc.

Life Cycle

The adult beetle lays eggs singly in the soil upto 10 cm deep. The incubation period range from 7 to 10 days. Eggs are laid with the onset of monsoon in July. There are three grub instars. The full grown grubs go deep into the soil to pupate. The pupal period takes 13 days. There is only one generation in a year. The beetle hibernates in soil and emerges out of it with pre-monsoon shower in June. The beetle emerges from the soil between 7.30–8.00 PM and immediately flies to their host plants.

Management

1. Plough the field twice during May-June. It would help in exposing the beetles resting in the soil.
2. Wherever possible, sow the crop early *i.e.*, between June 10 to 20.
3. Treat the seed before sowing with 12.5 ml of chlorpyriphos 20 EC per kg of kernels.
4. Pre-sowing soil treatment with phorate 10G at 25 kg per ha or quinalphos 5G at 30 kg per ha effectively protects the crop from the attack of grub.

Castor

Castor Semilooper, *Achaea janata* (Lepidoptera: Noctuidae)

Marks of Identification

The moths are stoutly built with smoky grey or brown forewings. Hindwings are dark with a white band in the middle and three to four white spots at the anal margin. The larva is a semilooper. Body is grey or black in colour, with red or whitish side strips. A full grown larva measures about 60-70 mm in length. The pupa is ashy grey. The colour pattern of third and fourth instar larvae is variable.

Nature of Damage

The first instar larvae nibble the outer tissues of the leaves while the second stage larvae bite holes on castor leaves. The later instar larvae eat the leaves completely leaving behind only veins and petioles. Normally, it does not attack stem and branches, however, under severe infestation, the larva damages even inflorescence and young capsules. The moths punctures citrus fruits at night and suck the juice.

Life Cycle

The female moth lays about 450 eggs singly and scattered over the lower surface of leaves. The eggs are round, concave from the under side, convex from below and pale green in colour. The incubation period is 3-4 days. The newly hatched larva starts feeding on leaves and full grown in 15-20 days after 4-5 moultings. Pupation takes place in dried leaves either on plant or fallen in soil in a cocoon. The pupal period lasts 10-25 days, the longevity of adults range between 7-19 days. There are 5-6 generations in a year. The infestation starts from July and continues till December.

Management

1. Light ploughing after harvesting of castor crop helps in reducing the pupal population.
2. The crop should be seeded during first fortnight of August so that the crop may escape from the incidence of this pest.
3. Light trap is one of the best tools to attract and destroy the adult moths.
4. Hand collection of big sized larvae helps in large scale reduction of the pest.
5. Egg parasitoids *viz.*, *Trichogramma chilonis*, *T. achae*, *Telenomus* spp. and *Trissolcus* are the most effective egg parasitoids of semilooper egg. Therefore, Inundative release of *Trichogramma* @ 100000 per ha per week should be made after the onset of moth emergence. *Microplitis maculipennis* is also effective larval parasitoid.
6. The crop should be sprayed with quinalphos 25 EC 0.05 per cent or endosulfan 35 EC 0.07 per cent based on ETL (4 larvae per plant).

Plate 13.15: Groundnut Jassid

Plate 13.16: Red Hairy caterpillar

Plate 13.17: Castor Semilooper

Plate 13.18: Groundnut Leafminer Adult

Plate 13.19: Castor Hairy Caterpillar

Plate 13.20: Mustard Saw Fly

Plate 13.21: Groundnut Aphid

Plate 13.22: Mustard Aphid–Adult

Plate 13.23: Mustard Aphid Infested Crop

Plate 13.24: Mustard painted bug

Shoot and Capsule Borer, *Conogethes (=Dichocrocis) punctiferalis* (Lepidoptera: Pyralidae)

Marks of Identification

The adult is a small sized bright-yellow coloured moth with numerous black spots on the wings. Full grown larvae measures about 25 mm in length. It is brownish in colour with a pinkish tinge and numerous spiny warts on the body. The pupa is reddish brown measuring about 11 mm in length and 3 mm in breadth.

Nature of Damage

Pest appears from the flowering stage of the crop till maturity. The freshly hatched larva initially feeds on the greenish coat of capsules in between the warts. The larva then bores the capsule destroying the seeds inside. It webs together capsules along with excreta and frass. The larvae may also bore into the tender shoots and attached inflorescence killing the terminal shoots. The borer incidence usually starts from early September reaching its peak in November. The incidence declines in January.

Life Cycle

The female moth lays eggs on tender shoots and capsules. The egg period is about 6-7 days. The larva bores into the shoots and capsules. It passes through 4-5 instars and become full fed in 2-3 weeks. Pupation takes place in a silken cocoon in shoots and capsules. The pupal stage lasts for 7-10 days. Entire life is completed in 4-5 weeks with 5-6 generations in a year. It is active throughout the year on perennial castor.

Management

1. Remove and destroy the infested shoots and capsules
2. Dusting of methyl parathion 2 per cent or chlorpyriphos 1.5 per cent or quinalphos 1.5 per cent @25 kg per ha should be done.

3. Spraying of monocrotophos 0.04 per cent or chlorpyriphos 0.05 per cent should be done at an interval of 15 days for effective management of this pest.

Hairy Caterpillar, *Euproctis* Sp. (Lepidoptera: Lymantriidae)

Marks of Identification

The adult has brown forewings with dark scales and their colour extends to two spurs across yellow margin area below the apex and to the centre of the wing margin. The hindwings are yellow. Caterpillar is hairy dark brown with a wide yellow band dorsally on abdominal segments. A medium orange red line runs along the yellow band and a fine yellow band occurs on each side of larva above the spiracular line.

Nature of Damage

On hatching, larvae feed gregariously for sometime on host. Under severe infestation, they also feed on capsule and reduce the crop yield.

Life Cycle

The eggs are laid in masses covered with buff-coloured hairs. The egg stage lasts for 4 days. Larval development is completed in 16-20 days. Pupal stage lasts for about 8 days. The pupa is enclosed in a filmy brownish-white silken cocoon. Adult period is 2-8 days.

Management

1. Destruction of first instar gregarious larvae helps in large scale reduction in population.
2. Dusting of methyl parathion 2 per cent or chlorpyriphos or quinalphos 1.5 per cent @ 25 kg per ha should be done for the effective control.
3. Spraying of monocrotophos 0.04 per cent or dichlorvos 0.05 per cent or chlorpyriphos 0.05 per cent should be done at an interval of 15 days for effective management.

Sesame or Til

Sesame Gall Fly, *Asphondylia sesame* (Diptera: Cecidomyiidae)

Marks of Identification

The adults is mosquito-like and small fly. The maggots are whitish in colour.

Nature of Damage

The maggots feed on the young capsules and cause irritation to the plants, as a result, capsules are stunted, twisted and malformed. Occasionally, buds and flowers are also attacked in similar manner. The infested flowers do not develop into capsules, the pest causes maximum damage during September- October. The intensity of pest damage varies from 1-21 per cent.

Life Cycle

Female fly lays eggs in the young flower buds. The incubation period lasts for 2-4 days. Young maggots feed on the ovary of the flower. Larval period lasts for 14-24 days. The maggots pupate inside the gall. The pupal period lasts for 7-8 days. Total life cycle is completed in 23-26 days. There are 4 overlapping generations during the season.

Management

1. The adult flies can be killed using light traps. The flies are also attracted in day-time to molasses or *gur* added in water.
2. Timely sown crop can escape from the attack of the pest.
3. Infested galls should be collected and destroyed away from field to reduce further infestation.
4. Dusting the crop with methyl parathion 2 per cent or malathion 5 per cent or spray crop with endosulfan 0.07 per cent or monocrotophos 0.04 per cent is found effective.

Shoot and Leaf Webber of Sesame, *Antigastra catalaunalis* (Lepidoptera: Pyralididae)

Marks of Identification

The moth is a small insect with a wing span of about 2 cm having dark brown markings on the wing-tips. The young larvae are pale yellow and gradually become green and develop black dots all over the body.

Nature of Damage

Larvae start attack from 15th day of sowing. In 1st instar stage, acts as leaf miner, in later stage, comes out of the mine and acts as webber by webbing the top leaves or tip of shoot, by remaining within epidermis results in drying of webbed portion. When the flowers are formed, it bores into flowers and feeds on reproductive parts. When capsule formed, it acts as capsule borer. By suitably adjusting its feeding habit, it attacks throughout the crop period. Pest is active during rainy season.

Life Cycle

Female lays up to 140 eggs singly on the tender portion of plants at night. The eggs are shiny, pale-green and hatch in 2-7 days. The larvae feed on leaf epidermis or within the leaf tissue. It become full grown in 10-33 days and pupates in the soil in a silken cocoon. Pupal development is completed in 4-20 days. A generation is completed in about 23 days and there are nearly 14 generations in a year.

Management

1. Early sown crop *i.e.*, sown in June, is infested less than late sown crop.
2. Ichneumonids, *Temelucha biguttula*, *Eriborus* sp., *Compoplex* sp. and *Trathala flavor-orbitalis* are known to parasitise the larvae. *T. biguttula* is a specific larval parasite.

3. Spray the crop twice (first at pest appearance and then at flowering stage) with 250 ml of fenvalerate 20 EC.

Hawk Moth, *Acherontia styx* (Lepidoptera: Sphingidae)

Marks of Identification

The full-grown caterpillar measures about 5 cm in length and 1 cm in width, often retracts some of its anterior body segments and looks like a sphinx. The adult is a large reddish brown, robust thick-set moth with a wing span of about 10 cm. The forewings are decorated with a mixture of dark-brown and grey patterns with dark or black wavy markings and prominent yellow spots on each wing. There is a prominent Death's head mark on the thorax. The moths are swift fliers and often make hawk like darts to a source of light at dusk.

Nature of Damage

The larvae feed voraciously on leaves and defoliate the plants. The insect is capable of inflicting heavy damage at times, but generally it is not a very serious pest in India.

Life Cycle

The female moth lays egg singly on the underside of leaves. The eggs are fairly large in size and hatch in 2-5 days. The pale-yellow larvae start feeding on leaves. The larval period is usually long and may last for 2 months or more. The pupal stage lasts for 2-3 weeks in summer and about 7 months in winter. The winter is passed in pupal stage and there are 3 generations in a year.

Management

1. The pest can be suppressed by hand-picking the larvae in the initial stages of attack and also by ploughing soil in winter.
2. The eggs are parasitized by *Agiommatus acheronitiae* (Hymenoptera: Pteromalidae) while larvae are parasitized by *Apanteles acherontiae* (Hymenoptera: Braconidae).
3. Dusting of methyl parathion 2 per cent or chlorpyriphos 1.5 per cent or quinalphos 1.5 per cent @ 25 kg per ha should be done.

Linseed

Linseed Gall Midge, *Dasineura lini* (Diptera:Cecidomyidae)

Marks of Identification

The adults are small, delicate, mosquitoes-like, orange coloured insects.

Nature of Damage

Damage is done by maggots. They feed on buds and flowers. In case of severe infestation, no pod formation takes place.

Life Cycle

The female lay whitish, minute curved eggs in clusters, underneath the sepals of compact flower buds. A single female lays 22-103 eggs and infests 8-17 buds. The egg, larval, pupal and adult stages last for 2-5, 5-14 and 1-3 days, respectively. One generation takes 14-27 days to complete. Full-grown maggots come out of the bud fall to ground for pupation and prepare silken cocoons about 5-7 cm below the surface of the soil. There are four generations of this pest per season.

Management

1. The adult flies should be killed by installing light trap.
2. Inter-cropping of wheat, mustard and lentil with linseed in 4:2 and 6:1 row ratio reduces the bud-fly infestation.
3. Linseed varieties *viz.*, R-552, R-958, LCK-38, EC-22582, LC-1013, LMH-412, LCK-88062, Neela and Jawahar are quite promising against the midge.
4. Spray deltamethein 0.002 per cent for effective control of this pest.

Safflower

Safflower Aphid, *Dactynotus carthami* (Hemiptera: Aphididae)

Marks of Identification

The adult is black in colour and one of the bigger aphid in shape and size.

Nature of Damage

In case of severe infestation, the whole plant is covered with aphids which causes excessive drainage of sap, leading to stunted growth. The yield of aphid infested crop is reduced by 20-25 per cent.

Life Cycle

The duration of the life cycle is two weeks in January and one week in February. One female lays upto 13 eggs in one day. The reproduction capacity of a female is 29.5, with a maximum of 56 young ones. The duration of the life cycle is about eight days.

Management

1. Spray methyl demeton (0.03 per cent) or monocrotophos (0.03 per cent) or cypermethrin (0.005 per cent) or dimethoate (0.03 per cent) for control of this pest.

Safflower Capsule Fly, *Acanthiophilus helianthi* (Diptera: Tephritidae)

Marks of Identification

The adult fly is ash coloured with light brown legs. The full grown maggots are 5 mm long.

Nature of Damage

The injury is caused by the maggots which feed upon the floral parts including thalamus. The infested buds begin to rot and an offensive smelling fluid oozes at the apices giving a soaked appearance to the buds. The pest causes reduction in the yield of safflower.

Life Cycle

The adults are active from March to May. The female lays eggs in clusters of 6-24 within the flower buds or the flower. The eggs hatch in about one day and the young maggot starts feeding on the florets and the thalamus. They become full-grown in one week. They pupate inside the buds. The pupal stage lasts 7 days. Three generations are completed during a crop season.

Management

1. The early removal and destruction of affected bud is helpful in checking the spread of the pest.
2. The pest can be effectively and economically controlled by three sprays of dimethoate (0.03 per cent) or monocrotophos (0.04 per cent) or cypermethrin (0.003 per cent).

Pests of Fiber Crop
Cotton

Jassid, *Amrasca biguttula biguttula* (Hemiptera: Cicadellidae)
Marks of Identification

Adults are about 3 mm long and greenish yellow during the summer, acquiring a reddish ting in the winter. The winged adults jump or fly away at the slightest disturbance and are also attracted to light at night.

Nature of Damage

Both nymphs and adults suck cell sap from the lower surface of leaves. Affected leaves at first show signs of yellowing and curling at the margins in the lower parts of plant. Later on, crinkling and curling of leaves spread to other parts of the plant followed by bronzing and drying of leaves. Cotton jassid also causes phytotoxaemia in plants.

Life Cycle

The pest is practically active throughout the year, but during winter, only adults were found on potato, brinjal and tomato. In spring, cotton jassid migrates to okra and then to cotton. Singh (1978) reported that adults copulated 2-10 days after maturation. They mated in a tail to tail position for 5-25 minutes. Thirty to 45 days old leaves were preferred for oviposition. Eggs are laid in the large veins on the lower surface of the leaves. A female, on an average lays 15 yellowish eggs singly during a oviposition period of 9-28 days. Egg stage lasts for 4-11 days. Nymphal stage in six instars is completed in 7-21 days. Adults live for 5-7 weeks. There are 7-11 generations in a year.

Management

1. Use jassid tolerant varieties like L-604, LRA-5166, L-603, Savitha, Narasimha (NA-1325), NHH-44, H-8, NHH-390 and Lam Hybrid.
2. Spray any synthetic insecticide.
3. Seed treatment with imidacloprid or carbosulfan protects the crop from all sucking pests including jassids for about a month.
4. Stem application with monocrotophos protects the crop from jassids for about a month.

Cotton Aphid, *Aphis gossypii* (Hemiptera: Aphididae)

Marks of Identification

Nymphs are light yellowish green or greenish black or brownish in colour. Adults are mostly wingless but few winged forms are also observed.

Nature of Damage

Adults and nymphs both suck the plant sap gregariously. They also excrete honey dew.

Life Cycle

Pest breeds practically throughout the year on different crops. It breeds parthenogenitically and produces both alate and apterae adults. Females are viviparous. Each female produces 8-22 nymphs/day. There are four nymphal instars. Total development is completed in 7-9 days. Aphids lay eggs during winter and overwinter in this stage.

Management

1. Spray dimethoate or methyl demeton for its control.
2. Seed treatment with imidacloprid (5 g per kg seed) keeps the crop free from the attack of this pest.
3. Paint on stem with a mixture of monocrotophos: water (1:4) or imidacloprid: water (1:20) at 20, 40 and 60 days age of crop.

Thrips, *Thrips tabaci, Scirtothrips dorsalis* (Thysanoptera: Thripidae)

Marks of Identification

The adults are slender, yellowish brown and measure about 1 mm in length. The males are wingless whereas females have long, narrow strap-like wings, which are furnished with long hairs along the hind margins. The nymphs resemble the adult in shape but are wingless and slightly smaller.

Nature of Damage

Thrips attack the cotton plants at young as well as at advanced stage. At young stage, the attacked leaves become small and silvery in appearance accompanied by upward curling and crinkling. At advanced stage, the affected leaves show browning

Plate 13.25: Whitefly

Plate 13.26: Red Cotton Bug

Plate 13.27: Cotton Jassid

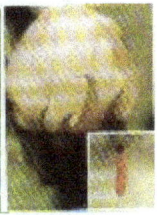

Plate 13.28: Pink Bollworm

Plate 13.29: Helicoverpa Affected Cotton

Plate 13.30: *Spodoptera litura*

Plate 13.31: Cotton Boll Affected by *H. armigera*

and blackening at lower surface followed by drying of leaves along the mid ribs and veins.

Life Cycle

Female lays 50-60 kidney shaped eggs singly in the tissues of tender leaves. Eggs hatch in 4-9 days. Fully fed nymphs descend to soil for pupation at depth of about 2.5 cm. The nymphal, pre-pupal and pupal periods last for 4-6, 1-2 and 2-4 days, respectively. Whole life cycle is completed in 11-21 days.

Management

1. Spray dimethoate (0.03 per cent) or methyl demeton (0.05 per cent) for its control.

2. Seed treatment with imidacloprid (5 g per kg seed) keeps the crop free from the attack of this pest.

Whitefly, *Bemisia tabaci* (Hemiptera: Aleyrodidae)

Marks of Identification

The louse-like nymphs clustered together on the under surface of the leaves and their pale yellow bodies make them stand out against the green background. In the winged stage, they are 1.0-1.5 mm long and their yellowish bodies are slightly dusted with a white waxy powder. They have two pairs of pure white wings and have prominent long hindwings.

Nature of Damage

Adults and nymphs suck the cell sap from lower surface of leaves and cause chlorotic yellow spots on upper surface of affected leaves. Whitefly also excretes honey dew, which make the leaves sticky. Sooty mould (*Cladosporium* Sp.) growth on such leaves interferes with photosynthesis of plants.

Life Cycle

A female can lays 150-200 eggs in her life span. On an average, 28-43 eggs are laid singly on lower surface of leaves in an oviposition period of 2-18 days. Eggs are stalked, light yellow in colour and measure 0.2 mm. They hatch in 3-5 days. The nymphs on emergence feed on cell sap and grown in three stages to form the pupae within 9-14 days. The life cycle is completed in 14-122 days and 11 generations are completed in a year.

Management

1. Whiteflies can be effectively attracted and controlled by yellow sticky traps, which are coated with grease or sticky oily material.
2. Use whitefly tolerant varieties such as LPS-141 (Kanchan), LK-861 and NA-1280.
3. Spray trizophos (0.05 per cent) or profenophos (0.05 per cent).

Red Cotton Bug, *Dysdercus cingulatus* (Hemiptera: Pyrrhocoridae)

Marks of Identification

The adult bugs are elongated slender insects, crimson red with white bands across the abdomen. The membranous portion of their wings, antennae and scutellum are black.

Nature of Damage

The damage is caused by both nymphs and adults. They suck the sap of the leaves as well as of green bolls and stain the lint by introducing a bacterium, *Nematospora gossypii*. That is way called as cotton stainer. The bugs are gregarious in habit.

Life Cycle

A female lays 100-130 eggs in loose masses in the cracks and crevices of the soil near the plant. Eggs are bright yellowish in colour. The eggs hatch in 4-6 days. The newly hatched nymphs are provided with a row of black spots and row of white spots on each side. There are 5-6 nymphal stages in the life cycle, which is completed in 19-32 days depending upon the environmental conditions. The adult lives for 12-42 days.

Management

1. Spray with any synthetic insecticide.
2. Seed treatment with imidacloprid (0.005 per cent) or carbosulfan (0.05 per cent) protects the crop from all sucking pests including jassids for about a month.
3. Stem application with monocrotophos (0.04 per cent) protects the crop from jassids for about a month.

Dusky Cotton Bug, *Oxycarenus hyalinipennis* (Hemiptera: Lygaeidae)

Marks of Identification

The adults are 4-5 mm in length, dark brown and have dirty white transparent wings. The young nymphs have a round abdomen and as they grow older, they resemble the adults, except for being smaller and having prominent wing pads instead of wings.

Nature of Damage

The nymphs and adults suck the sap from immature seeds, where upon these seeds may not ripen, may lose colour and may remain light in weight. The adults found in the cotton are crushed in the ginning factories, thus staining the lint and low its market value.

Life Cycle

The eggs are usually laid in the lint of half opened bolls, either singly or in small clusters of 3-18 each. The egg stage lasts 5-10 days and the nymphs on emergence pass through 7 stages, completing the development in 31-40 days. The life cycle completes in 35-36 days. There are many overlapping generations in a year.

Management

"Same as in case of red cotton bug."

Spotted Bollworm, *Earias vittella* and *Earias insulana* (Lepidoptera: Noctuidae)

Marks of Identification

Larvae are chocolate brown in colour and bluntly rounded. Adults of *E. vittella* are medium sized moths, head and thorax ochreous-white, forewings are pale white with a broad wedge shaped horizontal green patch in the middle, and the hindwings are cream white in colour *i.e., E. insulana* adults are smaller than *E. vittella*. Head and thorax are pea green in colour and forewings are uniformly pale yellowish green.

Nature of Damage

When the cotton plants are young, the larvae bore into the terminal portion of the shoots, which wither away and dry up. Later on, they cause 30-40 per cent shedding of the fruiting bodies. The infested bolls open prematurely and produce poor lint, resulting in lower market value.

Life Cycle

A female lays about 400 eggs. Eggs are spherical, light bluish green in colour and sculptures, are laid singly on shoot tips, buds, flowers and fruits. Egg period is 3-7 days. Larval period lasts for 10-18 days. Full grown caterpillars are enclosed in an inverted boat shaped cocoon. Pupation takes place in 8-12 days. Pest is active round the year and prefers high humidity and high temperature. It is more abundant in South India than North India.

Management

1. All withered and drooped top shoots should be clipped early during the cotton season and destroyed.
2. Growing of a trap crop like lady's finger helps in reducing the pest population on the cotton crop.
3. Eggs are parasitized by the hymenopteran, *Trichogramma evanescens*, and caterpillars by *Bracon* spp. and pupae by *Chelonus rufus* and *Chalcis tacharae*.
4. Application of neem products in the early stages and contact and stomach poison insecticides in the later stages particularly endosulfan (0.07 per cent) of chlorpyrifos (0.05 per cent) reduce the pest damage.

Pink Bollworm, *Pectinophora gossypiella* (Lepidoptera: Gelechiidae)

Marks of Identification

The caterpillars are pink in colour and found inside flower buds, panicles and the bolls of cotton. The adult moth is deep brown measuring 8-9 mm across the spread wings. There are blackish spots on the forewings, and the margins of the hindwings are deeply fringed.

Nature of Damage

The pink bollworm larvae do most spectacular damage to mature cotton bolls in which they enter as tiny just-hatched larvae, their entry holes blocked and they remain inside, devouring both seed and fibre forming tissues. The attacked bolls fall off prematurely and those, which mature do not contain good lint.

Life Cycle

The females lay whitish, flat eggs singly on the underside of the young leaves, new shoots, flower buds and the young green bolls. The eggs hatch in one week, soon after emergence caterpillar enters the flower buds, the flowers or the bolls. They become full-grown in about two weeks and come out of the holes for pupation on the ground, among fallen leaves, debris, etc. Within one week, the moths emerge to start the life cycle all over again. By October-November, 4-6 generations are completed. The last life-cycle is very long covering 5-10 months.

Management

1. The destruction of off-season cotton sprout, alternate hosts, minimize the incidence of the pest.

2. Deep ploughing by the end of February is helpful in reducing the carry over of this pest to the next season.

3. Pest is naturally regulated by many parasitoids *viz., Apanteles pectinophora, Bracon* spp. *Microchelonus* sp., *Elasmus* and certain predators including spiders.

4. Spray endosulfan 0.07 or monocrotophos 0.1per cent or cypermethrin 0.003 per cent for its control.

American Bollworm, *Helicoverpa armigera* (Lepidoptera: Noctuidae)

Marks of Identification

The moth is stoutly build and is yellowish brown. There is a dark speck and a dark area near the outer margin of each forewing. The forewings are marked with grayish wavy lines and black spots of varying size on the upper side and a black kidney shaped mark and a round spot on the underside. The hindwings are whitish and lighter in colour with a broad blackish band along the outer margin. The caterpillar when full grown is 3.5 cm in length, being greenish with dark broken grey lines along the sides of the body.

Nature of Damage

The larvae damages by boring into squares, flowers and bolls and feeds on inner contents, while feeding it thrusts its head inside the square/boll leaving the rest of its body outside. The entry hole is large and circular.

Life Cycle

The female lays eggs singly on tender parts of the plants. A single female may lays as many as 741 eggs in 4 days. They hatch in 2-6 days. The young larvae feed on the foliage for some time and later bore into the bolls and feed inside. They full fed in 13-19 days. The full grown larvae come out of the boll and pupate in the soil. The pupal period lasts 8-15 days. There may be as many as 8 generations in a year. The caterpillars feed on their fallows if suitable vegetation is not available.

Management

1. Collection and destruction of eggs and larvae on trap crop (marigold) as well as main crop.
2. Recommended doses of fertilizers should be applied.
3. Sow 3-4 lines of maize or jowar around the cotton crop to monitor the moth.
4. Sow trap crop like marigold at 125 plant per ha along with cotton.
5. Intercrop like green gram, black gram, groundnut, soyabean should be sown.
6. Use of pheromone traps (10 per ha) for pest monitoring as well as trap the male moths.
7. Arrange 15-20 bird perches per ha.
8. Spray 5 per cent neem oil before egg laying.
9. Spray monocrotophos 0.1 per cent or endosulfan 0.07 per cent or chlorpyriphos 0.05 per cent.

Leaf Roller, *Sylepta derogata* (Lepidoptera: Pyralidae)

Marks of Identification

Moths are yellowish-white, with black and brown spots on the head and the thorax. They measure about 28-40 mm across the spread wings and have a series of dark brown wavy lines on the wings.

Nature of Damage

The larvae feed on cotton leaves and in years of serious outbreaks, the cotton plants may be completely defoliated. American cotton is preferred over *desi* cotton by the pest.

Life Cycle

The pest is active from March-October and passes the winter as a full-grown caterpillar. The moths are active at night and they mate and lay 200-300 eggs singly on the underside of the leaves. The eggs hatch in 2-6 days. The young caterpillars feed on the lower surface of leaves. The older larvae roll leaves from the edges inwards upto the mid rib and feed on leaf tissues from the inside. The larvae grown through seven stages and full-fed in 15-35 days. They pupate either on the plant, inside the rolled leaves or among plant debris in the soil. They emerge as adult in 6-12 days and live for about 7 days. The life cycle is completed in 23-54 days and there are 5-6 generations in a year.

Management

1. The destruction of off-season cotton sprout, alternate hosts minimize the incidence of the pest.
2. Deep ploughing by the end of February is helpful in reducing the carry over of this pest to the next season.
3. Spray monocrotophos (0.1 per cent) or quinalphos (0.1 per cent) or methyl parathion (0.1 per cent).

Ash Weevil, *Myllocerus* spp. (Coleoptera: Curculionidae)

Marks of Identification

The weevils are grey and are 3-6 mm long. The grubs are white, legless, cylindrical and are about 8 mm in length.

Nature of Damage

Both adults and grubs cause damage. The grub feed underground on the roots of the cotton seedlings and destroys them. One grub can destroy 9 seedlings in 40 days. The adults feed on leaves, buds, flowers and young bolls cut prominent round holes.

Life Cycle

A female lays 120 eggs in 24 hours. The eggs hatch in 3-5 days and the young grubs feed on the roots of the cotton and other plants. The grub completes their development in 1-2 months. They pupate inside the soil inside an earthen cell. Pupal period last one week. The adult live for 8-11 days. During the active period, the life cycle is completed in 6-8 weeks. There are 3-4 generations in a year.

Management

1. The pest can be suppressed by disturbing the soil upto a depth of 7.5 cm and destroying the eggs, grubs and pupae.
2. Dust 2 per cent methyl parathion (@ 20 kg per ha) or 1.5 per cent quinalphos (20 kg per ha).

Pests of Sugarcane

Early Shoot Borer, *Chilo infuscatellus* (Lepidoptera: Crambidae)

Marks of Identification

The moths are light straw to brownish grey in colour. Females are slightly bigger than male moths. The wing span in male measures 19-26 mm and in female 23-35 mm. Antennae are lamellate and flat in males and filiform in females with 41 joints in both the sexes. Full grown larvae measure 20-25 mm in length and 4 mm in width. The head is dark brown and directed towards the anterior. Body is cylindrical and exhibits a dirty white colour with 5 violet dorsal strips from second thoracic to eight abdominal segments.

Nature of Damage

The borer larvae enter the plants laterally by one or more hole in the stalk and bore downwards as well as upwards killing the growing point, thereby cutting off the central leaf spindle which dries up forming a dead heart that can be pulled out easily. The cut off portion inside the bored plant rots and the dead hearts emits an offensive odour on being pulled out. The larvae feed on the soft tissues and make cavities extending to the setts. The injury to shoots after the internode formation seldom results in a dead heart. Borer infestation during the germination phase kills the mother shoots resulting in the drying up of the entire clump, creating gaps in the field.

Life Cycle

A female moth deposits about 400 eggs in one night in several egg masses. The moths live for 4-9 days. Hatching takes place in 4-6 days at sunrise or little letter. Freshly hatched larvae measure about 1.5 mm in length and have a black head and prothorax. The body is dirty grey and strips are not prominent. The larvae generally enter in between the first leaf-sheath and stem and feed on the soft inner tissues of the sheath like a leaf miner for few days. After that, the larvae enter into the stalk and kill the growing point in 7-8 days. Larval period lasts for 16-30 days, while pupal period lasts for 6-12 days. Life cycle is completed in 27-133 days. There are 5-7 generations in a year.

Management

1. Use of Trash traps for attracting and collecting moths.
2. Socking cane setts in water for 24-48 hrs for killing the larvae within and hastening germination.

3. Removal of first leaf-sheath.
4. Two to three light earthing-up during the early stages of crop growth.
5. Pulling out of dead hearts and killing the larva with a spoke.
6 Release *Sturmiopsis inferens* @ 125 gravid females per ha or innundative release of *Trichogramma chilonis* @ 1,25000 per hectare.
7. Spraying shoot borer GV @ 10^7-10^9 IB per ml.
8. Application of chlorpyriphos 20EC @ 50 ml/10 litre of water on the leaf whorls and collar region of the plant by means of knapsack sprayer will prevent the spread of infestation. In commercial fields, the 10 litre spray solution should be applied on shoots in 100 metre row length.

Internode Borer, *Chilo (Sacchariphagus) indicus* (Lepidoptera: Crambidae)

Marks of Identification

Moths are straw coloured with a slightly dark spot on each of the forewings. Male are smaller and darker in colour than females. They are sluggish and fly short distances when disturbed.

Nature of Damage

The neonate larvae feed on the leaf spindle or leaf sheath by scraping the tissues and characteristic white streaks are discernible on leaf lamina when it opens. The larvae bore into the tender cane top. In young and weak shoots, especially in the rations, the formative internodes are damaged badly, resulting in the formation of dead hearts. There is no offensive smell when the dead hearts are pulled out. The larvae tunnel upwards in a characteristic spiral fashion, sometimes feeding extensively and at other times superficially near the periphery, depending on the variety under cultivation. Rarely it feeds downwards also. The borer feeds on the inner tissues and the frass is pushed out to the exterior. The damaged internodes get hardened and crushing of these becomes difficult.

Life Cycle

The moth lays eggs in clusters in 2-3 parallel rows on both surface of the green leaves dorsally on the midrib and ventrally parallel to the midrib. The number of eggs in a cluster varies from a few to hundreds. The incubation period is 5-6 days. The newly hatched larva measures about 1.5-2.0 mm in length and is light orange in colour with a black head and a prominent prothoracic shields. The larvae moult 6-7 times and complete their larval period in 37-54 days. The full grown larvae comes out of the tunnel and selects a semi-dry leaf-sheath, spins a silken cocoon in about 10-19 hours and rests inside. The pupal period ranges from 7-10 days. Under tropical conditions, the borer remains active throughout the year and so all stages of the pest can be observed at any time. There are 4-5 overlapping generations in a year.

Management

1. Borer free setts may be planted
2. Detrashing of the crop may be done at fifth, seventh and ninth months.
3. Water shoots may be removed at eight or ninth months.

4. High dose of nitrogen may be avoided.

5. Water may be drained off in low lying areas.

6. Release *Trichogramma chilonis* @ 50000 per ha starting from the 4th to 11th month stage of the crop.

7. Install pheromone traps in water tray for mass trapping of the males..

Top Shoot Borer, *Scirpophaga excerptalis* (Lepidoptera: Crambidae)

Marks of Identification

The moth is silver white in colour. Both males and females may have one black spot on each of the forewings. The female have tuft of anal hairs of crimson red, orange or buff colour. The wing expanse is on an average 27 mm in males and 33 mm in females.

Nature of Damage

The pest occurs in all the cane growing areas in India, it is easy to spot a top bored tiller with reddish brown charred and sometimes curved dead heart, leaves with shot holes and galleries in the mid-rib of the leaves showing downward passage of the borer caterpillar. In grown up canes, owing to cessation of growth, top buds give rise to side shoots forming as bunchy top. The loss in yield amounts to about 20 per cent. Bored cane mature early, but with a rise in temperature from January a decrease of one to two units in sucrose per cent in juice is generally observed. Moderate and high humidity conditions favour profuse borer activity and multiplication.

Life Cycle

Eggs are laid in clusters of 2-5 overlapping rows mostly on the under surface of the leaves near the midrib and are covered by the hairs of the anal tuft. Generally, eggs are laid by a female in 3-6 masses, though rarely as high as 13 egg masses. An egg mass may have 10-80 eggs and a female lays 47-216 eggs during its life span of 1-4 days. The newly hatched caterpillars are grayish brown with black head and thorax with long hairs on the body. There are four larval instars. Upon the fourth instar, the larva deeds above the growing point. Then it cuts across the growing points, causes dead hearts, and enters the top internodes. The larvae are full-fed in 21-40 days. The pupal period lasts for 4-21 days and one life cycle is completed in 24-78 days.

Management

1. Collection and destruction of egg masses and moths and rouging of affected sugarcane tops before moth emergence.

2. Cut infested shoots from the top and destroy the caterpillar.

3. Carbofuran 1 kg a. i. per ha or phorate 3 kg a. i. per ha effectively control top borer and increases the yield.

Stem Borer, *Chilo tumidicostalis* (Lepidoptera: Pyrallidae)

Marks of Identification

The full grown caterpillar measures 25-30 mm in length and 3.5 mm in width. It is white in colour with four broad pink strips present sub-dorsally and laterally in pairs. In adult moths, the frons widely conical with a distinct corneous point at apex, ocelli present. Forewings are cinnamon brown, suffused with reddish to dark brown with number of scattered dark brown scales. Hindwings are whitish except for few light brown scales in the costal area in male moths.

Nature of Damage

Two types of infestation is observed: (*i*) **Primary infestation**: It is caused by the newly hatched larvae aggregating in the top 3-5 internodes of the cane. As the larvae feed inside the internodes, fresh wet frass shiny red in colour is pushed to the exterior through the boredholes in the top internodes. Top leaves of infested canes dry completely. Tunneling of internodes is so severe that the dried top portion of the cane easily breaks off at the slightest jerk. (*ii*) **Secondary infestation**: In the phase of secondary infestation, grown-up borer larvae migrate to adjacent canes or to the lower healthy portion of the canes showing primary infestation. One caterpillar may bore one to five internodes in a cane. However, cane tops do not dry up in this case. Losses in the yield and sucrose are more due to primary than secondary infestation.

Life Cycle

The moths emerge during night and are attracted to light. Mating and oviposition takes place only at night. The egg masses are deposited on the underside of the first, second and third leaves from the top in 2-4 tiers. The numbers of egg in an egg mass varies from 90-250. One female can lays as many as 800 eggs in 4-5 egg masses. The caterpillars from one egg masses hatch almost simultaneously and penetrate into one of the tender internodes. After a period of ten days, the borer larvae disperse to adjoining cans, each one boring into a separate internode. Life cycle is completed in 44-83 days with an incubation period of 7 days, larval period of 27-70 days and pupal period of 6-11 days. There is overlapping of broods and all stages of the pest are observed in the field simultaneously.

Management

1. Used light traps to collect moths.
2. Collection and destruction of egg masses
3. Removal and destruction of cane tops showing primary infestation.
4. Use of resistant varieties *viz.,* Co.-356 and Co.–513

Root Borer, *Emmalocera depresella* (Lepidoptera: Pyralidae)

Marks of Identification

Small moth with pink head, brown thorax and abdomen. The forewings and hindwings are light yellow in colour. Forewings possess light black longitudinal stripes. Adult moth is 27 mm across the wings.

Nature of Damage

Caterpillars bore at the base of a stem, which is very close to the root. Although they do not actually bore into the roots, but since they do so near the soil surface they are called root borers. It differs from the shoot borer damage in the way that the dead-heart when pulled out the whole plant comes out and it does not possess bad smell. Besides central leaf whorl, some side leaves also dry.

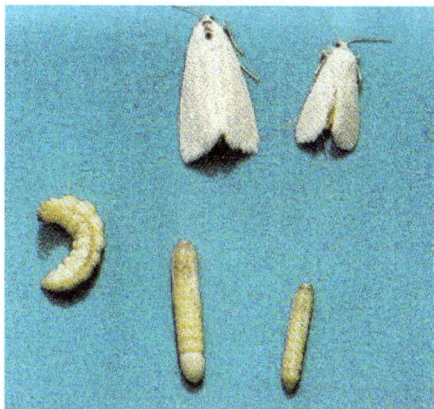

Plate 13.32: Sugarcane Top Borer

Plate 13.33: Sugarcane Internode Borer

Plate 13.34: Different Stages of Sugarcane Internode Borer

Plate 13.35: Sugarcane Shoot Borer

Plate 13.36: Sugarcane Shoot Borer Affected Crop

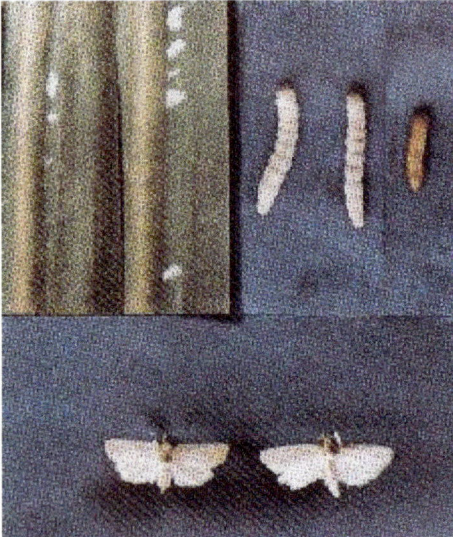

Plate 13.37: Sugarcane Stalk Bore-life Stage

Plate 13.38: Sugarcane Pyrilla

Plate 13.39: Sugarcane Mealy Bug

Life Cycle

Moths emerge during the early morning hours and remain hidden beneath leaf-sheaths during the day. Mating occurs in the early hours of night, copulation lasts for 20-30 minutes. On an average, 212 eggs are laid by one female. Egg period is 5-8 days. On hatching, the larva crawls to the base of the stem and enters it by making a single hole at or just below the ground level. The larvae feed in an irregular semicircular pattern. Larval period varies between 23-43 days during hot weather, 25-59 days during rainy season and 191-245 days during winter and post monsoon period. Pupation occurs in the silken tube outside the cane. The pupal period lasts for 7-11 days during hot weather and rainy season and 8-14 days during winter. Duration of one life cycle is completed 30- 61 days in different seasons. There are 3-5 generations in a year.

Management

1. Removal of dead hearts and killing the larvae within using a pointed cycle spoke.
2. Digging and destruction of shoots at weekly intervals.
3. Collection of moths using light traps.
4. Grow twelve feet *arhar* (*Cajanus cajan*) around cane field to repel the moths.
5. Deep harvesting of canes which are to be ratooned.
6. Digging and destruction of stubbles after harvest.
7. Application of endosulfan @1.0 kg a. i. per ha over cane setts.

The Green Borer, *Raphimetopus ablutellus* (Lepidoptera: Physitidae)

Marks of Identification

The moths are brown in colour with a reddish tinge. The hindwings are white. Antennae of females are filamentous and those of males are comb shaped. The female moths measures 23.5 mm and male 23.3 mm. Male moths are active, while female are sluggish. The full grown larva measures 16-19 mm in length and 2-2.7 mm in width. The abdominal segments are copper-green coloured above and bluish green below.

Nature of Damage

The pest generally infests the young crop, resulting in drying up of the central whorl to form a dead heart, which when pulled out gives an offensive smell.

Life Cycle

Eggs are deposited in a loose mass on soil and in crevices. On an average, one female lays 37 eggs. Incubation period is 3-8 days. There are five larval instars. Duration of the larval period is 20-25 days. Pupation takes place in the soil inside earthen cells. Pupal period lasts for 7-9 days. The duration of the life cycle is 30-43 days. There are three generations in a year.

Management

1. The larvae is parasitized by *Stenobracon deesae* the grub of the parasitoid feeds on the larvae.

The Gurdaspur Borer, *Bissetia steniella* (Lepidoptera: Pyralidae)

Marks of Identification

The moth is brownish in colour with a wing span of 35-40 mm. The forewings are pale grey brown and have several blackish spots along the outer margins. The hindwings are white in colour. The larvae are polymorphic and show three colour variations *viz.*, light violet, violet and light brown and their numbers occur in the ratio of 1:2:3, respectively.

Nature of Damage

This pest attacks the plants after the cane formation. The young larvae enter the stem *i.e.*, internode and feed gregariously on the internal contents. The larva move upward in a spiral manner making minute holes on the cane surface. This is the characteristic damage due to this pest. Later on the larvae feed deeper into the stem resulting in a single straight tunnel moving upward. As a result of infestation the leaves turn yellow. The damaged internodes become weak resulting in the breaking of the stem even with slight disturbance.

Life Cycle

The eggs are laid in clusters and arranged in 304 overlapping rows in an oval pattern. The egg masses are generally deposited on the upper surface of the leaf in and along the groove of the mid rib near the leaf-sheath. The duration of egg and larval stages varies from 4-11 and 21-42 days, respectively. Pupation generally occurs within 7-8 cm of the exit hole in a specially constructed cell. The pupal period lasts for 6-13 days. The total life cycle is completed in 35- 284. There are 2 or 3 broods in a year.

Management

1. Borer free sets may be used for planting.
2. Ratooning of heavily infested crops is to be avoided.
3. Adjusting the time of harvest also helps in checking this pest.
4. Socking setts in 0.2 per cent trichlorphon for two hours before planting kills the larvae within.

White Fly, *Aleurolobus barodensis* (Hemiptera: Aleyrodidae)

Marks of Identification

The adult flies are pale yellow insects exhibiting brisk fluttering movement. Females are larger and less active than males.

Nature of Damage

The nymphs of white flies suck the sap from the undersurface of the leaves. As a result, the leaves turn yellow and pinkish in severe cases of infestation and gradually dry up. Heavily infested leaves are covered by the sooty mould (*Capnodium* sp.) which adversely affects photosynthesis. White fly infestation not only causes retardation in plant growth, but also reduces the sugar content in canes. Average loss in sucrose per cent in juice amounts to 1.98 units in plant cane and 2.52-3.33 units in unmatured rations.

Life Cycle

Eggs are laid in linear rows of 2-15 mm long. Fresh egg is creamy yellow in colour, finally changing to black. Egg period lasts for 8-10 days. During winter it may prolong to 39 days. There are four nymphal instars. The duration of the first, second, third and fourth instar varies from 2-4, 4-5, 3-5 and 10-15 days, respectively. Total life cycle is completed in 25-47 days.

Management

1. Discouraging rationing in low lying areas, early harvesting of rations and adequate manuring of plant and ratoon crops with not less than 100 kg nitrogen per h are recommended.
2. Proper drainage is to be ensured to avoid waterlogging in low lying areas.
3. Spraying with 0.04 per cent monocrotophos or dichlorovos 0.1 per cent or endosulfan 0.07 per cent after removal of dried puparia bearing lower leaves reduces pest incidence, considerably.

Leaf Hopper, *Pyrilla perpusilla* (Hemiptera: Fulgoridae)

Marks of Identification

The adult is straw coloured, has a soft body and is very active. The head is prominently drawn forward into a rostrum or snout and measures 10 mm in length. Females are longer than the males. Wing expanse of male and female varies from 16-18 and 19-21 mm, respectively.

Nature of Damage

The adults as well as the nymphs suck the sap from the undersurface of the lower leaves near midrib, resulting in yellowish white spots. When the infestation is high, the leaves gradually turn pale and wither away. The hopper exude a sweet sticky fluid commonly known as *honey dew*, which promote quick and luxurious growth of the fungus, *Capnodium* sp., and as a result, the leaves get completely covered by the sooty mould. The black coating interferes with the photosynthesis of the leaves and crop growth is adversely affected. Due to continuous desapping by large number of hoppers, top leaves in the affected canes dry up and the lateral buds germinate.

Life Cycle

Eggs are laid in clusters of 30-50, covered with waxy threads on the ventral side of green leaves parallel to the midrib. Each female lays 600-800 eggs. The female generally prefers low, shady and concealed sites for oviposition. Incubation period during summer, monsoon and winter months ranges from 10-15, 6-10 and 15-18 days, respectively. There are five nymphal instars. The duration of each instar is 7-10 days. The total nymphal period during summer, monsoon and winter month ranges from 40-52, 34-40 nd 70-132 days, respectively. Females live longer (6-8 weeks) than males (4-6 weeks). The pest completes 4-5 generations in a year.

Management

1. Burning of trash after harvest should be carried out in the middle of March in order to destroy the unhatched egg masses and over-wintering nymphs.

2 Removal of sprouts from the stubble at least once by the end of April helps in reducing the pest build-up.

3 Removal of dry leaves from August onwards, effectively reduces pyrilla population. During pre-monsoon period, eggs are laid on the lowest two leaves may be removed and destroyed.

4. Spray *Metarhizium anisopliae* @10^7spores per ml.

5. Release of pyrilla adults along with *Metarhizium anisopliae* spores @ 250 ml per ha.

6. Redistribution and colonization of *Epiricania melanoleuca* cocoons/egg masses from heavily parasitized fields to low parasitized field.

7. Spraying with monocrotophos (0.04 per cent) or endosulfan (0.05 per cent) after removal of dried lower leaves reduces pest incidence considerably.

Insect-Pests of Vegetables
Brinjal
Shoot and Fruit Borer, *Leucinodes orbonalis* (Lepidoptera: Pyralidae)
Marks of Identification

Adults has brownish and red markings on the whitish forewing. Hindwings are opalescent with black dots.

Nature of Damage

It is the most important and destructive pest of brinjal. It starts damaging brinjal plant, a few weeks after transplanting. Larva bores into tender shoots causing withering of terminal shoots. It also bores into petiole of the leaves, flower buds and developing fruits causing withering of leaves, shedding of buds and making the fruits unfit for consumption and marketing. Attacked fruits show holes on them plugged with excreta. It causes 70 per cent loss.

Life Cycle

Eggs are laid singly on leaves, shoots and fruits. They hatch in 3-5 days. Larva is a borer within shoot, leaf midrib, petiole and fruit and feeds on the internal tissues. It becomes full fed in 10-15 days. It is stout and pink coloured with sparsely distributed hairs on the body and brownish head. Pupation takes place in a tough grayish cocoon on the plant itself for a period of 6-8 days. Adult lives for 2-5 days and female lays upto 250 eggs.

Management

1. Destroy Brinjal Crop Stubble

Many farmers store dried brinjal stubble from the previous season crop for use as fuel for cooking. Such stubble is heaped around the field or nearby dwelling. Fruit and shoot borer pupae can survive in this stubble for several weeks. When the new brinjal crop is planted, the moths from these pupae fly and lay eggs on the crop, starting a pest epidemic. This stubble should be destroyed, burnt or buried after harvest.

2. Use Healthy Pest Free Seedlings

Most farmers grow brinjal seedling in the open field, often near abandoned brinjal crops or heaps of dried brinjal stubble from the previous season. Adults from these old plants fly and lay eggs on seedlings. These eggs are very hard to see. The use of such contaminated seedlings spread the pest into the field. To avoid this, brinjal seedlings should be raised away from source of infestation. Also growing seedlings under nylon netting prevents borer moths from laying eggs on the plants.

3. Remove and Destroy Infested Shoots

Before initiation of fruiting, the larvae feed inside the tender shoots. These are visible as dried tips. Cut and destroy these larvae infested shoots immediately. This will be helpful in reducing the pest population. These shoots must be destroyed by burning or burying them at least 20 cm deep in soil. These pruning activities are very important in the early season. Once fruiting begins, most larvae will prefer to enter in fruits rather than shoots. In newly infested fruits, it is very difficult to detect the damage and the insect population multiply and spread the infestation. Always destroy damaged or infested fruits. After the final harvest, the old plants should be uprooted and destroyed promptly because they may harbor pest larvae, which could become a source of infestation in future.

4. Use of Pheromone Trap

The pheromone lures are commonly available. The pheromone traps are installed in the field. They attract male moths continuously up to six weeks. Traps should be installed in the field 2-3 weeks after transplanting and continue till the last harvest. 10-15 m distance should be maintained between traps in the field. The traps are hung in such a way that the lure is just above the plant canopy.

For this, install either funnel based or disc shaped (like WOTA) pheromone trap @ 30 per hectare to monitor and attract male moths.

5. Use of Chemical Pesticides

The indiscriminate use of toxic, broad spectrum insecticides should be avoided. This is because within hours of hatching from eggs, the larvae enter fruit or shoot. Once inside these plant parts, insecticides cannot reach larvae and kill them. Also due to frequent use of insecticides, the pest becomes tolerant and cannot be killed easily with these chemicals. These chemicals also kill the natural enemies such as spiders, mantis, earwigs, ladybird beetles and wasps that are naturally found in the field. These predators feed on pests and reduce damage to the brinjal crop. For this reason, the use of pesticides, especially the broad spectrum have to be used for successful control of this pest. If selective, preferably, botanical insecticides, such as neem are used, the natural enemies will survive and be able to help kill fruit borer larvae. If it is necessary, then always apply least toxic insecticides at the recommended dose.

Stem Borer, *Euzophera perticella* (Lepidoptera: Pyralidae)

Marks of Identification

The moths measures about 32 mm across the spread wings and have pale yellow abdomens. The head and thorax are grayish, the forewings are pale straw-yellow and the hindwings are whitish. The full grown caterpillar is creamy white and have a few bristly hairs. Their body taper posteriorily.

Nature of Damage

Larva bore into main stem of both young and old plants and move downward. Top shoots of young plants droop and wither and older plants become stunted and fruit bearing capacity is adversely affected.

Life Cycle

Eggs are laid singly or in batches on young leaves, petioles and tender branches. They hatch in 3-10 days. Yellowish larvae with red head feeds on the exposed parts for a few minutes after which it bores into the stem at leaf or branch axil and cover the holes with excreta and frass. It feeds on the internal tissues and becomes full grown in 26-58 days. Pupation is within a silken cocoon inside. Pupal period lasts for 9-16 days. Total life cycle ranges from 26-58 days.

Management

1. Collect and destroy affected plants.
2. Avoid continuous cultivation and ratooning of brinjal.
3. Spray endosulfan 35 EC 2 ml with neem oil 2 ml/liter starting from one month after planting at 15 days interval.

Hadda Beetle, *Epilachna viginitioctopunctata* and *E. dodecastigma* (Coleoptera: Coccinellidae)

Marks of Identification

Adults are hemispherical in shape, pale brown in colour and mottled with black spots. *Epilachna viginitioctopunctata* has 14 spots on each elytra, while *E. dodecastigma* has six spots.

Nature of Damage

Both adults and grubs cause considerable damage to the leaves by scraping away chlorophyll from epidermal layers of leaves which get skeletonized and gradually dry away.

Life Cycle

A female lays 120-180 eggs. Egg period is 2-4 days. Grubs are yellowish in colour, stout with spines all over the body. Larval period lasts for 10-35 days and pupation occurs on the leaves or stem. Pupa is yellowish with spines on the posterior part and the anterior portion being devoid of spines. Pupal period lasts for 3-5 days. Life cycle takes 20 to 50 days to complete on different hosts and weather conditions. There are 7 generations in a year.

Plate 13.40: Chili Thrips

Plate 13.41: Chill–White fly

Plate 13.42: Chill-Thrips Affected Crop

Plate 13.43: Cabbage Butterfly

Plate 13.44: Fruit Borer Affected Chili

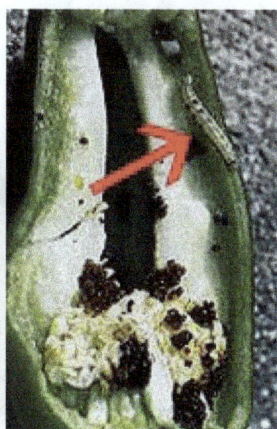

Plate 13.45: Chili Fruit Suffered by Borer

Plate 13.46: Chili Affected by Broad Mite

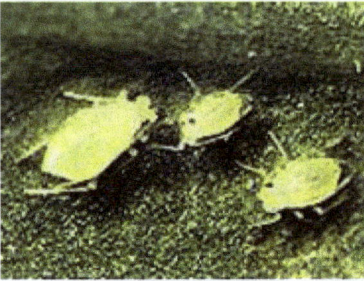

Plate 13.47: Tomato Aphid

Plate 13.48: Adult Fruit Borer

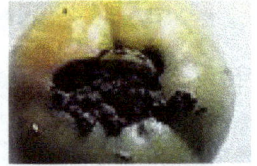

Plate 13.49: Borer Damaged Fruit of Tomato

Plate 13.50: Borer Damaged Fruit of Tomato

Plate 13.51: Borer Damaged Fruit of Tomato

Plate 13.52: Larvae of Fruit Borer

Plate 13.53: Brinjal Epilachna grub

Plate 13.54: Epilachna Beetle of Brinjal (*Hadda* beetle)

Plate: 13.55: Brinjal Stem Affected by Shoot Borer

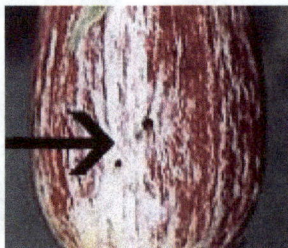

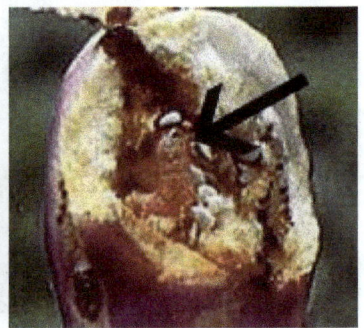

Plate: 13.56: Brinjal Fruit Affected by Shoot and Fruit Borer

Management

1. Collect damaged leaves and egg masses on the leaves and destroy them.
2. Shake the plants to dislodge grubs, pupae and adults in a pail of kerosenated water early in the morning or collect them mechanically and destroy.
3. Spray endosulfan 35 EC 2ml per liter or malathion 50 EC 2 ml per liter of water.

Tomato

Fruit Borer, *Helicoverpa armigera* (Lepidoptera: Noctuidae)

Marks of Identification

Adult are medium sized light brown stout moth with a "V" shaped speak on the forewing. Hindwings are with dull black border. The young caterpillars are yellowish-white in colour and gradually become greenish in colour.

Nature of Damage

The young larvae feed on tender foliage and full grown caterpillar attack fruits. They make circular bore holes and thrust only a part of their body inside fruit and eat inner contents. Caterpillar move from one fruit to another and one caterpillar consumes 2–8 fruits in its life span.

Life Cycle

Eggs are yellowish white, ribbed, dome shaped and are laid singly on leaves and flowers. Egg period lasts for 2-4 days. Larval period lasts for 15-24 days. Full grown larvae drop down from the plants and burrow into the soil and pupate therein. Pupal period is 10-14 days. Total life cycle is completed in 4-6 weeks. There are 5-8 generations in a year.

Management

1. Grow less susceptible genotypes *viz.*, Rupali, Roma and Pusa red palm.
2. Grow simultaneously 40 days old American tall marigold and 25 days old tomato seedlings at 1:16 rows to attract the female moth for egg laying.
3. Set up pheromone traps with *heli* lure at 15 per ha and change the lure once in 15 days.
4. Collect and destroy the damaged fruits and grown-up caterpillars.
5. Six release of *Trichogramma chilonis* @ 50,000 per ha per week coinciding with flowering time and based on ETL.
6. Spray *Helicoverpa armigera* NPV at 500 LE per ha along with cotton seed oil (300 g per ha), to kill larvae.
7. Spray endosulfan 35 EC 2 ml per liter or *B.t.* 2 g per liter or quinalphos 2.1 ml per liter.
8. Encourage activity of parasitoids, *Eucelatoria bryani, Campolites, Chelonus*, etc.

Tobacco Caterpillar, *Spodoptera litura* (Lepidoptera: Noctuidae)

Marks of Identification

Adults are stout, with wavy white markings on the brown forewings and white hindwings having a brown patch along its margin. Full grown caterpillar is pale brown with greenish to violet tinge. There are yellow and purplish spots present in the sub-marginal areas.

Nature of Damage

Freshly hatched larvae feed gregariously, scrapping the leaves from ventral surface. They feed voraciously during night and hide in the morning. In case of severe infestation, etire crop is defoliated overnight.

Life Cycle

Female lays dirty white coloured eggs in cluster on the under surface of the leaves and are covered with brown hair. Incubation period is of 3-5 days. Larval period is 20-28 days. Pupation takes place in the soil earthen cocoon for 7-11 days. Total life cycle is completed in 30-40 days during summer and 120-140 days in winter season. There are 7-9 overlapping generations in a year.

Management

1. Plough the soil to expose and kill pupae.
2. Grow castor along border and irrigation channels as indicator or trap crop.
3. Flood the field to draw the hibernating larvae.
4. Set up light or pheromone traps @ 15 per ha.
5. Remove and destroy egg mass in castor (trap crop) and tomato.
6. Collect damaged leaves and gregarious early instar larvae and destroy them.
7. Hand pick grown-up larvae and kill them.
8. Spray NPV for *S. litura* at 250 LE along with teepol 1 ml per ha in evening hours.
9. Spray chlorpyriphos 20 EC 2.0 liter per ha or DDVP 76 WSP 1.0 liter per ha or endosulfan 35 EC 1.25 liter per ha or NSKE 5 per cent.
10. Prepare poison bait with rice bran 5 kg jaggery (0.5 per cent), carbaryl 50 WP 0.1 per cent, water 3 liter per ha and spread the bait in the evening hours.

Hadda Beetle, *Epilachna viginitioctopunctata* and *E. dodecastigma* (Coleoptera: Coccinellidae)

"As described in brinjal pests."

Whitefly, *Bemisia tabaci* (Hemiptera: Aleyrodidae)

Marks of Identification

The adults are tiny white in colour, usually crowd in between the veins on ventral sides of leaves. Nymphs are oval, scale like and greenish white in colour.

Nature of Damage

Nymphs and adults suck the sap from the tender leaves. The affected leaves become yellow, the leaves wrinkle and curl downward and are ultimately shed. These insects also exude honey dew which favour the development of sooty mould. These insects also act as vector, transmit leaf curl virus disease and cause severe loss.

Life Cycle

Eggs are pear shaped, light yellowish in colour and are positioned upright on leaves by tail like appendages. Incubation period is 3-5 days. On hatching the nymphs suck the sap. Nymphal development takes place in 9-14 days in summer and 20-80 days in winter. Pupal period lasts for 3-8 days. Total life cycle is completed in 14-107 days. There are about 9-11 overlapping generations in a year.

Management

1. Uproot and completely destroy the diseased leaf curl plants.
2. Collect and destroy the damaged leaves along with eggs, nymphs, pupae and adults.
3. Remove alternate weed hosts.
4. Use nitrogenous fertilizers and irrigation, judiciously.
5. Use yellow sticky traps smeared with sticky substrates to attract and kill the adults.
6. Spray fish oil rosin soap (FORS) 2 per cent or neem oil 0.5 per cent along with teepol 1 ml per liter or methyl demeton 0.025 per cent or endosulfan 0.07 per cent along with FORS.
7. Apply systemic insecticides in early stage of the plant growth and contact insecticides in the later stages for vector control.
8. Encourage activity of parasitoids, *Eretmocerus masii* and predatory coccinellid *viz., Brumus* and *Chrysoperla*.

Okra

Shoot and Fruit Borer, *Earias vittella, E. insulana* (Lepidoptera: Noctuidae)

Marks of Identification

Larvae are chocolate brown in colour and bluntly rounded. Adults of *E. vittella* are medium sized moths, head and thorax ochreous-white, forewings are pale white with a broad wedge shaped horizontal green patch in the middle, and the hindwings are cream white in colour *i.e. E. insulana* adults are smaller than *E. vittella*. Head and thorax are pea green in colour and forewings are uniformly pale yellowish green.

Nature of Damage

Caterpillars bore into tender shoots and tunnels downwards. Shoots wither, droop and growing points are destroyed. Caterpillars also bore into buds, flowers and fruits and feed on inner tissues. Damaged buds and flowers wither and fall down. Damaged fruits become deformed in shape and remain stunted in growth.

Life Cycle

A female lays about 400 eggs. Eggs are spherical, light bluish green in colour and sculptures, are laid singly on shoot tips, buds, flowers and fruits. Egg period is 3-7 days. Larval period lasts for 10-18 days. Full grown caterpillars are enclosed in an inverted boat shaped cocoon. Pupation takes place in 8-12 days. Pest is active round the year and prefers high humidity and high temperature.

Management

1. Adopt clean cultivation and remove alternate host plants.
2. To check the population build-up of pests, do not cultivate cotton and okra together or one after other in a year.
3. Collect and destroy the affected fruits.
4. Grow resistant/tolerant varieties *viz.*, Kalyanpur Boni, Pusa Makhmali, HB-55, AC-218.
5. Use pheromone traps for *Earias* spp.
6. Release of egg parasitoids *Trichogramma chilonis* at 100000 per ha and egg larval parasitoids *Chelonus blackburni* at 10000 per ha.
7. Spray endosulfan 35 EC 750 ml per ha or *Bt* 2 g per liter of water.

Leaf Hopper, *Amrasca bigutulla bigutulla* (Hemiptera: Cicadellidae)

Marks of Identification

Adults are wedge shaped, pale green in colour with black dot on posterior portion of each forewing.

Nature of Damage

Both nymphs and adults suck cell sap from the ventral surface of leaves and inject their toxic saliva into plant tissues. Infested leaves turn yellowish and curl. In case of severe infestation, leaves turn dark brick red, become brittle and crumple.

Life Cycle

Pear shaped, elongated and yellowish white eggs are laid singly in the tissues of main veins on the undersurface of leaves. Egg period lasts for 4-10 days. Newly emerged nymphs suck the sap from tender portion and leaves of plant and become adult in 7-20 days. Pest population appears with the onset of cloudy weather and decreased after heavy monsoon showers.

Management

1. Uproot and completely destroy the diseased plants.
2. Collect and destroy damaged leaves along with eggs, nymphs and adults.
3. Remove alternate weed hosts.

4. Use yellow sticky traps smeared with sticky substances to attract and kill adults.

5. Grow resistant/tolerant varieties *viz.*, AE-30, Reshmi, AC-3375, Okra Red, Sel.- 6,8,11,HR-4,Punjab-7.

6. Spray fish oil rosin soap (FORS) 2 per cent or neem oil 0.5 per cent along with teepol 1 ml per liter or methyl demeton 0.025 per cent or endosulfan 0.07 per cent along with FORS.

7. Apply systemic insecticides in early stage of the plant growth and contact insecticides in the later stages for vector control.

Cucurbitaceous Vegetables

Fruit Fly, *Bactrocera cucurbitae, B. ciliatus* (Diptera: Tephritidae)

Marks of Identification

This fruit fly is distributed throughout the country and attacks almost all kind of cucurbits and is a very serious pest. Adult is reddish brown fly with lemon yellow surved vertical markings on thorax and fuscous shading on outer margins of wings. *B. ciliatus* it is a polyphagous attacking all kind of cucurbits. Adult fly is bright brown or ferrugineus brown with hyaline wings and two round dark brown spots on fourth abdominal segment.

Nature of Damage

Maggots tunnel and feed within fruits and cause damage. Infested fruits decay aided by bacterial action. They rot and drop down. Fly prefers tender fruits. Adult also causes injury by making ovipositional puncture on the fruits through which fruit juice oozes out. More than 50 per cent loss is caused to vegetables by this fly.

Life Cycle

Female flies puncture soft and tender fruits and lay eggs. Eggs are laid in the fruit in a cavity 2-4 mm deep, singly or in clusters of 4-10 and sealed with a gummy secretion from ovipositor. Oviposition takes place at intervals of 2-5 days, laying 1-26 eggs a day. Egg period lasts from 1 day in summer to 6-9 days in winter. Larva bores into fruit feeding on the internal contents. Larval period varies from 3 days to 3 weeks. Fullygrown larvae fall on to ground to pupate in soil. Pupal period is 6-15 days. Male fly lives up to 56 days and the female up to 66 days.

Management

1. Grow resistant or early maturing variety (Arka-tinda, a variety of round gourd, Arka-saryamakhia- pumpkin variety)

2. Change sowing dates as the fly population is low in hot dry conditions and its peak is noticed during rainy season.

3. Use attractants like citronella oil, eucalyptus oil, vinegar, dextrose and lactic acid to trap flies.

4. Use ridge gourd as trap crop and apply carbaryl 0.15 per cent or malathion 0.1 per cent on congregating adult flies on the undersurface of leaves.

5. Collect and destroy all fallen and damaged fruits by dumping in a pit and covering with a thick layer of soil to prevent carry over of pest.

6. Frequently rake-up soil under the vines or plough infested field to destroy puparia and apply endosulfan 4 per cent dust at 25 kg per ha on the soil.

7. Use cue lure (25 per ha) to monitor and kill male adults of fruit flies or prepare cue and malathion 50 EC mixture at 1:1 ratio and take 10 ml mixture with six holes and place 12 traps per ha or keep 5 g wet fish meal in a polythene bag with six holes and add 0.1 ml of dichlorvos and place 12 traps per ha.

8. Protect fruits with polythene bags.

9. Use bait spray combining molasses or jaggery 10 g per liter and one of the insecticides, fenthion 100 EC 1 ml per liter or malathion 50 EC 2 ml per liter or dimethoate 30 EC 1 ml per liter, two rounds at fortnight interval before ripening of the fruits.

Black Pumpkin Beetle, *Aulacophora foveicollis, A. cincta, A. intermedia* and Red Pumpkin Beetle, *Rhapidopalpa foveicallis* (Coleoptera: Chrysomelidae)

Marks of Identification

Adult is oblong. Dorsal surface is orange-red while ventral surface is black. Grub is creamy white with a silghtly darker oval shield at the back.

Nature of Damage

Damage is caused mainly by adult beetles, which feed extensively on leaves, flowers and fruits making holes and cause death or retardation of growth. Seedlings, when infested are totally destroyed. Damage done by grub to the seedling also is serious.

Life Cycle

Red coloured beetle lays spherical eggs singly or in batches in the moist soil around the base of the host plants. As many as 300 eggs are laid by a female. Egg period lasts for 6-15 days, larval period for 13-25 days and pupal period for 7-17 days. Grub undergoes four instars entering soil each time to moult. Mature larvae enter soil and pupate within a water proof cocoon. Five generations are completed in a year. Adults live for more than one month. They hibernate under old cucurbitaceous creepers, grasses and weeds and in soil.

Management

1. Ensure clean cultivation.

2. Adopt early sowing and cultivate less susceptible varieties.

3. Collection and destruction of adult beetles.

4. Spray malathion 50 EC or dimethoate 30 EC 1 ml per liter or fenthion 100 EC 1 ml per liter.

Cruciferous Vegetables/Cole Crops

Diamond Back Moth, *Plutella xylostella* (Lepidoptera: Plutellidae)

Marks of Identification

Adult is a small grayish moth, which when at rest shows a series of three yellowish diamond, shaped markings dorsally on the wings. Larvae feeds on foliage and grows to a pale-green caterpillar, brownish at the anterior.

Nature of Damage

Larva causes serious damage by defoliation. Leaves present a withered appearance or eaten up completely. Larva damages cabbage and cauliflower. It makes holes and soil them with excreta. It is very active in cold season.

Life Cycle

A female moth lays up to 57 eggs. Egg, larval and pupal period lasts for 3-6, 14-21 and 7-11 days, respectively. It pupates in a thin transparent cocoon on the foliage. It completes 8-12 broods in a year.

Management

1. Remove and destroy all debris, remnants and stubbles after crop harvest and plough fields.
2. Grow 20 days old mustard seedling as intercrop at the time of cabbage or cauliflower planting at 2:25 ratio to attract diamond back moths for oviposition. Periodically spray mustard crop with diclorovos 0.075 per cent at 10 or 15 days interval to avoid dispersal of larvae.
3. Install pheromone traps at 12 per ha.
4. Grow DBM resistant cauliflower variety *i.e.*, Nashik-1.
5. Spray cartap hydrochloride 0.05 per cent or *Bacillus thuringiensis* 2 g per liter or quinalphos 0.07 per cent at primordial or head initiation stage when ETL crosses 2 larvae per plant.
6. Spray NSKE 5 per cent along with teepol or sandovit 0.5 ml per liter after primodial stage.
7. Release parasitoids, *Diadegma semiclausum* or *Cotesia plutellae* for effective control.

Leaf Webber, *Crocidolomia binotalis* (Lepidoptera: Pyralidae)

Marks of Identification

The larva is green with a red head and it has longitudinal red strips on the body. It is 2 cm in length.

Nature of Damage

The caterpillar causes damage to the crop by webbing the leaves together and feeding on them. They also feed on flower buds and bore into the pods.

Life Cycle

Eggs are laid in mass of 40-100 on the underside of leaves. They hatch in 5 to 15 days in different conditions. Larval period lasts for 24-27 days in summer and up to 51 days in winter. It pupates in a cocoon within webbed up leaves and flowers or just below ground surface for 14 to 40 days depending on climate. More than one generation may be completed in the season.

Management

1. Remove and destroy the webbed leaves.
2. Use light trap at 1 per ha to attract and kill adults.
3. Spray carbaryl 0.2 per cent or malathion 0.15 per cent at fortnight interval.
4. Encourage the activity of parasitoid, *Apanteles crocidolomiae*.

Cabbage Head Borer, *Hellula undalis* (Lepidoptera: Pyralidae)

Marks of Identification

The adult moth is slender, pale yellowish-brown, having grey wavy lines of the forewings. Its hindwings are pale dusky. The larva is creamy yellow with a pinkish tinge and has seven purplish brown longitudinal stripes.

Nature of Damage

The caterpillar mine into the leaves, later on feed on the leaf surface, shelter and make within the skin passages. As they grow bigger they bore into the heads of cauliflower and cabbage. When attack is heavy, the plants are riddled with worms and outwardly the heads look deformed.

Life Cycle

The female moths lay eggs singly or in clusters on the under surface of the leaves or some other parts of the plant. The egg hatches in 2-3 days. The caterpillar feeds in the head of the cabbage and become full-grown in 7-12 days, after undergoing four moultings. The full grown caterpillar spins a cocoon among the leaves and pupate inside it. The pupal period is about 6 days and the life cycle is completed in 15-25 days.

Management

1. Collection and mechanical destruction of caterpillars in early stage of attack helps to check the infestation.
2. Grow resistant variety of cauliflower *viz.*, Early Patana, ES-96 and ES-97.
3. Spray quinalphos 0.05 per cent or endosulfan 0.1 per cent or malathion 0.1 per cent.

Cabbage Butterfly, *Pieris brassicae* (Lepidoptera: Pieridae)

Marks of Identification

Adult are a snow white butterfly having black marking on wings. The full grown larvae are 40-50 mm long. The young larvae are pale yellow, and become greenish-yellow later on. The head is black and the body is decorated with short hairs.

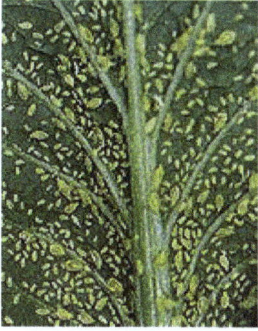

Plate 13.57: Cabbage Aphid

Plate 13.58: Heavily Aphid Infested Plant of Cabbage

Plate 13.59: Cabbage Semilooper

Plate 13.60: Semilooper Affected Cabbage Head

Plate 13.61: Cabbage Flea Beetle Affected Crop

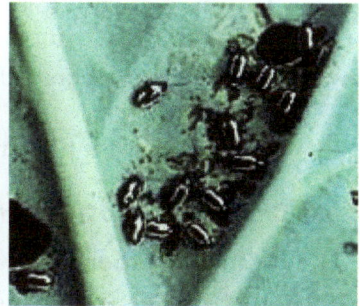

Plate 13.62: Flea Beetle of Cabbage

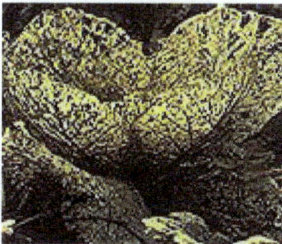

Plate 13.63: DBM Affected Leaves

Plate 13.64: Pupae of DBM

Plate 13.65: Adult of DBM

Plate 13.66: Diamondback Moth (DBM) Larvae

Nature of Damage

The caterpillar causes damage. The first instar caterpillar just scrape the leaf surface, whereas the subsequent instars eat up leaves from the margins leaving intact the main veins.

Life Cycle

Eggs are laid in cluster of 50-80 on leaf surface. The eggs are flask shaped. Egg hatch in 11-17 days. The caterpillar feeds gregariously during the early instars and dispers in latter instars. They pass through five stages and are full-fed in 14-22 days. The pupal period lasts for 7-15 days. There are four generations in a year.

Management

1. Collect and destroy mechanically caterpillars in the early stage of attack.
2. Grow resistant/tolerant varieties of cabbage *viz.*, Red rock mammoth, Red rickling, Red drum head.
3. Spray quinalphos 0.05 per cent or endosulfan 0.1 per cent or malathion 0.1 per cent.
4. Encourage the activity of *Apanteles glomeratus*.

Flea Beetle, *Phyllotreta cruciferae* (Coleoptera: Chrysomelidae).

Marks of Identification

Adult beetles are elongate oval in shape and metallic bluish green in colour. Head has impunctate vertex and black antennae.

Nature of Damage

The adults mostly feed on the leaves by making innumerable round holes in the host plants. The affected leaves dry up. A special kind of decaying odour is emitted by the cabbage plants after being attacked by this pest.

Life Cycle

A single female lays 50-80 eggs in soil. Incubation period lasts for 5-10 days. The grubs feed on roots and do not cause much damage. Grub period lasts for 9-15 days. Pupal period is of 8-14 days. There are 7-8 generations of this pest in a year.

Management

1. Hand pick and destroy adult beetles.
2. Spray quinalphos 0.05 per cent or endosulfan 0.1 per cent

Aphids, *Brevicoryne brassicae* and *Lipaphis erysimi* (Hemiptera: Aphididae)

Marks of Identification

Nymphs of *B. brassicae* are yellowish green while adults are long and darker in colour than nymphs. *L. erysimi* nymphs and adults are lighter in colour and slightly longer in size.

Nature of Damage

Nymphs and adults infest young shoots and leaves. They produce honeydew profusely and make plant parts sticky and leading to fungal growth on them.

Life Cycle

A female produces 26-67 young ones in her life span. The nymphal period lasts for 11-15 days. Life cycle is completed in 11-18 days and there can be 11-20 generations in a year.

Management

1. Remove and destroy damaged plant parts along with nymphs and adults.
2. Use yellow sticky traps at 12 per ha to attract winged aphids.
3. Grow resistant/tolerant varieties of cabbage *i.e.,* Red drum head, Early queen, Red rock mammoth, Glory, Red pickling, Express mail, etc.
4. Spray dimethoate 0.03 per cent or neem oil 2 per cent along with 0.5 ml teepol.
5. Encourage activity of natural enemies: predators, Coccinellids and Syrphids, parasitoids, *Diaeretiella rapae,* pathogens, *Entomophthora corcnata* and *Cephalosporium aphidocola.*

Pea and Beans

Pea Stem Fly, *Ophiomyia phaseoli* (Diptera: Agromyzidae)

Marks of Identification

Adults are metallic black flies, having hyaline wings. Females are slightly bigger than males. Maggots are yellowish in colour.

Nature of Damage

Maggots mine the leaves, bore inside the petioles and tender stems and tunnel downwards. Adults also puncture the leaves. Affected leaves turn yellow while the stems drop down and gradually wither away.

Life Cycle

A female lays 14-64 eggs. Egg hatches in 2-4 days. The maggot mine in the stem. The maggot is full fed in 9-12 days. Pupation takes place in the underground portion of affected stem. Pupae are barrel shaped and brown in colour. Pupal period lasts for 18-19 days. There are 8-9 generations in a year.

Management

1. Remove and destroy all affected branches during initial stage of attack.
2. Seed dressing with dimethoate 4 ml per kg of seeds at the time of planting.
3. Spray with 0.03 per cent dimethoate or endosulfan 0.07 per cent.

Spotted Pod Borer, *Maruca testulalis* (Lepidoptera: Pyraustidae)

Marks of Identification

Dark brown with a white cross band in the middle of the forewings and the hind wings are white with a darker border.

Nature of Damage

Presence of semi-solid excreta at the junction of the borehole. Young shoot with dried tip, large scale dropping of flowers. Larva present inside the webbing of leaves, flowers and young pods, faecal material accumulates outside the borehole. It feeds on the seeds by boring into the pods.

Life Cycle

Eggs are elongate oval in shape, light yellow in colour and are laid singly on or near flower buds of host plants. Young caterpillar feed on reproductive parts of flowers and move from one flower to another. Later they web inflorescences with adjacent leaves and developing pods and feed within by boring into the flowers and pods. Full grown caterpillars are light brown in colour with irregular brownish black dorsal, lateral and ventral spots. Incubation period is 2-3 days. Larval stage lasts for 8-14 days and pupal period lasts for 6-9 days.

Blue Butterfly, *Lampides boeticus* (Lepidoptera: Lycaenidae)

Marks of Identification

It is medium sized butterfly. The colour of the wings is violet metallic blue to dusky blue. The tail of hind wings is black and tipped with white. The female is slightly bigger than the male. In males, the abdomen is slender and tapering, while in female it is long and broader at the tip.

Nature of Damage

The larva bores into the buds, flowers and green pods just within couple of hours after hatching and feeds inside the developing grains.

Life Cycle

Eggs are laid on the buds, flowers, green pods and on shoot and leaves. Greenish white in colour, round in shape with a slight depression at the top. Incubation period lasts for 5-7 days. Newly hatched larva is yellowish green in colour with black head and a dark-brown patch on the prothorax and cylindrical body with scattered hair. Full-grown larva is yellowish green to yellowish red sometimes light purple in colour, ventral surface is light green. Whole larva is covered with small setae and marked with irregular black markings. It looks like a slug. Larval stage is completed in 10-27 days. Pupa are green in colour, later on it darkens and wings are also visible. Pupal development is completed in 10-15 days. The pest breeds throughout the year and passes through 5 generations in a year.

Management

1. Grow resistant or tolerant varieties.
2. Mechanical destruction of caterpillars in the initial stage of attack.

3. Spray quinalphos 0.05 per cent or endosulfan 0.07 per cent or *Bacillus thurginsis* 0.1 per cent during 50 per cent flowering stage.

Spiny Pod Borer, *Etiella zinckanella* (Lepidoptera: Phycitidae)

Marks of Identifications

Grayish brown moth, distinct pale-white band along the costal margin of the forewings. Hindwings are semi-transparent with a dark marginal line. Prothorax is orange coloured.

Nature of Damage

Entrance hole in the green pod disappears and leaves little evidence that the pod is infested. In pods, the larva devours many seeds. The pod always contains a mass of frass and held together by a loosely spun web. Young larva bores into floral parts, making rough and irregular incision.

Life Cycle

Eggs are laid singly or in small groups on immature pods either along the midrib or on the calyx. Freshly laid eggs are glistening white and adhere securely to whatever they touch. Incubation period lasts for 3-5 days. Larval period is 12-17 days. Pupa light green in colour changes to light brown or amber. Pupates in the ground at a depth of 2 to 4 cm. Pupal period lasts for 14-17 days. Total life cycle occupies 35-36 days.

Management

At flower initiation stage, spray the crop with endosulfan 35 EC 750 ml per ha and repeat the treatment after three weeks.

Bean Aphid, *Aphis craccivora* (Hemiptera: Aphididae)

Marks of Identification

Apterous females are shiny, dark brown or black. Alate forms are greenish black with transparent wings.

Nature of Damage

Colonies of nymphs and adults found on leaves, terminal shoots and pods and suck the plant sap. Serious pest when rainfall is low.

Life Cycle

The common mode of reproduction is through vivipary and parthenogenesis, though reproduction by ovipary has been also recorded. A single apterous, parthenogenetic female produced 29 nymphs. The nymphs generally undergo 4 moults before reaching the adult stage. The duration of each instar is usually one day, though in some cases it was even 3 days. Within a day after it becomes an adult the apterous female starts producing its brood. A female reproduces upto a maximum of 12 days.

Management

1. Groundnut intercropped with cereals (*e.g.* millets or maize) reduced the incidence of this pest.
2. Encourage the activities of the predators, *viz.*, *C. septumpunctata*, *M. sexmaculatus*, *B. suturalis* and *X. scutellarae*.
3. Spray demeton-o-methyl (0.05 per cent).

Potato

Potato Aphid, *Myzus persicae, Aphis gossypii, A. fabae* and *Rhopalosiphum rufiabdominalis* (Hemiptera: Aphididae)

Marks of Identification

Myzus persicae

It is light to dark green or pink with well developed frontal tubercles, which project inwards. The cornicles are long, cylindrical and slightly swollen in the middle. In the winged form, a dark patch can be seen on the abdomen.

Aphis gossypii

Colour variable, ranging from pale yellow to brown or grey black or light to dark green. The tips of the leg joints, eyes and cornicles are black.

A. fabae

Variable in colour, from black to olive green, often with irregular dark pigmented areas over the abdomen. Cornicles slightly black, imbricated, slightly tapering towards exterior. In the winged forms some black bars occur on the abdomen.

Rhopalosiphum rufiabdominalis

Olive green to almost black. Antennae 5 or 6 segmented. Media of forewing sometimes once branched. Usually with reddish blotches at the base of siphunculi.

The first two species *viz.*, *Myzus persicae* and *Aphis gossypii* mainly act as major pest.

Nature of Damage

Both adults and nymphs are destructive as they suck the sap from the plant. Beside this they act as potential vector of potato leafroll (PLRV) and Potato leaf yellow (PLY). They are not serious as pest on the crop but play a vital role in limiting seed production. The losses in yield by the aphid transmitted virus range between 40-85 per cent.

Life Cycle

The winged viviparous females start appearing on potato and other secondary host plants from middle of November onwards. These forms reproduce parthenogenetically and give birth to living young ones. These pass through four nymphal stages. Each stage is of about 1-5 days duration and one generation is completed in about 15 days. This type of asexual reproduction goes on for many generations on the secondary hosy plants.

Management

1. Remove all weeds, hosts prone to virus and for aphids especially those having yellow flowers, and volunteer (selfgrown potatoes) plants from within and around the vicinity of field.

2. The haulm cutting of seed crop should be done as soon as the aphid number crosses the critical level *i.e.*, 20 aphids per leaves.

3. Application of phorate 10 G @ 1.5 kg a. i. per ha on furrows at planting time will keep the pest under check upto 45-60 days. This should be followed by need based foliar application of any suitable systemic insecticides such as dimethoate 30 EC @ 0.03 per cent.

Cut Worm, *Agrotis ipsilon* and *A. segetum* (Lepidoptera: Noctuidae)

Marks of Identification

Adults are medium to heavy bodied moths. *A. ipsilon* moth is dark brown to grayish brown with large areas of black patches on the forewings and thorax. The moths of *A. segetum* are slightly smaller than *A. ipsilon*, measuring 20 mm in length and 35 mm in wing span.

Nature of Damage

Crop damage is caused by the caterpillars only. They feed at night on young shoots or under ground tubers. In the early stages of the crop, the caterpillars cut the stem of the young plants near the ground and feed on the shoots and leaves. After tuberization, they feed by bore and nibble into the tubers affecting both tuber yield and market value.

Management

1. Deep ploughing during hot weather reduce the population of immature stage. A number of birds, such as crow, mynah, starling etc feed on the insects that get exposed upon ploughing.

2. Natural enemies play an important role in the management of cutworms. Some of the important parasites of cutworms are *Broscus punctatus* and *Liogryllus bimaculatus*, *Auplopus hypsipylae*, *Ichneumon* sp., *Turanogonia chinensis*.

3. Spray chlorpyriphos 20 EC @ 2.5 liters per ha for spraying the foliage and drenching the ridges on noticing the cutworms attack.

White Grub, *Lachnosterna longipennis* and *L. coracea* (Coleoptera: Scarabaeidae)

Nature of Damage

The damage to potato crop is caused by the grubs in two ways *i.e.*, when they initially feed on rootlets or roots and after tuberization on tubers. The first stage grubs feed on the organic matter available in the soil but they prefer live roots. The damage to potato tubers is caused, mainly by the second and third instar grubs which make large, shallow and circular holes and render them unfit for human consumption.

Life Cycle

The beetles start coming out of the soil at dusk, soon after the first pre-monsoon shower in May end or early June and settle on nearby bushes or trees *viz.*, Acacia, neem, roses, etc., first for mating and afterwards for feeding. The beetles after feeding on the host foliage return to soil (5 to 10 cm deep) early in the morning (before sunrise) for egg laying. The eggs and the first instar grubs are found near the root zones of potato plants between June and August, while second and third instars attack tubers during August- October, following the harvest. These grubs move down into the soil, with fall in temperature. These grubs reach upto one meter depth and hibernate in earthen cells. They come up in March-April with the rise in temperature of atmosphere and soil. The full grown grub pupate during April-May for about a fortnight and the beetles emerge between late May and early June.

Management

A. Beetles

1. A majority of beetles are attracted to the light source hence electric or petromax light traps may be operated for mass-collection.
2. Wild shrubs and other hosts of beetles growing in or around the field should be removed, excepting a few which should be sprayed with insecticides before the emergence of beetles.
3. Adults may be killed by spraying the host trees or shrubs with insecticides like chlorpyriphos 20 EC or endosulfan 35 EC or quinalphos 25 EC at 0.1 to 0.2 per cent concentration.

B. Grubs

1. Repeated ploughing before monsoon expose the grubs and pupae for predation by natural enemies such as crows, mynahs, etc., or may be hand collected and destroyed.
2. Flooding of the field, wherever possible for 7-10 days.
3. Apply only well rotten FYM, compost in the field.
4. Apply phorate 10 G or carbofuran 3 G @ 2.5–3.0 kg a. i. per ha in furrows at planting or near plant base at earthing time. Application of granular insecticides at earthing time is more effective.

Potato Tuber Moth, *Phthorimaea operculella* (Lepidoptera: Gelechidae)

Nature of Damage

This is one of the most destructive pest of potato under the worm and dry environment of field and storage. The larvae attack potato in two ways *i.e.* by mining younger leaves and feeding on the tubers. Larvae penetrate the leaves and feed within leaf veins or stems of the plant and on tubers in storage deposit the eggs near the eye buds, causing irregular galleries or tunnels deep inside the tuber. Infested tubers are seen at maturity in the field and top of the storage. Pest infestation in field and storage varies from 0-84 and 0-52 per cent in different parts of country.

Life Cycle

There are four stages of this pest namely egg, larva, pupa and adult. On an average, 60-70 eggs are laid on leaves, eye buds of the tubers and soil around the plant. Eggs are laid either singly or in groups of 6 to 7 on an eye bud. The incubation of eggs, larval and pupal period varies from 2 to 28, 6 to 50 and 6 to 33 days, respectively, depending on the environmental conditions. Life cycle of PTM is usually completed within 21-30 days. Several generations are completed throughout the year.

Management

A. Field Control Measures Prior to Harvest

Measures to protect the potato crop from planting to harvest are primarily cultural control methods. These are as follows:

Good soil preparation: Adequate soil preparation not only ensures vigorous plant growth but also helps to destroy the resting stages, mostly pupae, of the tuber moth before planting.

Deep planting: Placing tuber seed to a depth of 5-10 cm prevent female moths from ovipositing in seed tubers and keeps larvae from migrating to tubers from infested aboveground sprouts. Newly emerged larvae of PTM can burrow to a depth of 10 cm to the seed tubers.

High hilling: High hilling of growing plant protects the developing tubers from ovipositing female and reduces the possibility of larvae reaching the bulking tubers.

Frequent irrigation: Adequate watering and cultivation prevent cracks from forming in the soil. Soil cracks allow female moths to reach the potato tubers for oviposition, and provide shelters to adult moths.

Pheromone traps: Commercial pheromones are available for PTM. Mass trapping of male moths reduces the probabilities of moth mating, thus causing a drop in egg fertility.

Pesticide application: Spray monocrotophos 40 EC @ 1.5 litre in 1000 liters of water on 30 days old crop. It may be repeated two weeks before harvesting.

B. Control Measures at Harvest

The two most important control measures at harvest are protecting harvested tubers from ovipositing females and removing crop residues from the field. The following practices should be adopted:

Timely harvesting: During the last phase of the crop *i.e.* tuber filling and plant senescence, the infestation rate accelerates. Delaying harvest by one or two months can increase damage as much as 70-80 per cent

Storing healthy tubers: Only healthy tubers should be stored. Infested tubers should be buried under at least 10 cm soil.

Covering tubers: Female moths become active in the evening and most eggs are laid at that time. Harvested tubers should not remain exposed to ovipositing females overnight. If they cannot stored immediately tubers should at least be covered.

Destruction of harvested residues: Insect pupates in tubers and dry stems left in the field. Moths from these pupae infest the crop. The tubers left in the field become volunteer plants for the pest. For these reasons, all harvest residues must be destroyed.

C. Measures to Avoid Damage in Storage

Cleaning stores: Cleaning floors, walls, and ceilings of stores before storing healthy tubers destroys pupae and other life stages of the moth.

Storing healthy tubers: Tubers should be sorted and infested ones discarded before storing. Tubers exposed to moth oviposition should not be stored, as eggs are commonly overlooked during sorting. Storing infested tubers or those that have been exposed to moth oviposition along with healthy tubers may result in infestation of the entire store with in 3 to 4 months.

Use of repellents: The foliage of some plants, rich in essential oils, such as neem, eucalyptus and *Lantana camara*, repels PTM. The leaves are dried under the shade, crushed, and then used to cover the tubers in stores.

Use of pheromone traps: This reduces the further multiplication of the PTM and reduces the population (@ 4 traps per 100 cu m).

Microbial pesticides: Use of microbial pesticides *viz.*, *Bacillus thuringiensis* (B.t.) and Granulosis virus (GV), are also very effective in reducing the population of PTM.

Sweet Potato

Sweet Potato Weevil, *Cylas formicarius* (Coleoptera: Curculionidae)

Marks of Identification

Adult weevils are ant like, slender bodied having elongated snout like bluish brown head with non-geniculate antennae, bright red thorax and legs and brownish red abdomen. Females make small cavities on the tubers or stems and lay eggs singly.

Nature of Damage

It is a specific pest of sweet potato. Grubs bore into stems, tunneling inside and feed on soft tissues. Grubs and adults bore into tubers both in the field and storage godowns. Affected tubers develop dark patches, which later start rotting. Pest is disseminated from field to field through infested vines and is carried over from season to season by breeding in damaged tubers left in the field after harvest.

Life Cycle

Each female lays 100-200 eggs. Grubs are fattish, legless, pale-yellowish white in colour. Pupation takes place in larval burrows. Incubation, grub and pupal stages last for 5-10, 16-20 and 4-8 days, respectively. Life is completed in 4-5 weeks. More than one generations are completed in a year.

Management

1. Remove previous sweet potato crop residues and alternate host, *Ipomoea* sp., and destroy them.
2. Use pest free planting material.

3. Use deep rooting varieties like White Star, CL-44 and Pusa Red to avoid the attack of this pest.

4. Use cut sweet potato tubers (100 g) as trap during 50-80 days after planting (DAP) at 10 days intervals. Set the traps at 5 m apart at 4 PM and collect and destroy adult weevils at 6 AM next day.

5. Dip planting materials in fenthion 0.05 per cent or monocrotophos 0.05 per cent.

6. Rake up soil and earth up at 50 days after planting.

7. Drench soil with endosulfan 0.05 per cent, or spray it if needed.

8. Harvest immediately after maturity and destroy the crop residues.

9. Install yellow sticky trap at 12 per ha or pheromone trap at 10 per ha.

10. In godowns, treat out side of bags containing tubers with malathion 5 per cent.

Onion and Garlic

Onion Thrips, *Thrips tabaci* (Thysanoptera: Thripidae)

Marks of Identification

The adults are slender, yellowish brown and measure about 1 mm in length. The males are wingless whereas females have long, narrow strap-like wings, which are furnished with long hairs along the hind margins. The nymphs resemble the adult in shape but are wingless and slightly smaller.

Nature of Damage

Damage is done by adults and nymphs. In onion and garlic, the leaves of attacked plants become curled, wrinkled and gradually dry up. The plant do not form bulbs nor do the flowers set seed.

Life Cycle

The pest is active throughout the year. The adult female lives for 2–4 weeks and lays 50-60 eggs singly in slits which are made in leaf tissues with its sharp ovipositor. The eggs hatch in 4-9 days. The nymphs pass through four stages and are full fed in 4-6 days. The full grown nymph pupate in soil at a depth of about 25 mm. The pre pupal and pupal periods last 1-2 and 2-4 days, respectively. There are 7 generations in a year.

Management

1. Grow resistant varieties *viz.*, White Persian, Grano, Bombay White, Sweet Spanish, Crystal Wax.

2. Soil application of phorate 10 G @ 2.0 kg a. i. per ha before sowing has been found effective to control onion thrips.

3. Spray 0.05 per cent malathion or 0.075 per cent acephate.

Onion Maggot, *Delia antique* (Diptera: Anthomyiidae)

Marks of Identification

The adult flies are about 6-8 mm in length, slender, large winged, rather bristly. Maggots are small white and about 8 mm long.

Nature of Damage

Only maggots cause damage. Maggots tunnel through bulbs and leaving only the outer sheath. The leaves of the infested plants turn brown from the tip down wards. The attack does completely destroy the bulbs but it causes subsequently rotting in storage.

Life Cycle

Eggs are laid by female near the base of the plant or cracks or in soil. The incubation period is 2-7 days. After hatching, the maggots mine through bulbs and cause damage to bulb. Maggots attain its full growth in 14-21 days and then crawls out the bulb to pupate inside the soil. Pupal period lasts for 14-21 days.

Management

Soil application of phorate 10 G at 15 kg per ha or carbofuran 3 G at 25 kg per ha.

Waternut or Singhara

Singhara Beetle, *Galerucella birmanica* (Coleoptera: Chrysomelidae)

Marks of Identification

Adult beetles on emergence are bright yellow but soon become reddish brown. Antennae is brown-black. The beetles are 6-8 mm long. Grubs are yellowish-brown. The beetles are sluggish and not active fliers.

Nature of Damage

Grubs and adults feed usually on leaves and sometime also on petioles and integument of the fruits. Full grown grubs are voracious feeders and more destructive than adult beetles.

Life Cycle

A single female lays on an average 101 eggs in clusters of 5-8 eggs, glued firmly to the upper surface of leaves. Pre-mating period is about 2 days, during July and extends upto 7 days in November. Pre-oviposition period ranges from 15 hours to 4 days. Oviposition period lasts for 6-19 days. Incubation, grub and pupal periods are 4-9, 9-20 and 5-10 days, respectively. Total life cycle from egg to adult stage is completed in 18-32 days. Longevity of male and females is 11-28 and 13-33 days, respectively. The entire life cycle of the beetle and its immature stages is passed on the leaves of water nut plant. The pest is active during August-September.

Management

1. Mechanical collection of adults and egg-masses from the leaves and their destruction in early stages of crop is quite effective.

2. The pest can be controlled by dusting malathion 2 per cent @ 25 kg per ha.

Pests of Fruits

Apple

San Jose Scale, *Quadraspidiotus perniciosus* (Hemiptera: Diaspidae)

Marks of Identification

The adults are grayish tiny insects showing sexual dimorphism, the female being rounded, 2 mm across and wingless while the males are elongated and winged, bearing 2 well developed forewings whereas hindwings are rudimentary.

Nature of Damage

The damage is caused by nymphs and female scales which suck the sap from the twigs, branches and fruits. All parts of the plant above the ground are attacked and the injury is due to loss of the cell-sap. At first, the growth of the infested plants is checked, but as the scales increase in number, the infested plants may die.

Life Cycle

The pest is active from March to December and passes the winter in nymphal stage. The female gives birth to young ones, which hatch from the eggs developed within the body of the female. Each female may give birth to 200-400 nymphs. They become full grown in 3-40 days and the females again start giving birth to young ones within the next 50-53 days. There are four to seven overlapping generations in a year.

Management

1. For effective management of this pest, orchard sanitation should be given priority. Infested pruned material should be collected immediately and burnt.

2. The parasite, *Encarsia perniciosa* may be released to check the overwintering population on wild host plants growing around.

3. Spray diesel oil emulsion + Bordeaux mixture is to be emulsified and diluted 5-6 times before spraying or spray 7.5 liter of ESSO tree spray oil emulsion in 250 liters of water per ha during the winter season when the trees are in dormant stage and completely defoliated.

4. In case of severe infestation spray methyl demeton 25 EC (1.25 liter per ha).

5. To protect the plants in the nursery, apply carbofuran granules 0.75–1.0 g a. i. per plant.

Apple Wooly Aphid, *Eriosoma lanigerum* (Hemiptera: Aphididae)

Marks of Identification

The infested plants have pale green leaves and whitish cottony patches on the stems and branches. Characteristic galls or knots are formed on roots and other underground portions of the plants.

Nature of Damage

The pest is active throughout the year. It attacks primarily the underground roots but alate forms also attack trunks, branches, stems, twigs, leaf petioles and fruit stalks. Upward and downward migrations are accentuated during hottest and coldest season, respectively. Due to desaping caused by these pests, the affected trees present a sickly appearance, lose vigour and the growth of these trees as also their fruiting capacity are adversely affected, in case of young tree, the roots disintegrate to such an extent that these trees are easily blown over by the even moderately strong winds.

Life Cycle

The pest overwinters either as egg or young nymph on the roots of the host tree. The eggs hatch and the nymphs mature during spring. There are four nymphal instars and the total duration of nymphal stage lasts for 32-43 days. There may be 13 generations in a year.

Management

1. To prevent damage by this pest, use resistant root-stock like Golden delicious, Northern spy and Morton stocks 778, 779 and 793.
2. Release exotic parasitoid, *Aphelinus mali.*
3. Select healthy plant material from nursery and then before planting in the orchard, treat them with chlorpyriphos 0.05 per cent.
4. Spray 500 ml of nicotine sulphate 40 EC or 800 ml of malathion in 500 liter of water per ha.

Codling Moth, *Cydia pomonella* (Lepidoptera: Tortricidae)

Marks of Identification

The adult moth is small, about 12-14 mm in wing span and is 6-8 mm long. The forewings are dark grayish and are marked with wavy lines and a copper coloured metallic eye like circle towards the outer margin. The hindwings are pale grey. The full grown larvae are 16-22 mm long and are pinkish or creamy-white in colour with a brown head. The larvae have eight pairs of legs.

Nature of Damage

Damage is caused by larvae which burrow into the fruit and feed on the pulp. The infested fruits cannot be marketed for table consumption.

Life Cycle

A single female may lay about 100 eggs in her life time. The egg hatch in 4-12 days. The young larvae after emergence enter into the fruit through the calyx. The larval period lasts for 3-4 weeks. At this stage, the larva burrows its way out of the apple fruit and falls to the ground and spins a silken cocoon in which it transforms into a yellowish-brown pupa. The pupal period is completed in 8-14 days. There are two generations in a year.

Management

1. The orchard should be kept clean of all the debris and weeds and remove loose barks from the old trees to prevent the hibernating larvae to find shelter.
2. Pluck and collect fallen infested fruits to burry or burn them to destroy the hiding larvae.
3. Enforce quarantine rules strictly to prevent spread of this dangerous pest.
4. Spray endosulfan 0.07 per cent to protect the fruits from this pest.

Gypsy Moth, *Lymantria obfuscate* (Lepidoptera: Lymantriidae)

Marks of Identification

The female moths are dark grey and have atrophied wings. The males are comparatively more active in moving around and in mating. Female moths are apterous type and are unable to fly. The caterpillar is 40-50 mm long and is clothed in tufts of hairs.

Nature of Damage

The caterpillars are gregarious and feed voraciously at night time. Their habit to defoliate the host trees completely results in the failure of fruit formation.

Life Cycle

The eggs are laid under the loose bark and covered over with yellowish-brown hairs. The eggs over-winter as such and hatch in March-April. The larval period is completed in 66-100 days. The pupal formation takes place in soil among debris and pupal stage lasts 9-21 days. The male moth lives for 4-10 days and the females for 11-31 days. One generation is completed in a year.

Management

1. Collect and destroy sluggish females and the egg-masses that are so conspicuous are easy to locate.
2. Spray endosulfan 0.07 per cent to protect the fruits from this pest.

Mango

Mango Hopper, *Amritodus atkinsoni, Idioscopus clypealis, I. nitidulus* and *Amrasca splendens* (Hemiptera: Cicadellidae)

Marks of Identification

Freshly hatched nymphs are wedge shaped and whitish in colour with two small red eyes, gradually with each moulting the colour changes to yellow, yellowish-green, green and ultimately to greenish-brown. Adults are also wedge shaped having greenish brown body and pale yellow vertex. Forewings are thicker than hind wing, bronzy sub-hyaline with veins pale yellow and a white line along the costal margin forming distinct mid longitudinal line when the insect is at rest. Adult of *A. atkinsoni* are darker in colour and bigger in size, male being 4.2 to 4.8 mm long and females 4.7 to 5.1 mm those of *I. clypealis* are smaller and narrower, males being 3.4 to 3.7 mm long and females 3.6 to 3.9 mm. *I. nitidulus* is dirty brown in colour and is medium sized. *A. splendens* adults are pale yellow or greenish yellow in colour.

Nature of Damage

Enormous number of nymphs are found clustering on the inflorescences sucking sap. The infested flowers shrivel, turn brown and ultimately fall off. On attaining maturity the hoppers leave the blossoms and move on leaves and trunks of the trees. Swarms of adults are commonly seen hovering in mango groves and sitting on all plant parts. *A. atkinsoni* adults take shelter in cracks of crevices of mango tree trunk bark. Both nymphs and adults suck the sap usually from the ventral surface of leaves. As a result, growth of the trees is stunted. The hoppers also exude honey dew which encourages the development of fungi, resulting in growth of sooty moulds on dorsal surface of leaves, branches and even fruits. The characteristic fungi are *Capnodium mangiferae* and *Meliola mangiferae.*

Life Cycle

The pest is active throughout the year but during the hot months of May-June and the cold months of October-November, only the adults are found sitting in thousands on the bark of trunks and branches. A female deposits 100 to 200 eggs. The eggs hatch within 4-7 days. The nymphs become full grown in three stages, in 8-13 days. The life cycle from the time eggs laid to the time the adult appear, takes 15-19 days.

Management

1. High density planting should be avoided as it provides favourable condition for hopper multiplication.
2. Regular irrigation should be given in mango orchard to prevent intermittent flushes. Split doses of nitrogen fertilizers should be applied.
3. Avoid waterlogging or damp conditions.
4. In case of dense orchards, prune some of the branches during early winter to have better light interception.
5. Spray endosulfan 0.07 per cent or phosalone 0.075 per cent or monocrotophos 0.04 per cent or deltamethrin 0.0075 per cent. The insecticides should be sprayed alternatively whereas the hopper population exceeds the count of 5 nymphs per inflorescence of adults or both.

Stem Borer, *Batocera rufomaculata* (Coleoptera: Cerambycidae)

Marks of Identification

The adults are longicorn beetles, well built, large and pale grayish, measuring about 5 cm in length and 2 cm in breadth. The beetle is provided with long legs and antennae and a dirty white band, extending from head to tip of the body on each side. A number of dirty yellowish spots are present on the elytra. Head is distinct with large prominent eyes and the pronotum is ornamented with two crescent orange yellow spots. Full grown larva is a stout yellowish or white fleshy grub, measuring about 6 cm in length. Its head is dark with strongly developed mandibles.

Nature of Damage

This pest is quite serious when it appears since it attacks the main stem and the branches of the trees. When the stem or any branch is attacked, the sap and masses of frass exude from the bored hole. Often the damage may be visible by falling off the leaves of the attacked branches and sudden collapse of the branches. In case of severe damage, the entire tree is often killed. The damage is done either to the roots or the stems. As the grub feeds, a harmonious sound is produced by its mandibles. On this account, the insect is popularly known as violin beetle.

Life Cycle

Copulation generally commences 2 days after the adult emerged. The pre-oviposition and oviposition period are on an average 1-1.5 and 22 days, respectively. Female lays eggs parallel to the bark layers in depth of 2.5 to 4.0 mm. Eggs are laid mostly on trunks and primary branches. The incubation period is 7-13 days. Larval period is 140-160 days. The pupal period varies from 19-36 days. Adult live for 25-26 days. The total life cycle completed in 170–190 days.

Management

1. Cut and destroy the infested branches with grubs and pupae within.
2. Beetles wherever found in the garden should be collected and destroyed.
3. The grub should be extruded through hooked wires or destroy by injecting dilute kerosene or petrol or 0.02 per cent dichlorvos @ 5 ml per hole and then the holes be sealed with the mud.
4. To kill effectively, the stem borer larva hidden in up-down tunnel through insecticide, apply insecticide solution with the help of syring (10 ml).

Fruit Fly, *Bactrocera dorsalis* (Diptera: Tephritidae)

Marks of Identification

The adult fly is stout and measures 14 mm across the wings and 5 mm in maximum length. The flies are strong fliers and can fly up to two kilometers in search of food. The fly is brown or dark brown in colour with hyaline wings and yellow legs. The thorax is ferruginous without yellow middle stripe. The abdomen is conical in shape and dark brown in colour. The young maggot is white and translucent.

Nature of Damage

The dark puncture caused by the oviposition of adult fly is not very conspicuous as its colour blends with the dark green colour of the fruit. It is very clearly visible in some yellow and pale brown varieties. The maggots on hatching feed on the pulp of the fruit for few a days and a brown rotten parch appears on the fruit surface. The mesocarp becomes dirty brown. Infested fruits finally fall on the ground. The fruit is affected from late April to June.

Life Cycle

A female on an average lays 50 eggs in one month. Eggs hatch in 3-10 days. As maggots develop, they pass through 3 stages in the ripening pulp and are full-grown in 6-29 days. They leave fruit and move away by jumping in little hope. On reaching a suitable place, they burry themselves into soil and pupate. In 6-44 days, they emerge as flies and reach ripe fruit for further multiplication. Life cycle is completed in 2-13 weeks and many generations *i.e.*, 10-12 are completed in a year. Flies are present in the field all through the year.

Management

1. Collect and destroy fallen and infested fruits by dumping in a pit and covering with a thick layer of soil.
2. Use parasitoids *Opius compensatus* and *Spalangia philippinesis*.
3. Use methyl eugenol lure trap (10 per ha) to monitor and kill male adults or prepare methyl eugenol and malathion 50 EC mixture at 1:1 ratio and take 10 ml mixture per trap.
4. Prepare methyl eugenol impregnated plywood based fruit fly trap. For this purpose, take plywood block measuring 5x5 mm size and 12 mm thickness. Prepare a solution of ethyl/methyl alcohol (6 ml) + methyl eugenol (4 ml) + DDVP (1 ml) and pour block in it for 24-48 hours. Thereafter, use them for direct nailing on tree trunk @ 10/ha or use in a discarded through away mineral water bottle @ 10/ha.
5. Use bait spray combining molasses or jaggery 10 g per liter and one of the insecticides, malathion 50 EC 2 ml per liter, dimethoate 30 EC 1 ml per liter two rounds at fortnight interval before ripening of the fruit.

Shoot Borer, *Chlumetia transversa* (Lepidoptera: Noctuidae)

Marks of Identification

Young caterpillars are yellowish-orange in colour with characteristic dark brown prothoracic shield. Full grown caterpillars are dark pink with dirty spots and measure 20 to 24 mm in length. Adults have thorax and abdomen clothed with rufous, fuscous and gray scales. Forewings are dark gray beautifully patterned with wavy design. Hindwings are fuscous, apical side being darker than proximal side. Wing expanse is 15-20 mm.

Nature of Damage

The newly hatched larvae bore into midribs of the leaves and feed therein for 2-3 days, and thereafter, they come out and bore into the tender shoots. They make tunnels downwards up to 100 mm to 150 mm in length and expelled out excreta through the entrance hole and the shoot becomes hollow. The affected shoots show dropping of leaves and give a wilting look. The attack is noticed during the period when there is new flush on the trees and saplings. The young caterpillars attack during the earlier part of April with the commencement of hot winds. The larvae also eat young leaves and inflorescence.

Life Cycle

The adult female lays eggs singly on the tender leaves and hatch within 2-3 days. The caterpillar takes 10-12 days to mature and then it leaves the tunnel and enters into the cracks, and crevices of bark of the tree, dried malformed panicles and also in the soil for pupation. It makes a silken cocoon and within that it pupates. The pupae are unaffected till the onset of monsoon. Generally from these pupae the moths emerges out in 15-18 days. The total life cycle completed in one and half month's period.

Management

1. To control this pest, clip off and destroy promptly affected shoots in the initial stage of attack.
2. Spray carbaryl 0.2 per cent or endosulfan 0.07 per cent or monocrotophos 0.04 per cent 2-3 times at three weeks interval commencing from initiation of new flush of leaves.

Leaf Gall Midge, *Procontarinia matteiana* (Diptera: Cecidomyiidae)

Marks of Identification

The adult is a minute, yellowish colour with grayish black. The males are larger than the females. The head bears two large compound eyes and long antennae. There are two tiny transparent wings. The legs are slender and long each bearing strongly arched claw with additional dent.

Nature of Damage

The pest lays its eggs on ventral surface of leaves. On hatching the maggots bore inside the leaf tissues and feed within, resulting in formation of small raised wart-like galls on the leaves. The affected leaves get badly deformed and drop prematurely.

Life Cycle

The eggs are laid in tender leaves and life embedded within leaf tissue. The oviposition sites are marked by small reddish spots. Within a week after oviposition, the individual galls having eggs show symptoms as lenticular thickenings. The incubation period lasts for 3-4 days. The larval period lasts for 2-12 months. There are three overlapping generations in a year. The adults live for about two days.

Management

1. Cultural practice such as summer ploughing has been found useful in reducing the midge population in the following years.
2. Cut the heavily infested leaves at the early stage of infestation and destroy them.
3. Foliar application of monocrotophos 0.04 per cent or dimethoate 0.03 per cent or methyl-o-demeton 0.03 per cent should be applied at 15 days interval during new flush of leaves.

Mealybug, *Drosicha mangiferae* (Hemiptera: Margarodidae)

Marks of Identification

This pest is univoltine and commonly called as mealybug. The nymphs and adult females are flat, oval, waxy whitish insects, sometimes mistaken for fungal

Plate 13.67: Mango Hopper

Plate 13.68: Mango Hopper Infested Fruits

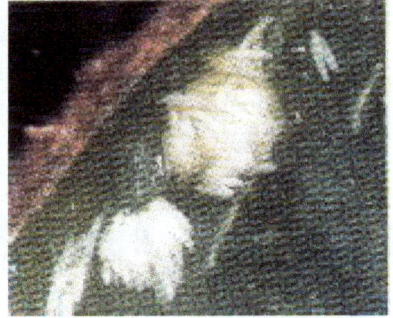

Plate 13.69: Mango Fruit Fly–Eggs

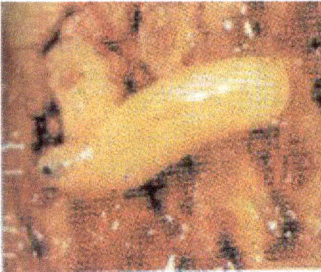

Plate 13.70: Mango Fruit Fly–Maggots

Plate 13.71: Mango Fruit Fly–Pupae

Plate 13.72: Mango Fruit Fly Adult

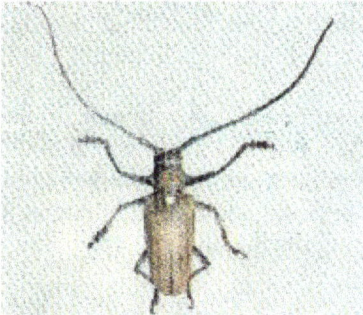

Plate 13.73: Mango Stem Borer

Plate 13.74: Mango Stem Borer–Grub

Plate 13.75: Mango Stem Borer Affected Stem

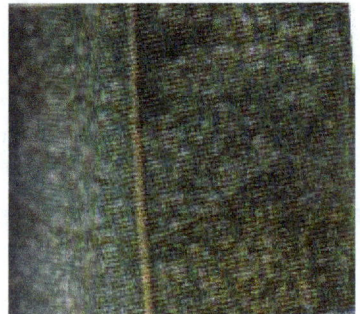

Plate 13.76: Mango Scale

growth. Adult females are wingless while males are crimson coloured bugs with two dark brownish black wings and cause no damage, except fertilizing the female.

Nature of Damage

Nymphs and adult females suck the sap from inflorescence, tender leaves, shoots and fruits. Excessive desaping of infested tissues leads to their wilting, drying and ultimately fruit setting is affected. They secrete honeydew over which sooty mould (*Capnodium mangiferae*) develops, as a result of which leaves and inflorescence become shining black and sticky. Due to the growth of sooty mould on the leaves, photosynthetic activity is adversely affected.

Life Cycle

This pest is active from December to May and spends rest the year in egg stage. Eggs are generally deposited in ovisacs in soil under the trees. They hatch at the end of December or in January and nymphs feed for a month on low-growing vegetation and then ascend the trees and cluster on the growing shoots and suck the sap from them. Nymphal duration is 76- 134 days. Adult duration is 22-27 days.

Management

1. Remove weeds from orchards which act as additional hosts for mealybug.
2. Ploughing or digging of the soil at depth of 15 cm around the tree during summer exposes eggs to natural enemies and extreme sun heat.
3. Rake the soil around the tree trunk and mix methyl parathion 2 per cent dust @ 200 g per tree.
4. Nymphs should be prevented from crawling up the trees by applying 400 gauge alkathene or plastic sheet 25 cm wide around the trunk about one meter above the ground level. Before tying the band, paste a layer of mud or wet soil around the tree trunk, to make that portion smooth and then wrap the alkethene or plastic sheet and tie it on both sides with sunhemp thread.
5. Nymphs found congregating below the lower edge alkathene band should be killed mechanically or by spray of methyl parathion 0.05 per cent.
6. Foliar application of methyl parathion 0.075 per cent or monocrotophos 0.05 per cent or diazinon 0.1 per cent should be applied as and when the pest appears on mango trees.

Mango Stone Weevil, *Sternochetus mangiferae* (Coleoptera: Curculionidae)

Marks of Identification

It is a short stoutly built, ovoid, dark-brown weevil and found inside the stone of fruit or in its pulp.

Nature of Damage

The insect attacks mango varieties with a relatively soft flesh. The injury caused by the grub feeding in pulp sometimes heals over but a certain number of fruits always get spoiled when the weevils make an exit through ripe or near ripe mango fruits.

Life Cycle

The weevil lays eggs in the skin or ripening fruit. On emergence from the egg, the grub enters further inward, eating its way through the unripe tissue until it bores into the embryo of the mango-stone. It pupates inside the stone and the weevil cuts its way through the stone. The life cycle is completed in 40-55 days and there is only one generation in a year.

Management

1. Field sanitation including collection and destruction of all fallen fruits at weekly interval destroys adult weevil.
2. Regular fallen fruit sampling, starting from pea size of fruit helps in forecast of the pest and helps in timing of first spray.
3. Racking of soil below the tree in October-November and March.
4. At pea to marble size, spraying of fenthion 0.08 per cent followed by monocrotophos 0.05 per cent after two weeks checks the pest population.

Citrus

Citrus Psylla, *Diphorina citri* (Hemiptera: Aphalaridae)

Marks of Identification

The adult is small 3-4 mm in length, active in habits and rests on the leaf surface with closed wings, the tail end of the body being turned upward. The insect is brown with its head lighter brown and pointed. The wings are membranous, semi-transparent with a brown bend in the apical half of the forewings. The hindwings are shorter and thinner than the forewings. The nymphs are flat, louse like and orange yellow creatures, and are seen congregated in large numbers on young leaves and buds.

Nature of Damage

Both nymphs and adults suck the plant sap but the nymphs are more destructive to the plants. They suck the cell sap from tender shoot, leaves and flowers causing curling of leaves, defoliation and drying of twigs. It is also reported that the insect produces a toxic substance in the plant as a result of which the fruit remains under sized and poor in juice. The insect is also responsible for spreading the greening virus.

Life Cycle

A single female lays about 800 eggs on tender leaves and shoots after the pre-oviposition period of about 24 hours. The eggs are laid either singly or in groups of two or three. The egg hatch in 10-20 days in winter and 4-6 days in summer. There are five nymphal stages and the development is completed in 10-21 days. There are great variations in the longevity of adults in different seasons. There are 8-9 overlapping generations in a year.

Management

1. Spray suspension of neem or *datura* (*Ipomea fistullosa*) leaf extract (10 per cent) or azadirachtin 0.03 per cent or neem oil 0.5 per cent is suggested to control the pest.

2. Spray methyl-o-demeton 0.025 per cent or dimethoate 0.03 per cent or malathion 0.05 per cent or monocrotophos 0.04 per cent or phosalone 0.07 per cent or endosulfan 0.07 per cent at peak activity of this pest is suggested.

3. The optimum dose of fertilizers should be applied in citrus orchards as higher NPK dose fertilizer results in higher infestation of psylla than lower NPK dose of fertilizers.

4. Trees should be planted in wider spacing (6.0 m x 6.0 m and 6.0 m x 4.5 m) to minimize the infestation of this pest.

Leaf Miner, *Phyllocnistis citrella* (Lepidoptera: Phyllocnistidae)

Marks of Identification

The adult is a tiny moth measuring 4.2 mm across the wings. On the front wings there are brown strips and prominent black spots along the tips. The hind wings are pure white and both pairs are fringed with hairs. The full grown larva measures 5.1 mm in length, apodus and is pale yellow or pale green with light-brown well developed mandibles.

Nature of Damage

The caterpillar attacks only young and tender leaves, whereas older leaves are avoided. The larvae make serpentine mines in the leaves feeding on epidermal cells of the leaf leaving behind the remaining leaf tissue quite intact. The mining larvae feed actually more on sap than on solid tissue of the leaf. The mined leaves turn pale, set destroyed and may dry up. The attacked leaves remain on the plant for a considerable long time and the damage gradually spreads to fresh leaves. Heavily attacked plants can be spotted from a distance and young nurseries are most severely affected, the young plants of orange and grape fruit may not even survive. In large trees, the photosynthesis is adversely affected, vitality is reduced and there is an appreciable reduction in yield. Attack by this pest also encourages the development of citrus canker.

Life Cycle

The pest is active throughout the year and breeds on young growth. The duration of various stages depends upon the prevalent temperature. The moth lays about 83-127 eggs singly in 3 days on young leaves or tender shoots, usually on the lower surface, particularly near the midrib. The eggs hatch in 2–10 days. Larval duration is 5-30 days. They settle down in enlargements of the galleries near the leaf margin, and spin cocoons for pupation. The pupal stage lasts for 5-25 days. The life cycle is completed in 12-55 days and about 16 overlapping generations are produced in a year.

Management

1. Prune the heavily affected parts during winter and burn the same for effective control.

2. Spray dimethoate 0.03 per cent on need basis (one larva per 15 cm twig length) for effective control of the pest.

3. Spray suspension of neem or datura (*Ipomea fistullosa*) leaf extract (10 per cent) or neem seed extract 2 per cent for the control of the pest.

4. Spray methyl-o-demeton 0.025 per cent or dimethoate 0.03 per cent or monocrotophos 0.04 per cent at peak activity (May-June) of the pest.

5. The varieties, Adinima lime and Zumukhiya of acid lime are found tolerant to the attack of leaf miner.

6. Spray nursery plants 5 ml of fenvalerate 20 EC or 10 ml of cypermethrin 10 EC or 3.5 ml of deltametherin 2.8 EC or 15 ml of monocrotophos 36 EC in 10 liter of water at fortnightly intervals.

Lemon Butterfly, *Papilo demoleus* (Lepidoptera: Papilionidae)

Marks of Identification

The full grown caterpillar is yellowish green and has a hornlike structure on the dorsal side of the last body segment and is 40 mm long and 6.5 mm wide. The black and white markings make the larvae look like bird dropping. The adult is a large beautiful butterfly,28 mm in length and 94 mm in wing expanse. Its head and thorax are black, being a creamy yellow coloration on the underside of the abdomen. Its wings are dull black, ornamented with yellow markings. The general coloration on the underside of the wings is slightly paler and the markings are also larger. The antennae are black and have club like structures at their ends.

Nature of Damage

The young larvae feed only on fresh leaves and terminal shoots. Habitually, they feed from the margin inwards to the midrib. In later stage, they feed even on mature leaves and sometimes the entire plant may be defoliated. The pest is devastating in nurseries and its damage to foliage seems to synchronize with fresh growth of citrus plants. Heavily attacked plants bear no fruits.

Life Cycle

The female butterfly lays eggs on tender shoots and fresh leaves, mostly on the under surface. A female lays on an average 75-120 eggs. The eggs are placed singly or in groups of 2-5 eggs. Eggs hatch in 3-4 days during summer and 5-8 days in winter. The larval period lasts 8-16 days in the summer and about 4 weeks in winter. The mature larva spins a supporting girdle around its body and pupates on a twig. The pupal stage lasts about 8 days in summer and 9-11 days in winter. A male lives for 3-4 days and female for about a week. They pass through three or four generations in a year.

Management

1. Hand picking of various stages of the pest and their destruction especially in nurseries and new orchards helps to suppress the population of the pest.

2. Spraying of entomogenous bacteria, *Bacillus thuringiensis* @ 2 kg per ha and neem seed extract 3 per cent gives high mortality of caterpillar.

3. Spray endosulfan 0.07 per cent or quinalphos 0.05 per cent during severe infestation *i.e.* September.

Fruit Sucking Moth, *Ophideres spp.* (Lepidoptera: Noctuidae)

Marks of Identification

The freshly hatched caterpillars are slender, thread like 3-4 mm long and pale greenish in colour. Full grown caterpillars are 50-60 mm long, stouty, velvety blue with yellow pattern on dorsal and lateral sides and having a hump at anal end. The moths are large and stoutly build and their prominent palpi are turned upwards. The head and thorax of the moths are greenish-gray and abdomen is orange. Forewings are greenish-grey with numerous faint striated reddish lines and three rufous spots, hind wing have apical area blotched with rufous, a round black spot in the center and a marginal black band. Wing expanse is 80-110 mm.

Nature of Damage

The moth punctures the ripening fruit for sucking juice with the help of its strong, piercing mouthparts. The damaged fruits soon start rotting as the puncture regions are easily infested with bacteria and fungi and ultimately the fruits drop prematurely. If the damaged fruit is squeezed, the juice spurts from the hole. In severe cases of infestation, almost all the fruits are lost.

Life Cycle

The moths are nocturnal and are not seen during the day. They lay eggs on a number of wild plants and weeds which are often found growing near citrus orchards. The eggs hatch in about 8-10 days within 24 hours of emergence, the young larvae start feeding on the foliage of host plants. A larva passes through five instars in 28-35 days. When full grown it makes a pupal case by webbing together pieces of leaves and soil particles. The pupa is thick and is dark reddish brown. This stage lasts about 14-18 days. The moths on emergence fly to nearby orchards for feeding on fruit-juice.

Management

1. Destruction of alternate host plants in the vicinity of the orchard is advised to minimize the pest infestation.
2. Dispose the fallen fruits which attract the moths.
3. Creating smoke in the orchards after sunset may keep the pest any from orchards.
4. Install light trap to attract the moths and destroy the attracted moths.
5. Bagging the fruits to protect them from the moth attack.
6. Kill moths with a bait containing jaggery 100 g+ vinegar 6.0 g+ malathion 10 ml+ water 1 liter. Wide-mouthed bottles (1 bottle per 10 trees) containing bait solution should be tied to the plants when the fruits are in unripe condition.
7. Spray tree with carbaryl 0.2 per cent at the time of maturity of fruits.

White Fly, *Dialeurodes citri* (Hemiptera: Aleurodidae)

Marks of Identification

The eyes are oval and pale yellow. The adult is a minute insect, measuring 1.02 to 1.52 mm. The males being smaller than the females. The antennae are six-segmented. The eyes are transparent, red and kidney shaped, with the lower half covered over with bristles. The head is somewhat pointed. The wings are more than twice in length of the body and extend beyond the tip of the abdomen. Both the wings and the body are completely covered with white waxy powder. The nymphs are pale yellow, with purple eyes and its body is marginally fringed with bristles.

Nature of Damage

The nymphs and adults reduce the plant vigour by sucking large quantities of cell sap. Severely infested leaves turn pale green and gradually become pale brown, get badly curled and are even shed. The nymphs also produce copious quantity of honeydew on which sooty mould develops covering the foliage with superficial black coating and interfering with photosynthesis of the plant. As a result, the growth of the tree is stunted and the affected trees produce few blossoms, most of which are shed and even the fruits that are formed are insipid.

Life Cycle

The female lays 200 or more eggs singly on the underside of soft young leaves during its life of 7-10 days. It may take 10-20 days for the eggs to hatch. The young nymphs on emergence crawls for few hours and then settles on the succulent portion of the twig. The nymph is full-fed in 25-71 days and then change to pupa. The pupal period is long in summer and cold season when it lasts 114-159 days.

Management

1. For effective management of flies, close planting, water logging or any other stress condition should be avoided.
2. In case of localized infestation, the affected shoots should be clipped off and destroyed.
3. Spray methyl-o-demeton 0.025 per cent or dimethoate 0.03 per cent or malathion 0.05 per cent or monocrotophos 0.04 per cent or phosalone 0.05 per cent or endosulfan 0.07 per cent at peak activity of this pest.

Citrus Black Fly, *Aleurocanthus woglumi* (Hemiptera: Aleurodidae)

Marks of Identification

The male and female adults are 0.8 and 1.2 mm long, respectively. When freshly emerged, head and thorax are bright red, eyes reddish brown while antennae and legs are whitish. Within 24 hours the adults become slaty-bluish. Wing has black patches or whitish background. The full grown nymphs are 1-2 mm long and dark brown to shiny black in colour and conspicuously spiny, bordered by a white fringe of wax.

Plate 13.77: Citrus Aphid

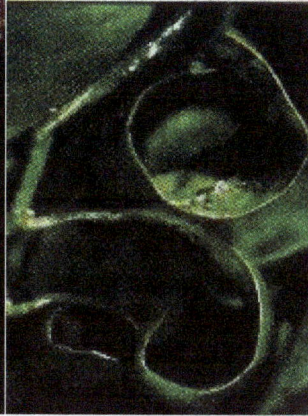

Plate 13.78: Citrus Aphid Affected Leaf

Plate 13.79: Citrus Black Fly

Plate 13.80: Citrus Blackfly Affected Fruit

Plate 13.81: Citrus Leaf Miner Affected Leaf

Plate 13.82: Citrus Leaf Miner Damage

Plate 13.83: Citrus Psylla–Adult

Plate 13.84: Citrus Mealy Bug

Plate 13.85: Citrus Fruit Sucking Moth

Plate 13.86: Fruit Sucking Moth Punctured Citrus Fruits

Plate 13.87: Lemon Butter Fly Caterpillar

Papilio demoleus
(Linnaeus)

Plate 13.88: Lemon Butter Fly Adult

Plate 13.89: Citrus Mite

Nature of Damage

Both adults and nymphs suck plant sap, reducing the vitality of trees. It results in curling of leaves and also the premature fall of flower buds and developing fruits.

Life Cycle

A single female may lay more than 100 eggs in spiral pattern on ventral side of leaves in her lifetime. There may be 16-22 eggs in a cluster. The eggs hatch in 7-14 days. The nymphs on emergence start feeding on cell sap. They pass through four nymphal instars and the nymphal stage is completed in 38-60 days. They pupate in leaf surface and the stage lasts 100-131 days. There are two distinct broods in a year.

Management

1. For effective management of flies, close planting, water logging or any other stress condition should be avoided.
2. In case of localized infestation, the affected shoots should be clipped off and destroyed.

3. First and second nymphal instars of the pest are more vulnerable to control hence the spray schedule should be undertaken at 50 per cent egg hatching stage that ensures maximum coverage of younger stage of the pest in two sprays at 15 days intervals with any of the insecticides *viz.*, monocrotophos 0.04 per cent, acephate 0.04 per cent, phosalone 0.05 per cent, dimethoate 0.06 per cent till drenching that ensure thorough coverage of the underside of leaves.

Bark Eating Caterpillar, *Inderbela quadrinotata* (Lepidoptera: Metarblidae)

Details given in the pests of guava.

Mealy Bug, *Planococcus citri* (Hemiptera: Pseudococcidae)

Marks of Identification

The adult is a minute insect, eyes are transparent, red and kidney shapes with the lower half covered with bristle. The wings are more than twice the length of the body. Both the wings are covered with white waxy powder. The nymph is flat, oval, pale yellow in colour with purple eyes. The body of the nymph is marginally fringed with bristles and scale like in appearance.

Nature of Damage

The pest cause heavy damage to nursery and grown up plants. Damage is severe in spring and autumn. The leaves and tender shoots get deformed and twisted into knot and loops. The leaves become curled up. It feeds at stem ends of fruits and often causes heavy fruit drop. Clusters of white formation of bugs are found at the joints of twigs. It excretes plant sap and reduces tree vigour. It also excrete honeydew which invites sooty mould.

Life Cycle

This pest breeds continuously. Yellow eggs are produced in a loose colony of waxy filaments. About 600-800 eggs covered in an ovisac are deposited in a fortnight which hatche within 7 to 15 days. There are 3 moults of female nymphs and 4 moults of winged male nymphs completing 3 overlapping generations in a year.

Management

1. Chemical control of the pest includes spraying of dimethoate 150 ml + kerosene oil 250 ml in 100 liters of water or carbaryl 10 ml + oil 10 ml or malathion 20 ml in 10 liters of water.

2. The possibility of the use of nymphal and adult parasitoid, *Leptomastrix dactylopii* and the predatory coccinellid, *Cryptolaemous montrouzieri* and the Chrysopid, *Mallada boninensis* needs to be explored for the control of the pest.

Citrus Aphid, *Toxoptera citricida* and *T. aurantii* (Hemiptera: Aphididae)

Nature of Damage

Adults and nymphs suck the sap from tender leaves and shoots devitalizing the plants. The pest secretes copious sugary solution on which sooty mould grows. Brown citrus aphid *Toxoptera citricida*, is responsible for vectoring citrus disease ' Tristeza' virus.

Life Cycle

Citrus aphid produces about 5 young ones daily for a period of 1-3 weeks parthenogenetically which attain maturity in about 6 days. Daytime warm temperature during February- March favours the pest population buildup. Winged colonizing adults called stem mothers give birth directly to nymphs with no egg stage and no mating is required. Nymph of black aphid gives out red hoemolymph and brown aphid, the yellow on squasing.

Management

1. An Ichneumouid parasitoid, *Lipolexis scutellaris* and the predators like chrysopids, coccinellids, syrphids etc., feed on the pest.
2. Spray quinalphos or dimethoate or malathion 0.05 per cent at weekly interval to control the pest, effectively.

Citrus Thrips, *Scirtothrips* spp. and *Heliothrips haemorrhaeodalis* (Thysanoptera: Thripidae)

Marks of Identification

The adults are slender yellowish brown in colour having apically pointed wings, measuring about 1 mm in length. The females possess long narrow wings with the fore margin fringed with long hairs. The nymphs resemble the adults in shape and colour but are wingless and smaller in size.

Nature of Damage

The nymphs and adult rasp and suck the sap from fully developed flower and leaf buds, young and grown-up fruits and also the leaves. The leaves become cup shaped and leathery. Two white lines running parallel to leaf midrib and a whitish silvery ring around the fruit neck are characteristics of thrips infestation.

Life Cycle

Males are practically unknown and the reproduction is by parthenogenesis. The female inserts eggs in the incisions made in leaf tissues which hatch in 2-7 days. The nymphal duration varies between 30 to 60 days depending upon the climatic conditions. It undergoes 3-4 instars to reach maturity. There are several generations in a year.

Management

1. Dimethoate or monocrotophos @ 1 ml per liter of water should be sprayed at bud burst stage and on berries. The surrounding vegetation should also be sprayed as the pest thrives on it.

Citrus Mite, *Eutetranychus orientalis* (Acarina: Tetranychidae)

Marks of Identification

The adult is small, plump and orange with thick deep brown patches on the dorsal side of its body and measures 0.33 mm in length. Its body is covered with prominent bristles, each born on a whitish tubercle. The antennae of the females are three segmented and are bright carmine pink. The newly hatched larva is light yellowish brown and has only three pairs of legs. The protonymph is orange brown and the duetonymph is orange brown with greenish tinge.

Nature of Damage

The mite extracts cell sap from leaves and fruits. Mite feeding causes pale stripping on the upper surface of leaves which are not seen on the lower surface. In severe infestation, the stripping enlarges to dry necrotic areas. But the stripping or silvering of green fruits disappears when the fruit changes colour. When large population feeds on maturing fruits the silvering may persist. Generally leaves drop and die-back of twigs starts. Mottling of leaves and chlorotic appearance due to multitude of grey unpleasant spots on fruits affecting the quality which is a serious concern.

Life Cycle

The pest is active throughout the year but has prominence during November-May having peak of 15 mites per leaf during February- June. Female lays upto 8 eggs per day mainly along the midrib of the leaf. It takes 10 to 15 days and 3 months for younger stages to develop fully and 17-20 days to complete entire life cycle. The pest has number of overlapping generations in a year.

Management

1. Water stress often aggravates mite problem. Trees are well irrigated particularly during the stress in late summer.
2. The most important natural enemies of citrus mite is a predacious mite, *Euseius hibisci* and the predators, *Agistemus* sp., and *Ambylesius hibisci*.
3. Foliar spray of dicofol @ 1.5 ml, or monocrotophos 1 ml or wettable sulphur 3 g per liter of water should be given as and when incidence is noticed.

Guava

Bark Eating Caterpillar, *Indarbela quadrinotata* (Lepidoptera: Metarblidae)

Marks of Identification

Freshly hatched larvae are dirty brown while the full grown larvae have pale brown bodies with dark brown heads. The adults are pale brown moths with rufous head and thorax. The forewings are pale rufous with numerous dark rufous bands. The hindwings are fuscous.

Nature of Damage

On hatching, the larvae bore into the bark or main stem. The larvae remain within the bored holes during day and come out at night to feed on the bark. They remain sheltered under the silken galleries. The attacked trees show the presence of winding silken galleries full of frass and facial matter on the bark surface extending from the bored holes downwards. Generally one larva is seen in each hole but in case of severe infestation 15-30 larvae bore the same tree. A severe infestation may result death of the attacked stem but not of the main trunk. There may be interference with the translocation of cell sap and thus arrest of tree growth is noticed with the resultant reduction in its fruiting capacity.

Life Cycle

The moth starts laying eggs in clusters of 15-25 eggs each, under the loose bark of tree. A single female lays about 300-400 eggs. The egg hatches in 8-10 days. Larvae takes as many as 9-11 months to complete development. Full grown larvae make a hole into the wood and pupate inside. The pupal stage is 3 to 4 weeks. Moths are short lived. There is only one generation in a year.

Management

1. Keep the orchard clean to prevent the infestation of this borer.
2. Kill the caterpillars mechanically by inserting an iron spike into the holes made by these caterpillars.
3. In case of severe infestation, clean the affected portion of the trunk or main stem and insert into the hole swab of cotton wool soaked in a 0.05 per cent emulsion of monocrotophos or dichlorvos and seal the hole with mud paste for killing the larvae in tunnels.

Fruit Fly, *Bactrocera dorsalis*, *B. correcta* and *B. zonata* (Diptera: Tephritidae)

The former two species of fruit fly are very serious pest. The details are given under mango fruit fly.

Grapevine

Grapevine Leafhopper, *Erythroneura* spp. (Hemiptera: Cicadellidae)

Marks of Identification

The grapevine leafhopper is a slender yellow-coloured insect with red markings and is about 3 mm long. It feeds on developing leaves and overwinters among fallen green leaves.

Nature of Damage

Both adults and nymphs suck the cell sap from leaves causing the foliage blotched with tiny white spots. Under heavy infestation the leaves turn yellow or brown and fall from the vines.

Life Cycle

During the spring season, the adults become active. Eggs are laid in leaf tissues and they hatch in about14 days. The nymphs are wingless and pale in colour and feed on the lower surface of the leaf. They moult five times and become full developed in 3-5 weeks. There are 2-3 generations in year.

Management

When the infestation level increases, spray fenitrothion 625ml after rainy season.

Grapevine Thrips, *Rhipiphorothrips cruentatus* (Thysanoptera: Heliothripidae)

Marks of Identification

The adults are minute, being 1.4 mm long, blackish brown, with yellowish wings. The nymphs are yellowish brown in colour.

Nature of Damage

They feed in large number on plant leaves. The attacked leaves take a whitish hue, acquire a withered appearance, and then turn brown. The affected leaves ultimately curl up and drop off the plant. Such vines either do not bear fruit or the fruit drops off prematurely. Sometimes the fruits are also attacked resulting in scab formation on the berries. The peak infestation is during hot weather-May to July in South India, March to April and August to October in North India.

Life Cycle

Both sexual and parthenogenetic reproduction occur side by side, the progeny of the latter being always males. 2-6 eggs are laid per day. Incubation, pre-imaginal, pupal and total life cycle durations required 3-8, 11-22, 2-5 and 14-33 days, respectively. There are 5-11 generations in a year. The pest overwinters in pupal stage in soil.

Management

1. Remove grasses from orchard and prune infested leaves.
2. Spray malathion 500 ml in 500 liters of water per 100 vines, once before flowering and repeat after the fruit set.

Grapevine Girdler, *Sthenias grisator* (Coleoptera: Cerambycidae)

Marks of Identification

Adult beetles are about 24 mm long, grayish-brown with white and brown irregular markings resembling the bark colour, elytra have an elliptical grayish median spot and an eye-shaped patch.

Nature of Damage

The adult beetles emerge usually around July to August and maximum damage is caused during August to October. The beetles with their strong mandibles, ring or girdle the young green branches of the host plant. The girdle's branches dry up.

Life Cycle

The female deposits eggs in clusters of 2-4 underneath the bark of girdled branches. Incubation period is 8-10 days. The grub stage lasts for 7-8 months. The life cycle is completed in more than a year. There is only one generation in a year.

Management

1. Cut the affected branches below girdling point and burn the same.

2. Hand collection and destruction of beetles also helps in mitigating the pest.

3. Spray the vines with 1.5 liters of monocrotophos in 1250 liters of water.

Pomegranate

Anar Butterfly, *Virachola isocrates* (Lepidoptera: Lycaenidae)

Marks of Identification

Full grown caterpillars are stout, 17 to 20 mm long, dark brown in colour and have short hairs and whitish patches all over the body. Adult butterflies are medium sized, glossy-bluish-violet (male) to brownish-violet (females) with an orange patch on each wing, the wing expanse is 40-50 mm.

Nature of Damage

The caterpillars damage the fruit by feeding inside and ridding through the ripening seeds of pomegranate. As many as eight caterpillars may be found in a single fruit. The affected fruits ultimately fall off and give an offensive smell. This pest may causes 40-90 per cent damage in fruits.

Life Cycle

The pest breeds throughout the year on one fruit or the other. The female butterfly lays shiny white, oval shaped eggs singly on the calyx of flowers and on small fruits and leaves. The eggs hatch in 7-10 days and the young larvae bore into the developing fruits. They feed there for 18-47 days till they are full grown. Then they pupate inside the fruit but sometime may pupate outside. The pupal period lasts 7-34 days. There are four overlapping generations in a year.

Management

1. Collect and destroy the fallen infested fruits to prevent build up of the pest population.

2. If number of fruit tree is limited, bagging of the fruit before maturity is advised to reduce the fruit damage.

Plate 13.90: Anar Butterfly–Larva and Damage

Plate 13.91: Anar Butterfly–Adult

3. Spray endosulfan 0.07 per cent or monocrotophos 0.04 per cent 2 to 3 times at 15 days interval commencing from infestation of fruit borer.

Ber

Fruit Fly, *Carpomyia vasuviana* (Diptera: Tephritidae)

Marks of Identification

The adults, smaller than the housefly, are brownish yellow, with brown longitudinal stripes on the thorax, being surrounded on the sides and the back with black spots. Wings are hyaline, transparent with grayish brown spots and it possesses bristly hairs on the tip of the abdomen. Full grown maggots are amphineustic, cretaceous or creamy white and about 6 mm long.

Nature of Damage

The female flies puncture the ripening fruits and lay their eggs inside the epidermis. The development around the punctures is arrested and this causes the deformation of fruit. On hatching, maggots feed on fleshy and juicy pulp. The infested fruit turn dark brown, rotten near the stones and emit a strong smell. One maggot is sufficient to destroy the entire fruit. Fleshy varieties of ber more seriously damages than the less fleshy once. Late maturing fruits are destroyed almost entirely.

Life Cycle

The female fly make cavity in the skin of fruit and lay 12 to 18 eggs singly or in groups of 2 to4. The eggs hatch in 2-3 days and the maggots feed on the flesh of the fruit, making galleries towards the center. The larvae are full grown in 7-10 days and come out of the fruit by cutting one or two holes in the skin. They move away, making jumps of 15-26 cm and reach a suitable place to pupate generally 6-15 cm below the soil surface. The pupal stage lasts 14-30 days and the shortest life cycle from egg to the adult emergence is completed in 24 days. There are 2-3 broods in a year.

Management

1. Collect and destroy the fallen infested fruits at alternate days.
2. Rake the soil around the trees during summer to expose the pupae to sun heat and natural enemies.
3. To escape egg laying on fruit, do not allow the fruits to ripe on the tree, harvest at green and firm stage.
4. Grow resistant/tolerant varieties *viz.*, Safada, Sanaur-1, Rohtaki Gola, Katha, Desi Alwar, Tikadi, etc.
5. Install methyl eugenol traps (0.056 ml or 4 drops each of methyl eugenol and dichlorvos to be recharged at weekly interval) in ber orchard to trap and kill the male fruit flies.
6. Three sprays of monocrotophos 0.03 per cent control the infestation effectively. The first spray applied when ber fruit attain peanut size.
7. The insecticides *viz.*, malathion 0.05 per cent, neemark 0.1 per cent, dichlorovos 0.05 per cent and quinalphos 0.05 per cent are also effective against fruit fly.

Banana

Banana Corm or Rhizome Weevil, *Cosmopolites sordidus* (Coleoptera: Curculionidae)

Marks of Identification

Adult weevils are 10-13 mm long shiny black in colour, having fairly long and curved snout and short elytra striated longitudinally. Though these wings are functional the weevils seldom fly. The grubs are creamy white, stout, fleshy, highly wrinkled and legless with spiny-shaped, 8-12 mm long body.

Nature of Damage

Infestation at early stage reduces the plant vigour. Early stage damage includes sick appearance and yellow lines on the leaf whereas in the advanced stage of infestation, plant show tapering of stem at the crown region, reduction in leaf size, poor bunch formation and choked throat appearance of bunch due to riddling by the grubs inside the corm.

Life Cycle

The adult weevil lay eggs singly into the corm. Eggs are laid superficially in the surface and crevices of the corm. Incubation period is 7-14 days. Larval stage lasts for two to six weeks. Pupation is completed in a week. Adult lives upto two years and can remain without food for 6 months.

Management

1. Weed free cultivation is essential to avoid the spread of weevils.
2. Remove the old and dried leaves.
3. Remove the pseudostem after harvest and treat it with carbaryl (2 g per liter)
4. Apply carbofuran 3 G (40 g per plant) before planting.
5. Before planting, treat the suckers with monocrotophos (14 ml in one liter of water) for 30 minutes.
6. Collect adult weevils using banana pseudostem traps both- longitudinally split (30 cm length) and disc-on-stump trap (100 per ha).

Banana Pseudostem Weevil, *Odoiporus longicollis* (Coleoptera: Curculionidae)

Marks of Identification

Adult weevils are chocolate to black in colour, 23-28 mm long with pointed head. Grubs are apodous, soft, fleshy, wrinkled and sparsely covered with brown hairs of varying lengths. Pupae are exarate and pale yellow in colour.

Nature of Damage

Presence of small holes and jelly exudation on the stem indicates the grub activity inside the stem. The apodous larvae are responsible for riddling of the pseudostem thereby cause serious damage. In the advanced stage of infestation, the severely affected plant breaks or topples along with the bunch.

Life Cycle

Eggs are laid inside the air chambers of the leaf sheath through the slits cut on leaf sheath. Incubation period ranges from 5-8 days. Larval period lasts for 26 days and the pupal period including pre-emergence resting period of adult lasts for about 20-14 days.

Management

1. Weed free cultivation helps in reducing the spread of infestation.
2. Remove the old and dried leaves.
3. After harvest remove the pseudostem and treat it with carbaryl (2 g per liter) to kill the egg laying weevils.
4. Collect adult weevils by using banana pseudo stem traps both-longitudinally split (30 cm length) and disc-on-stump trap (40 per ha).

Jack Fruit

White Tailed Mealy Bug, *Ferrisia virgata* (Hemiptera: Pseudococcidae)

Marks of Identification

The adult females are dark castaneous and 4.0-4.5 mm long. They are covered with a sticky cretaceous white ovisac. The nymphs are deep chocolate in colour, having their dorsum covered thinly with a whitish mealy material.

Nature of Damage

This pest is very active during dry season. It feeds on leaves and tender shoots and during dry weather it moves down and inhabit the roots. A prolonged period of drought may result in a severe outbreak of this pest. The peak period of its activity is August to November when the affected parts turn yellow, wither and ultimately die away.

Life Cycle

The eggs are enclosed in ovisac. The female lays 200-400 eggs which are cylindrically rounded, flat at both ends and chestnut in colour. The eggs hatch in 7-10 days. Male and female nymphs take 15 and 20 days to complete development, respectively.

Management

Spray 0.05 per cent monocrotophos or diazinon.

Leaf Webber, *Perina nuda* (Lepidoptera: Lymantriidae)

Marks of Identification

Full grown caterpillars are 22-25 mm long, having short erect tufts of dusky grey to brownish hairs. In the moths, there is extreme sexual dimorphism. Male moths are smaller than the female moths. The male moths have half ochreous and half transparent forewings with a wing expanse of 33-36 mm and the female is dull ochreous white in colour having 38-42 mm wing expanse.

Nature of Damage

The caterpillar folds or webs the leaves together and feed on the green matter causing extensive defoliation.

Life Cycle

A single female lays 60 to 400 eggs. They hatch in 4-6 days. Larval and pupal periods are 16-20 and 5-9 days, respectively. The adult longevity is 3–11 days and the total life-cycle is completed in 27-39 days.

Management

1. Hand picking and mechanical destruction of caterpillars in the initial stage of attack prevent further build up of population of leaf webber.
2. In case of severe infestation spray 750 ml monocrotophos or 1.50 liter chlorpyriphos 20 EC in 625 liter of water per ha. Spray should be directed on the upper surface of foliage.

Pineapple

Pine Apple Thrips, *Thrips tabaci* (Thysanoptera: Thripidae)

Marks of Identification

Adults are small, about one mm long, yellowish-grey to brown in colour with darker transverse bands across the thorax and abdomen. Forewings are shaped and possess short setae.

Nature of Damage

Nymphs and adults rasp the epidermis of leaves and imbibe the oozing sap, thereby devitalizing and distorting the leaves and affecting the growth adversely. In case of severe infestation, the leaves show silvery sheen and bear small spots of faecal matter. In many parts of the world this thrips has also been recorded as a vector transmitting yellow spot virus disease of both plant and fruits.

Life Cycle

Reproduction appears to be thelytokous and parthenogenetic. Males are extremely rare, the ratio being 3000 females to one male. Eggs are inserted into tissues of green leaf through a cut made by the ovipositor. A single female lays 40-50 eggs. These hatches in 4-9 days and the entire life cycle is completed in 10-13 days in summer and 21 days in winter season.

Management

1. Removal of alternate host plants near pineapple orchards.
2. Spraying 0.04 per cent diazinon.

Slug Caterpillar, *Latoia lepida* (Lepidoptera: Limacodiadae)

Marks of Identification

The full grown larvae is flat, fleshy and greenish in colour with white lines on the body. It is curved with spines having red or blue tips. The adult moth is short and stout with green and brown forewings.

Nature of Damage

The young caterpillars feed gregariously by scrapping the under surface of the leaves. The continuous feeding leads to loss of sap from plant tissues ultimately reduce the vitality of plants and the affected leaves dry up.

Life Cycle

The female moths lay eggs in batches of 15-35 on the under surface of the leaves. The eggs hatch in 7 days. On hatching, the young larvae start feeding in clusters on the under surface of leaves. The larva passes through five instars and is full-developed in 40-47 days. It pupates in silken cocoon in stems. The pupal stage lasts for 20-28 days and the life cycle is completed in about 10 weeks.

Management

1. Hand picking and mechanical destruction of caterpillars at early stage of infestation.
2. Dusting with methyl parathion (5 per cent) is effective and should be done when the caterpillars are still feeding gregariously.

Date Palm

Rhinoceros Beetle, *Oryctes rhinoceros* (Coleoptera: Scarabeidae)

Marks of Identification

The adult beetles are robust, 35-50 mm long, glossy-dark-brown to black dorsally and reddish brown ventrally. The pygidium in males is rather long, smooth and rounded while in case of females, it is short hairy and relatively pointed. Grubs are 80-100 mm long, soft, fleshy and wrinkled. Pupae are uniformly yellowish-brown.

Nature of Damage

Beetles are the destructive stage of this pest. They are nocturnal in habit. At night these beetles feed on the crowns of palms by boring into the unopened tender fronds and tunnel downwards feeding on soft tissues of the growing point. As a result, the growth of the tree is arrested. Young trees are more prone to attack by this pest and these trees when attacked ultimately wither and die away. The pest causes maximum damage during monsoon.

Life Cycle

Eggs are laid in rotting vegetation, manure or compost heaps. These hatch in 8-14 days. Grub period is 47-191 days and pupation takes place in the soil, 150-600 mm deep. Pupal period is 14-29 days and the adult remain in cocoon for 5-26 days till they become sexually mature. The life cycle is completed in 100-260 days and the adult live for 76-219 days.

Management

1. Keep the orchards clean and adopt all sort of sanitary practices.
2. All potential breeding sites *viz.,* compost heaps, refuse dumps, dung hills etc., must be destroyed constantly in and around the orchards by deep burying or burning. These breeding sites may be dusted with methyl parathion (5 per cent).
3. Use light trap @ 4/ha.
4. Filling up the crown portion with 5 per cent methyl parathion + saw dust or sand (1:1 ratio) three times in a year. This will kill the female weevils that usually rest in leaf axils.

Red Palm Weevil, *Rhynchophorus ferrugineus* (Coleoptera: Curculionidae)

Marks of Identification

Adult weevils are reddish brown in colour with red spots on thorax, cylindrical, 32-36 mm long, flattened with long and slightly curved snout. The grubs are fleshy, wrinkled, transversely apodous, pale yellow when freshly formed, later become light brown and full grown grubs which are 50-60 mm long.

Nature of Damage

The grubs are voracious feeders, once enter inside the trunk they seldom come out. They burrow in hard woody portion and tunnel through the tissues in all directions and feed on soft succulent tissues, discarding all fibrous material. In case of severe infestation, inside of trunk is completely hollowed and filled with decaying rubbish. As a result of this infestation, in case of young palms, the tops wither, rot and give out offensive smell, while in older trees, the tops wither, rot and give out offensive smell, the top portion of trunk bend and ultimately breaks at the bend. A thick reddish brown viscous fluid and chewed up and discarded fiber pieces are thrown out from the holes.

Life Cycle

The pest breeds throughout the year and all the stages are passed on the host tree. A female lays 76-355 eggs during its life span of 2-3 months. Eggs are laid singly in soft tissues at the base of leaf sheaths or in the cuts or wounds on the tree trunk. These hatch in 2-5 days. Grub and pupal period are 24-61 and 18-34 days, respectively and the entire life cycle is completed in 50-90 days. After emergence the weevils remain in cocoons for 4-17 days, till they become sexually mature.

Management

1. Maintenance of orchard sanitation, prompt destruction of dead and badly infested plants is advised.
2. Filling up the crown portion with 5 per cent methyl parathion + saw dust or sand (1:1 ratio) three times in a year. This will kill the female weevils that usually rest in leaf axils.

3. Prevent the occurrence of cuts, scars or wounds on tree trunks. As and when they appear, paint the same immediately with lime or coal tar. This will prevent the weevils from laying eggs on these trees.

4. If the attack of the pest is detected in the initial stage, remove and kill the grubs, mechanically.

Date Palm Scale, *Aspitous destructor*

Marks of Identification

The male scales on an average are 1.2 mm long and white in colour and the nymphs below are pale yellow and invisible. The female scales are brown white margin, more or less oval, flat and about 1.5 mm long. The mature females are creamy yellow, when gravid, these gradually become darker and darker till wine-red. Crawlers are also rose coloured and latter become light violet.

Nature of Damage

The pest prefers shade and invariably infests shaded pinnules and the main veins of the fronds. In severe cases even the fruits are attacked. Incidence is higher in trees where fruits have been wrapped with cloth as protection against solar radiation. Young trees grown in shade or those that are irrigated regularly suffer comparatively more from ravages of this pest. The continuous yellow incrustations of this scale on the leaflets interfere with photosynthetic, respiration and transpiration activities of the trees which ultimately affect adversely the normal growth of the trees and development of fruits, thus fruit become dry, deformed and do not mature.

Life Cycle

An oviparous female may lay 6-10 eggs in her life time. The eggs remain under the female scales and hatch in 8-18 days. The nymphs become mature females normally in 30-40 days. Male nymphs have to pass through pre-pupal and pupal stages before becoming winged or micropterous males. The life cycle varies from 50-60 days in summer to 75-85 days in spring.

Management

1. Good crop management restricts the pest development.

2. Remove and burn all the infested leaflets in initial stage of attack. This is possible only when the trees are young. In case of older trees, remove all the fronds except those at the very top and pour hot salty water over the crown and remaining leaves.

Jamun

Bark Eating Caterpillar, *Indarbela quadrinotata* (Lepidoptera: Metarblidae)

Marks of Identification

Freshly hatched larvae are dirty brown while the full grown larvae have pale brown bodies with dark brown heads. The adults are pale brown moths with rufous head and thorax. The forewings are pale rufous with numerous dark rufous bands. The hindwings are fuscous.

Nature of Damage

On hatching, the larvae bore into the bark or main stem. The larvae remain within the bored holes during day and come out at night to feed on the bark. They feed sheltered under the silken galleries. The attacked trees show the presence of winding silken galleries full of frass and faecal matter on the bark surface extending from the bored holes downwards. Generally one larva is seen in each hole but in case of severe infestation 15-30 larva may bore the same tree. A severe infestation may result death of the attacked stem but not of the main trunk. There may be interference with the translocation of cell sap and thus, arresting of tree growth is noticed with the resultant reduction in its fruiting capacity.

Life Cycle

The moth starts laying eggs in clusters of 15-25 eggs each, under the loose bark of tree. A single female lays about 300-400 eggs. The egg hath in 8-10 days. Larvae take as many as 9-11 months to complete development. Full grown larvae make a hole into the wood and pupate inside. The pupal stage is 3 to 4 weeks. Moths are short lived. There is only one generation in a year.

Management

1. Keep the orchard clean to prevent the infestation of this borer.
2. Kill the caterpillars mechanically by inserting an iron spike into the holes made by these caterpillars.
3. In case of severe infestation, clean the affected portion of the trunk or main stem and insert into the hole swab of cotton wool soaked in a 0.05 per cent emulsion of monocrotophos or dichlorvos and seal the hole with mud paste for killing the larvae in tunnels.

Tamarind

Chafer Beetle, *Holotrichia insularis* (Coleoptera: Scarabeidae)

Marks of Identification

Adults are brownish-black, convex beetles, 18-20 mm long and 8-10 mm wide. Full grown grubs are dingy white, fleshy, curved, 38-44 mm long and 6-9 mm wide.

Nature of Damage

Adult beetles feed on foliage and in case of severe attack the seedlings and young trees may be completely defoliated. If the attack is heavy the saplings are killed outright whereas the young plants wither and gradually dry up.

Life Cycle

Mating usually takes place after the females have fed for about a fortnight. A single female lays 40-65 eggs. Eggs are laid in moist sandy soil, 30-150 mm deep. On hatching, the grubs feed on roots. Pre-oviposition and incubation period last for 4-6 and 8-12 days, respectively. While grub and pupal stages occupy 11-16 and 2-3 weeks, respectively. There is only one generation in a year. Pupae and adults hibernate in the soil from November to June.

Management

1. Plough around the trees, during winter to expose and kill the hibernating pupae and adults.
2. Soil treatment with phorate 10G 50-100 g per tree.
3. To kill the adults, spray 0.1 per cent fenitrothion especially during breeding season.

Tamarind Fruit Borer, *Phycita orthoclina* (Lepidoptera: Phycitidae)

Marks of Identification

The moths are small, delicate insects, having elongate forewings. The hindwings are broad and bear hairs on the dorsal side. The full-grown larva is cylindrical, pink and measure 14mm long.

Nature of Damage

The larvae feed on the pulp and cause, excrements and webbing and render the fruit unfit for human consumption.

Life Cycle

The pest is active during December–April. The female moth lays flat, oval shaped and white eggs, singly on the pulpy portion inside the round shelled pods, through cracks and crevices found on their surface. A female on an average lays 190 eggs. Egg hatch in 4-5 days and larvae enter into the fruit pulp and feed thereby making silken web. Larval stage is completed in 27-40 days. The larva makes a silken cocoon inside the infested pod and pupate therein. Pupal stage lasts for 6-8 days.

Management

Spray monocrotophos 0.04 per cent to control the pest.

Papaya

Ak Grasshopper, *Poekilocerus pictus* (Orthoptera: Acrididae)

Marks of Identification

Nymphs are yellowish-white with orange and black strips and dots all over their bodies. Adults are stout, yellowish with broad bluish-green strips on head and thorax, antennae bluish black with yellow rings, abdomen yellowish with transverse bluish-black bands. Forewings are bluish-green with yellow veins and reticulations. Hindwings are hyaline.

Nature of Damage

Both adult and nymphs feed voraciously on leaves and skeletonize them. In case of severe infestation even the bark of the trees is not spared.

Life Cycle

Mating lasts for 5-7 hours and pre-oviposition period is 3-4 weeks. The female thrusts its abdomen into the soil and lays about 150-180 eggs at a depth of 120-200 mm depending upon the texture of the soil. There are atleast two broods in a year- a short one with incubation period of one month (June- July) and nymphal period two months and a long generation when eggs laid during September to November over winter and hatch around the end of March or early April and become adult in another 2.5 months.

Management

Grasshoppers can be managed by spraying 0.1 per cent malathion or dusting malathion 5 per cent @ 50 kg per ha.

Sapota

Bud Borer, *Anarsia achrasella* (Lepidoptera: Gelechiidae)

Marks of Identification

The forewings of the adult moth are dark ashy gray in colour while hindwings are fringed with yellowish colour. A tuft of yellowish silky hairs, presents on the last abdominal segment in case of male and projections on labial palp in female. The full grown larva is pink in colour.

Nature of Damage

The larvae bore into the tip of flower buds and feed on ovary and petals, adversely affecting the production. Sometime they feed on leaves also. The pest is active throughout the year.

Life Cycle

Female lays eggs on bud surface and stock of bud. Incubation period is 4-6 days. The larvae pass through four instars. Larval period is 11-13 days. The freshly formed pupa is brick red in colour. The pre-pupal and pupal periods are 1-2 and 5-6 days, respectively. The female lays on an average 39-63 eggs with an average oviposition period of 2-4 days. The longevity of male and female moths are 4-5 and 4-6 days, respectively.

Management

1. Remove and destroy all the infested leaves, buds and fruits to reduce the infestation.
2. Spray monocrotophos 0.04 per cent or dichlorovos 0.03 per cent or endosulfan 0.07 per cent or fenthion 0.1 per cent as and when flush of flower bud is set on the chiku trees.
3. Installation of traps containing black *tulsi* (*Ocimum sanctum*) leaves extract along with dichlorovos @ two Sapota trees during April to September to attract the male moths of bud borer.
4. The black *tulsi* leaf extract poison bait trap installed at 4.5 cm height from ground level of peripheral canopy of chiku tree followed by three sprays of endosulfan (0.07 per cent) at 15 days interval starting from bud initiation is recommended for effective control of this pest.

5. Two sprays of nimbecidine 0.3 per cent or any readymade neem based formulation at 20 days intervals starting from March are also found effective for the control of chiku bud borer.

Chiku Moth, *Nephopteryx eugraphella* (Lepidoptera: Pyralidae)

Marks of Identification

It is a major pest of chiku and occurs widely in India. The moth is gray in colour, forewings slightly suffused with brown and irritated with black, Hindwings are semi-hyaline. The destructive larva is 25 mm long, slender in body shape, and is pinkish with few longitudinal lines on the dorsal surface.

Nature of Damage

The larva causes considerable damage to young terminals affecting basal portion of young buds and either stigma and style or corolla of the chiku bud and feeds on ovary and inner parts of the bud, and then it move on the next buds thus, damaging many of them. Generally first larval instar completes within the flower bud and thereafter it comes out of the bud and produces a parchment. Bigger larva feeds voraciously on tender leaves, buds and young fruits causing considerable losses. The infestation is easily detectable by the presence of cluster of dried leaves hanging on webbed shoot.

Life Cycle

With the onset of spring season, the female moths start laying pale-yellow, oval shaped eggs singly or in batches of 2 or 3, on leaves and buds of young shoots. A female may lay as many as 374 eggs in 7 days. The eggs hatch in 2-11 days. The larvae feed for 13-60 days and complete development. They undergo pupation in the leaf webs and this stage is completed in 8-29 days. The life cycle is completed in 26-92 days depending upon the varying environmental conditions. There are 7-9 generations in a year.

Management

1. Remove and destroy all the infested leaves, buds and fruits to reduce the infestation.
2. Installation of traps containing black *tulsi* (*Ocimum sanctum*) leaf extract along with dichlorovos @ two Sapota trees during April to September is advised to attract male moths of the pest.

Plate 13.92: Chiku Bud Borer–Damage

Plate 13.93: Chiku Moth–Damage

3. The black *tulsi* leaf extract poison bait trap installed at 4.5 cm height from ground level of peripheral canopy of chiku tree followed by three sprays of endosulfan (0.07 per cent) at 15 days interval starting from bud initiation is recommended for effective control of this pest.

4. Spray monocrotophos 0.04 per cent or endosulfan 0.07 per cent following economic threshold level (one larva per twig) of this pest.

Sapota Fruit Fly, *Bactrocera spp.* (Diptera: Tephritidae)

Marks of Identification

The adult fly are stout and measures 14 mm across the wings and 5 mm in maximum length. The flies are strong fliers and can fly up to two kilometers in search of food. The fly is brown or dark brown in colour with hyaline wings and yellow legs. The thorax is ferruginous without yellow middle stripe. The abdomen is conical in shape and dark brown in colour. The young maggot is white and translucent.

Nature of Damage

The dark puncture caused by the oviposition of adult fly is not very conspicuous as its colour blends with the dark colour of the fruit. It is very clearly visible in some pale brown varieties. The maggots on hatching feed on the pulp of the fruit for few a days and a brown rotten parch appears on the fruit surface. The mesocarp becomes dirty brown. Infested fruits finally fall on the ground. The fruit is affected from late April to June.

Life Cycle

A female lays, on an average 50 eggs in one month. Eggs hatch in 3-10 days. As maggots develop, they pass through 3 stages in the ripening pulp and are full-grown in 6-29 days. They leave fruit and move away by jumping in little hope. On reaching a suitable place, they burry themselves into soil and pupate. In 6-44 days, they emerge as flies and reach ripe fruit for further multiplication. Life cycle is completed in 2-13 weeks and many generations *i.e.*, 10-12 are completed in a year. Flies are present in the field all through the year.

Management

1. To prevent the attack of fruit fly, harvesting of the fruits before ripening, and collection and destruction of infested fruits have been advised.

2. Install six methyl eugenol traps (0.056 ml or 4 drops each of methyl eugenol and dichlorovos to recharge at weekly interval) per hectare at a height of four feet placed at equal distance in Sapota orchard is advised to trap and kill the male fruit flies.

3. To control fruit fly besides fenthion 0.05 per cent spray, fixing of fruit fly traps containing methyl eugenol (1 trap per 2 trees) as suggested in case of mango fruit flies is also recommended.

Sapota Leaf Webber, *Nephopteryx eugraphella* (Lepidoptera: Pyralidae)

Marks of Identification

The full grown larva is cylindrical, of 2 cm length and pinkish with few longitudinal close set lines on the dorsal surface and hairs throughout the body. The moth is grey in colour.

Nature of Damage

The larvae constructs tunnel of webs and frass and feeds on tender leaves, leaf, flower buds and tender fruits throughout the year in serious proportions. The infested tree shows numerous webbed shoot, dried clusters which subsequently turn into dark brown patches on the leaves.

Life Cycle

The mean egg, larval, pupal periods and total life-cycle last 3-5, 17-32, 7-11 and 45 days, respectively.

Management

1. Removing and destroying the infested parts of the tree.
2. Spraying endosulfan (0.07 per cent) or monocrotophos (0.05 per cent) in conjunction with amino acid-based bio stimulant.

Phalsa

Plum Hairy Caterpillar, *Euproctis fraterna* (Lepidoptera: Lymantriidae)

Marks of Identification

Adults are yellow moths with pale transverse lines on forewings, wing expanse is 24-28 mm in male and 30-38 mm in female. Full grown caterpillars are 35-40 mm long and have red head and darkish-brown body with which hairs on the head and a tuft of long hairs at anal end.

Nature of Damage

The caterpillar feeds gregariously on leaf lamina skeletonizing the leaf completely. Later the caterpillars segregate and gnaw the leaves. In case of severe infestation, the entire tree may be defoliated.

Life Cycle

Eggs are laid in clusters on ventral surface of leaves and are covered with yellow hairs. A female lays 150-300 eggs. Incubation, larval and pupal period last for 4-10, 13-29, and 9-25 days, respectively. There are three generations in a year.

Management

1. The freshly hatched larvae feed gregariously so, hand picking on small scale is quite effective.
2. Dust methyl parathion (5 per cent) for the control of this pest.

Custard Apple

Mealy Bugs, *Ferrisia virgata, Maconellicoccus hirsutus* (Hemiptera: Pseudococcidae)

Marks of Identification

The eggs are buff to light yellow in colour and oval to cylindrical in shape having round end, measured 0.34 mm in length and 0.17 mm in width. Female is apterous, soft bodied, elongate oval in shape and deep brown in colour with a layer of prominent caudal filaments at posterior end and body is covered with a number of waxy filaments. Adult males are slender, delicate, elongate in shape and smoky brown in colour with a single pair of mesothoracic wings and two long waxy anal filaments. The freshly hatched adult female measures 1.96 mm in length and 0.90 mm in breadth while male measures 1.16 mm in length from head to tip of abdomen and 2.34 mm in breadth with wing expansion.

Nature of Damage

The infestation of this pest first appears on the seedlings grew up in the orchard. They infest leaves, shoots, buds and mainly fruits and suck the cell sap there from. As a result of heavy infestation development of fruits are reduced in size or get dropped. These insects also secrete honeydew on which a non-parasitic fungus, *Capnodium* species grows very rapidly, covering the fruits and plant parts with sooty mould. This black coating in turn interferes with the photosynthetic activities of the tree and as a result, its growth and fruiting capacity are adversely affected. The honeydew also attracts the ants which help in dissemination of the pest from tree to tree. Infested mature fruits also reduce their market value.

Life Cycle

They are active and mobile throughout the life. Reproduction is sexual as well as parthenogenetic, the later being more common. Mating takes place only once and a fertilized female lays 100-300 eggs in 3-4 weeks. The egg masses remain under the females till the young ones hatch out. Incubation period is about 3-4 hours. The development period of male and female nymphs varies from 31 to 57 and 26 to 47 days, respectively. Longevity of males is only 1-3 days while that of females extends from 36-53 days.

Management

1. To prevent the carry over of the pests, collect and destroy all the fallen infested fruits periodically and cut the branches portion being attacked between tree and also ground level.

2. Remove the seedlings and weeds, supporting the initial infestation, from the orchard.

3. Plough around the trees during summer to expose and kill the hibernated insect to natural enemies and sun heat.

4. Apply methyl parathion 2 per cent dust in the racked soil around the tree trunk to destroy the emerging crawlers.

5. Spray methyl parathion 0.05 per cent on stem to kill the climbing crawlers.
6. Remove and destroy the affected fruits and twigs to minimize the pest population.
7. Foliar application of the chlorpyriphos 0.05 per cent or quinalphos 0.05 per cent should be done as and when pest appears on tree.

Coconut

Leaf Eating Caterpillar, *Opisina arenosella* (Lepidoptera: Xylorycitidae)

Marks of Identification

The moth is ash grey in colour. It is medium sized, measuring 10-15 mm, with a wing expanse of 20-25 mm. The caterpillar feed hidden inside silk galleries on the underside of leaves.

Nature of Damage

As a result of the numerous galleries made by the feeding caterpillars, the foliage dries up. Infested trees can be recognized from the dries up patches in the fronds.

Life Cycle

A female moth lays 125 scale like eggs in small batches on the underside of the tips of the old leaves. The incubation period lasts about 3-5 days in summer and 10 days in winter. The larvae is full-grown in 40 days and pupate inside the galleries. The pupal period lasts for 12 days. The adult moths survive for 2-3 days. The pest has 7 generations in a year.

Management

1. Remove and destroy the infested leaves along with the caterpillar.
2. The infestation can be minimized by giving regular irrigation.
3. The coconut cultivar *viz.*, hybrid T x D, and west coast tall are tolerant to the pest attack.
4. To minimize the pest population in the coconut nursery, spraying of endosulfan 0.07 per cent or monocrotophos 0.05 per cent is recommended.
5. Monocrotophos 36 per cent WSC @ 5 ml per tree (below 15 years aged tree) and @ 10 ml per tree (15 years and above age) with equal quantity of water applied by root application method has been recommended for effective control of this pest.
6. The larval parasitoid, *Goniozus nephantidis* can be utilized for the biological control of this pest. The pupal parasitoid, *Brachymeria nephantidis* is also proved to be more potential bioagents to suppress this pest.

Rhinoceros Beetle, *Oryctes rhinoceros* (Coleop[tera: Scarabaeidae)

Marks of Identification

The adult is a stout-built black beetle, measured 35-50 mm in length, 14-21 mm in breadth and with a cephalic horn which is longer in male. The pygidium is densely and almost rudimentary in female separates the two sexes. The full grown grub is of stout, cylindrical but strongly arched convexly above and concavely beneath.

Nature of Damage

The adult beetle injures the trees by boring into the growing spear leaf cluster, spathes and petioles. The boring beetle chews the internal tissues and after imbibing the juicy part, throws out the fibrous part which protrudes out of the holes. The damaged leaf when emerges out is appeared like leaflets cut by scissor. The attacked central shoot topples down and repeated attack may even kill young palms. The damage to the spathes causes reduction in yield up to 10 per cent.

Life Cycle

A female may lay 100-150 eggs which hatch in 8-18 days and the grub start feeding on the decaying matter found in the vicinity. The grub pass through three instars to complete their development in 99-182 days, pupation takes place in chamber at a depth of about 30 cm and the beetle emerge after 10-20 days. They remain in the pupal cell for about 11-20 days before coming out of the soil and on emergence they are soft bodied creatures. The beetles are active at night and may be attracted to a source of light. The adults can live for more than 200 days. Generally one generation is completed in a year.

Management

1. Avoid the manure pits in orchard or nearly the orchard to prevent the egg deposition and feeding of this pest.

2. Filling up the crowns, particularly the inner most two-three leaf axils with a mixture of methyl parathion dust and sand (1 : 1) at least twice in a year before and after the rainy season is an excellent prophylactic measure against the beetles.

3. Rotting castor cake is recommended to attract and trap the adult beetles. For this water soaked castor cake and treated with carbaryl 0.1 per cent on a w/w basis is exposed in orchard in a small mud pot.

4. Floor of manure pits/compost pits should be treated with contact insecticide (methyl parathion dust) to control the full grown grubs which reach the floor of the pits for pupation in the soil below.

**Plate 13.94: Coconut
Rhinocerous Beetle Grub**

**Plate 13.95: Coconut
Rhinocerous Beetle (Male adult)**

5. Insecticidal treatment of the cattle manure or compost with methyl parathion 2 per cent dust should be given to control the immature stages of this beetle in such media.

6. Mechanical control involves monitoring the crowns periodically and extracting the adult beetles by means of a hooked pointed metal rod about 0.5 m long with hook at one end, and fill up the hole with a mixture of methyl parathion 2 per cent dust and an equal volume of fine sand to prevent the reinfestation of this pest.

Pests of Spices
Chilli

Chilli Thrips, *Scirtothrips dorsalis* (Thysanoptera: Thripidae)

Marks of Identification

Insects have modified piercing and rasping type of mouthparts. These insects are very small, about the size of flea. They are visible to the naked eyes. The young ones are yellow or white. Adults are darker and brownish with or without stripes on their back.

Nature of Damage

Both the nymphs and adults lacerate the tissues and suck the sap from tender leaves, growing shoots, flower buds and developing fruits. The leaves of affected plant curl and shed while the buds become brittle and drop down. They are also associated with *murda* disease.

Life Cycle

The pest is active throughout the year except during the rainy season. The female thrips lays 45-50 eggs inside the tissues of the leaves and shoots. The eggs hatch in 5 days. The larvae is full-fed in 7-8 days and pupate in 2-4 days. The adult live for 31 days. There are several overlapping generations in a year.

Management

1. Grow resistant/tolerant varieties *viz.*, Bonapari, LIC-13, 36, 45.
2. Parasitoids like *Orius maxidentex*, *Chrysoperla carnea*, *Amblyseius longispinosus*, *Hauptamannis* sp. and *Cheiracantius* sp. are key enemies of thrips.
3. Spray monocrotophos (0.05 per cent) or methyl-o-demeton (0.025 per cent)
4. Soil application of carbofuran @ 0.5 kg a. i. per ha or phorate 0.7 kg a. i. per ha 15 days after transplanting control the pest, effectively.

Whitefly, *Bemisia tabaci* (Hemiptera: Aleyrodidae)

Marks of Identification

The louse-like nymphs clustered together on the under surface of the leaves and their pale yellow bodies make them stand out against the green background. In the winged stage, they are 1.0-1.5 mm long and their yellowish bodies are slightly dusted with a white waxy powder. They have two pairs of pure white wings and have prominent long Hindwings.

Nature of Damage

Adults and nymphs suck the cell sap from lower surface of leaves and cause chlorotic yellow spots on upper surface of affected leaves. Whitefly also excretes honey dew, which make the leaves sticky. Sooty mould (*Cladosporium* Sp.) growth on such leaves interferes with photosynthesis of plants.

Life Cycle

A female can lay up to 150-200 eggs. On an average, 28-43 eggs were laid singly on lower surface of leaves in a oviposition period for 2-18 days. Eggs are stalked, light yellow in colour and measure 0.2 mm. They hatch in 3-5 days. The nymphs on emergence feed on cell sap and grow in three stages to form the pupae within 9-14 days. The life cycle is completed in 14-122 days and many generations are completed in a year

Management

1. Whiteflies can be effectively attracted and controlled by yellow sticky traps, which are coated with grease or sticky oily material.
2. Use whitefly tolerant varieties such as LPS-141 (Kanchan), LK-861 and NA-1280.
3. Spray trizophos (2.5 ml per liter of water) or profenophos (2 ml per liter of water).

Chilli Fruit Borer, *Helicoverpa armigera* (Lepidoptera: Noctuidae)

Marks of Identification

The moth is stoutly built and is yellowish brown. There is a dark speck and a dark area near the outer margin of each forewing. The forewings are marked with grayish wavy lines and black spots of varying size on the upper side and a black kidney shaped mark and a round spot on the underside. The hindwings are whitish and lighter in colour with a broad blackish band along the outer margin. The caterpillar when full grown is 3.5 cm in length, being greenish with dark broken grey lines along the sides of the body.

Life Cycle

The female lays eggs singly on tender parts of the plants. A single female may lay as many as 741 eggs in 4 days. They hatch in 2-6 days. The young larvae feed on the foliage for some time and later bore into the bolls and feed inside. They are full fed in 13-19 days. The full grown larvae come out of the boll and pupate in the soil. The pupal period lasts 8-15 days. There may be as many as 8 generations in a year. The caterpillars feed on their fellows, if suitable vegetation is not available.

Management

1. Grow less susceptible genotypes.

2. Grow simultaneously 40 days old American tall marigold and 25 days old tomato seedlings at 1:16 rows to attract the female moth for egg laying.

3. Set up pheromone traps with *heli* lure at 15 per ha and change the lure once in 15 days.

4. Collect and destroy the damaged fruits and grown-up caterpillars.

5. Six release of *Trichogramma chilonis* @ 50,000 per ha per week coinciding with flowering time and based on ETL.

6 Spray *Helicoverpa armigera* NPV at 500 LE per ha along with cotton seed oil (300 g per ha), to kill larvae.

7. Spray endosulfan 35 EC 2 ml per liter or *B. t.* 2 g per liter or quinalphos 2.5 ml per liter.

8. Encourage activity of parasitoids, *Eucelatoria bryani, Campolites, Chelonus,* etc.

Tobacco Caterpillar, *Spodoptera litura* (Lepidoptera: Noctuidae)

Marks of Identification

Adults are stout, with wavy white markings on the brown forewings and white hindwings having a brown patch along its margin. Full grown caterpillar is pale brown with greenish to violet tinge. There are yellow and purplish spots present in the sub-marginal areas.

Nature of Damage

Freshly hatched larvae feed gregariously, scrapping the leaves from ventral surface. They feed voraciously during night and hide in the morning. Entire crop is defoliated overnight.

Life Cycle

Female lays dirty white coloured eggs in cluster on the under surface of the leaves and covered with brown hair. Incubation period is 3-5 days. Larval period is 20-28 days. Pupation takes place in the soil earthen cocoon for 7-11 days. Total life cycle is completed in 30-40 days during summer and 120-140 days in winter season. There are 7-9 overlapping generations in a year.

Management

1. Plough the soil to expose and kill pupae.

2. Grow caster along border and irrigation channels as indicator or trap crop.

3. Flood the field to draw the hibernating larvae.

4. Set up light or pheromone traps as 15 per ha.

5. Remove and destroy egg mass in castor and tomato.

6. Collect damaged leaves and gregarious early instar gregarious larvae and destroy.

7. Hand pick grown-up larvae and kill them.

8. Spray NPV based on *S. litura* at 250 LE along with teepol 1 ml per ha in evening hours.

9. Spray chlorpyriphos 20 EC 2.0 liter per ha or DDVP 76 WSP 1.0 liter per ha or endosulfan 35 EC 1,25 liter per ha or NSKE 5 per cent.

10. Prepare poison bait with rice bran 5 kg, jaggery 0.5 per cent, carbaryl 50 WP 0.5 per cent, water 3 liter per ha and spread the bait in the evening hours.

Broad Mite, *Polyphagotarsonemus latus* (Acarina: Tarsonemidae)

Marks of Identification

Individuals are extremely small, about the size of a grain of sand and not clearly visible to the naked eye. They are found in groups hidden around the mid-vein on the undersides of the leaves. They appear crab like and are yellow or white.

Nature of Damage

Damage is usually confined to undersides of leaves, where areas between veins are brownish and dried out and brittle in severe cases. Young leaves are cupped downward and narrower than normal.

Life Cycle

Female lays eggs singly on the surface of the leaves. These eggs are white, oval and extremely large compared to adults that lay them. Populations are continual but appear to be limited at high temperatures. This mite also feed on tomato, potato, beans and pepper. Only males are produced from eggs of unmated females. The duration of the life cycle from egg to egg is approximately vary from 74 hrs.

Management

Application of sulphur @ 2.5 kg or dicofol 0.1 to 0.3 kg per ha controls this pest effectively.

Black Pepper (*Piper nigrum*)

Pollu Beetle, *Longitarsus nigripennis* (Coleoptera: Chrysomelidae)

Marks of Identification

The adult is small shiny yellow and blue flea beetle with stout hind legs. The full grown grub is yellowish with a black head and it measures 5 mm in length.

Nature of Damage

The grub causes damage by boring into the berries and eats the contents completely in about 10-11 days. Each grub destroys at least 3-4 berries during the larval period. The attacked berries appear dark in colour and are hollow inside and crumble when pressed. The grubs may also eat into the spike and cause the entire distal region to dry up. The adult feed voraciously on tender leaves and make holes in them.

Life Cycle

The female makes holes on the berries and lay 1-2 eggs in each hole. A female on an average, lays about 100 eggs. The eggs hatch in 5-8 days and the young grubs bore into the berry and feed for 20-32 days. Then they drop to the ground and pupate in an earthen cell in the soil at 5-7.6 cm deep. The pupal period is of 6-7 days. The total life cycle is completed in 30-50 days. There are four overlapping generations in a year.

Management

1. Tilling the soil in the base of vines at regular interval reduces the population considerably.
2. Spray quinalphos (0.1 per cent) or dimethoate (0.03 per cent) in late July and repeat in early October.

Leaf Gall Thrips, *Liothrips karnyi* (Thysanoptera: Thripidae)

Marks of Identification

The adults are black and measure 203 mm in length. The larvae and pupae are creamy-white in colour.

Nature of Damage

Both nymphs and adults infest leaves of black pepper and reduce the formation of tubular marginal leaf galls within which they live. The pest infestation is generally serious at higher altitudes and in nurseries in the plains. Apart from the formation of marginal leaf galls, the pest infestation results in reduction in size, crinkling and malformation of the infested leaves. In severe case of infestation, the growth of the vine and sometimes the formation of spikes may be adversely affected.

Life Cycle

The eggs are laid within the galls and they hatch in 6-8 days. The two larval stages *viz.*, pre-pupal stage and pupal stage last for 4-7, 4-6, 2, 2-3 and 2-3 days, respectively.

Management

1. *Montandoniola moraguesi* and *Androthrips flavipes* are the common predators of this pest.
2. Spray monocrotophos (0.05 per cent) or dimethoate (0.05 per cent) for controlling the pest.

Top Shoot Borer, *Cydia hemidoxa* (Lepidoptera: Tortricidae)

Marks of Identification

Adults are small with a wing expanse of 10-15 mm, the forewings being crimson-red and yellow and the gindwings are grey. Fully grown larvae are grayish-green and measure 12-14 mm in length.

Nature of Damage

The larvae cause damage, the earlier instars of the larvae live within the silken webs on the tender shoots and scrap and feed on them. Later, the larvae bore into the tender shoots and feed on the internal contents resulting in drying up of the tender shoots. Repeated infestations of new shoots affect the growth of the vine.

Life Cycle

The female lays eggs on the tender portion of the plant. After hatching from eggs the earlier larvae live in silken webs on the tender leaves and feed them by scraping green matter. The larval period lasts for 14 days. Pupation generally occurs within the infested shoots and sometimes outside. The pupal period lasts for 8-10 days. This pest is more active in the field during August- December.

Management

1. The larval population is parasitized by hymenopteran parasitoid *viz.*, *Apanteles* sp., *Eudederus* sp. and *Goniozus* sp.
2. Spray endosulfan (0.07 per cent) or monocrotophos (0.05 per cent) during June and September is very effective in controlling the pest infestation.

Black Pepper Mussel Scale, *Lepidosaphes piperis* (Hemiptera: Diaspididae)

Marks of Identification

Full grown females are small dark brown and boat shaped measuring about 3-5 mm in length. The size of males is 2 mm.

Nature of Damage

Mussel scale encrust the main stem of young veins and lateral branches, mature plants, spikes, berries and petioles of pepper veins. The female scale sucks the sap causing serious injuries on feeding parts. Except for roots, all other parts are terribly damaged. The infestation results in chlorotic spots, or patches, yellowing and necrosis of leaves due to desaping. Young veins succumb to the infestation whereas older infested lateral branches wilt and dry in patches resulting in necrotic vacant spaces in the canopy. Infestation by mussel scale causes significant loss of production as it affects all parts of plant including berries.

Life Cycle

The eggs are laid by the female under the scale cover and crawlers that hatch out from the eggs move about and fix themselves on the host plant at suitable places and thereafter become sedentary during the remaining part of their life. Early instar nymphs are spindle shaped. Once feeding is commenced, white silken threads are being secreted and cover the posterior part of abdomen. After a period of 5 days white coloured silken threads become melanised and turn darker, and harder. The total life cycle lasts for about a month.

Management

1. Pruning and destroying severely affects lateral branches.
2. Number of parasitoids and predators (*Chilocoris circumdatus, Sarajiscymnus dwipakalpa, Pseudoscymnus* sp.) are effective in regulating the pest population so their activities should be encouraged.
3. Apply neem formulations 0.5 per cent and fish oil insecticidal soap (3 per cent) four times at fortnight intervals after the harvest of berries suppress the pest very effectively.
4. Spray dimethoate (0.03 per cent) or monocrotophos (0.05 per cent) two times at 15 days interval control the pest effectively.

Ginger and Turmeric

Shoot Borer, *Conogethes punctiferalis* (Lepidoptera: Pyralidae)

Marks of Identifications

The adult is a small moth with orange, yellow wings with small black spots. Full grown larvae are light brown and 16-26 mm in length.

Nature of Damage

The larvae bore into pseudostems and feed on the growing shoot resulting in yellowing and drying of infested shoots. The presence of bore holes in the pseudostem through which frass is extruded and the withered central shoots are characteristic symptoms of pest infestation.

Life Cycle

Adult female lays 30-60 eggs during its life span. The egg period lasts for 3-4 days. There are five larval instars and they last for 3-4,5, 3-7, 3-8 and 7-14 days, respectively. The pre pupal and pupal period last for 3-4 and 9-10 days, respectively.

Management

1. Grow resistant/tolerant varieties of ginger- Rio-de-Janerio and turmeric, Dindigam Co-69 and Mannuthy local.
2. Spray malathion (0.1 per cent) during July- October is the most effective way to control the pest.

Leaf Roller, *Udaspes folus* (Lepidoptera: Hesperidae)

Marks of Identification

Adults are medium sized with brownish black wings having large white spots. The mature larva is dark green and is about 36 mm in length.

Life Cycle

The egg, larval and pupal periods lasts for 4-5, 13-25 and 6-7 days, respectively on ginger. On turmeric, the various periods in life cycle lasts for 3-4, 2-21 an d 6-7 days, respectively.

Management

Spray dimethoate (0.03 per cent) for the control of the pest.

Rhizome Scale, *Aspidiella hartii* (Hemiptera: Diaspidae)

Marks of Identifications

The adult females are minute, circular and light brown to grey.

Nature of Damage

The rhizome scale infests rhizomes of ginger and turmeric, both in field and in storage. In the field in severe cases of infestation, the plants wither and dry. In storage the pest infestation results in shriveling of buds and rhizomes. When the infestation is severe, it adversely affects sprouting of rhizomes.

Life Cycle

Females are ovo-viviparous and also reproduce parthenogenetically. About 100 eggs are laid by a single female.

Management

1. Dipping turmeric seed rhizomes in monocrotophos (0.1 per cent) for 5 minutes after harvest and before planting is effective in controlling the pest infestation.
2. The natural enemies *Physcus comperei* and *Adelencyrtus moderatus* are effective in regulating the pest population under the field conditions.

Cardamon, (*Ellettaria cardomomum*)

Cardamom Thrips, *Sciothrips cardamomi* (Thysanoptera: Thripidae)

Marks of Identification

The head and abdomen of this thrips are green brown and thorax and legs are pale yellowish brown. Adult measures approximately 1.2-1.5 mm in length.

Nature of Damage

Injury is caused by feeding in young leaf sheaths and basal ends of unripened flower bracts. Nymphs and adults feed with their piercing–raspim mouthparts, which lacerate young tissues and suck the juices that ooze from ruptured cells. First the injured area develops a silvery sheen because of air occupying the emptied cell cavities. Later on this area become whitish, yellow, and brown patches and streaks. Because of the secretive nature of feedings within the sheaths and unopened bracts, its early presence usually goes undetected. However, close observation in opened inflorescence confirm the feeding damage at the bottom of the flower bracts. This thrips is suspected to be a vector of a mosaic disease in cardamom.

Life Cycle

The pest is active throughout the year except during the monsoon season. The female lays 5-71 eggs at random on all the feeding area of the plant. The young nymphs emerge from the eggs in 9-12 days. The first two nymphal instars are active and grow by feeding on the plant sap. Pupal period lasts for 10-15 days. Life cycle is completed in 25-30 days.

Management

1. Clean cultivation practices *viz.*, sanitation such as the removal of alternative hosts and plant debris within and outside the field is advised.

2. Removal of dried shoot peels at the base of the stem results in one third reduction of thrips population.

3. Regularly spraying the cardomon crop either with fenthion 0.025 per cent or phenthoate 0.05 per cent or phosalone 0.05 per cent or methyl parathion 0.05 per cent or chlorpyriphos 0.05 per cent once in 3 days during summer months and once in 45 days during the winter months controls the pest, effectively.

Banana Aphid, *Pentalonia nigronervosa* (Hemiptera: Aphididae)

Marks of Identification

The wingless aphid is dark brown pyriform measuring 1.34 mm in length and with six segmented antennae which are longer than its body. Abdomen is dark brown, shining and slightly bulged. The winged forms are dark brown, elongated and pyriform. They are longer than the wingless form but body width is less.

Nature of Damage

The aphids feed on the leaf sheath and pseudostem. The insect causes little direct damage, but is of considerable significance being vector of cardamom mosaic in small cardamom.

Life Cycle

The reproduction takes place parthenogenetically. The longevity of adult ranges for 8-26 days with an average of 14 days. A single female lays 8-28 off-springs with an average of 14. The development is completed through three to four moults in 12-15 days, respectively. There are 21-24 generations in a year.

Management

Spraying of dimethoate or methyl-o-demeton (0.05 per cent) is effective for the control of this thrips.

Castor Capsule Borer, *Dichocrosia punctiferalis* (Lepidoptera: Pyralidae)

Marks of Identification

This is a serious pest of nursery plants and young green pods of cardamom. In nursery and main plants, it bore into the stem and causes death of the central core. It also eats away the tender seeds of the young barriers.

Nature of Damage

The full-grown caterpillar measures 25-30 mm in length and is reddish brown with black blotches all over the body and pale strip on the lateral side. The moths are orange yellow, with black markings in both the sexes.

Life Cycle

The female moths lay eggs on leaves and other safe parts of the plant. The eggs hatch in about a week. The larvae pass through 4-5 instars and are full-fed in 2-3 weeks. Pupation takes place inside the shoot or sometimes in the frass that is collected after feedings. The pupal stage lasts about one week. The life cycle is completed in 4-5 weeks and there are about 3 generations in a year.

Management

1. It is advisable that the infested shoots and barriers may be collected and destroyed.
2. Spray quinalphos or methyl parathion (0.05 per cent) during early blooming period. It helps in reducing the pest damage.

Cardamom Hairy Caterpillar, *Lenodera vittata* (Lepidoptera: Arctiidae)

Marks of Identification

The adult moth is stout and fairly big and densely covered with scales. The larvae are clothed with a dense tuft of capitate hairs and measure 106-110 mm in length.

Nature of Damage

The caterpillars are voracious feeders and cause extensive damage to capsule and the cardamom plants by feeding on leaves. They are active from August to December. Only pseudostems and midribs remain un-eaten in case of severe infestation.

Life Cycle

The moth lays cream coloured dome shaped eggs in rows on both the upper and the lower surface of leaves. A single female lays 100-130 eggs during an oviposition period of 6-9 days. The young larvae emerge from the eggs in 10-13 days. The larvae start feeding on leaves and other tender parts of the plant. They moult 6 times during the larval period of 110-120 days. Pupation takes place in the soil in earthen cell, where in stays for 5-7 months. There is only one generation in a year.

Management

1. The larvae is parasitized by a tachinid fly, *Carcelia kockrana*.
2. Spray 0.05 per cent malathion or 0.05 per cent monocrotophos to protect the crop from this pest.

Hairy Caterpillar, *Eupterote cardamomi* (Lepidoptera: Arctiidae)

Marks of Identification

The adults are large moths, ochreous in colour, with post-medial lines on the wings. They measure 70-80 mm in wing expanse. The larvae are hairy, dark grey in colour with pale brown head, bearing conical tufts of hairs on the dorsal side of the body. Full-grown caterpillar measures 90 mm in length.

Nature of Damage

The larvae feed on leaves of the shade trees up to the 6th or 7th instar and then they drop down on the cardamom plants growing underneath, with the help of silken threads. They start feeding on the leaves of cardamom voraciously and defoliate the cardamom plants causing heavy reduction in the yield.

Life Cycle

The moths emerge with onset of monsoon rains in June and July. The female moths lay 400-500, yellowish and dome-shaped eggs in the flat masses on the undersurface of the leaves. Each mass contains about 50-60 eggs. The hatching occurs in 15-17 days. The larva passes through 10 instars in 140-150 days. It pupates in the soil in a silken cocoon at a depth of 5-8 cm. The pupal period lasts for 7-8 months. The adult moths live for about 20 days. There is only one generation in the year.

Management

1. Larvae are parasitized by *Stuemia sericariae* and *Apantelis eupterote* under field condition.
2. Spray 0.05 per cent malathion or 0.07 per cent endosulfan for its management.

Rhizome Weevil, *Prodioctes haematicus* (Coleoptera: Curculionidae)

Marks of Identification

The adult is a brown coloured weevil measuring 12 mm in length.

Nature of Damage

The grub causes tunnels in the rhizome by feeding resulting in the death of entire clumps of the cardamom plant.

Life Cycle

The female weevil lays egg in cavities made on rhizomes. The young grubs emerge out of the eggs in 8-10 days and bore into rhizome making tunnels. They feed inside the rhizome and full-fed in 3 weeks. They pupate within the feeding tunnels. The pupal period lasts for 3 weeks. The adult weevil lives for 7-8 months. There is only one generation in a year.

Management

1. Destroy affected plants and seedlings.
2. If the grub population is more in soil, drench the base of the clump with 1.25 liter of malathion.

Cardamom Root Grub, *Basilepta fulvicorne* (Coleoptera: Scarabaeidae)

Marks of Identification

The adults are shiny metallic blue, green brown beetles measuring 4-6 mm. Males are smaller than the female. The full grown grubs are pale white, stout and 'C' shaped.

Nature of Damage

Root grub is a major pest of cardamom in nurseries and main field. The pest is generally serious in primary and secondary nurseries. The grubs feed and damage roots and portions of rhizomes, sometimes the entire root system is eaten away. The infested plants turn yellow and are stunted. Severely infested plants dies.

Life Cycle

The female lays eggs in dried leaves around the base of clumps. The egg, larval and pupal stages lasts for 3-19, 45-60 and 10-17 days, respectively.

Management

1. Collection and killing of adult beetles during March- April and August-September during peak period of emergence.
2. Apply chlorpyriphos (0.04 per cent) or phorate 10 G (20-40 g per clump) during May and October.

Coriander

Coriander Aphid, *Hydaphis coriandri* (Aphididae: Hemiptera)

Marks of Identification

Nymphs and adults are pear shaped, light green in colour. Their blue white look is due to powdery substances present on the body. Nymphs are 11 mm and adults are 12 mm long.

Nature of Damage

Both nymphs and adults are found to suck the sap from leaves, flowers and immature seeds mainly attacking umbels. Beside, they also excrete honeydew which favours the growth of sooty mould, as a result, the growth of plant retards, and quality and quantity of seeds are also affected. In case of severe infestation at early stage of development, tender leaves, growing points, flower stalks withers and dries up. When aphid infestation occurs at flowering and seed setting stage, the seeds are not formed and if formed they are shriveled and of poor quality. In case of severe infestation there is complete failure of crop.

Life Cycle

The aphid breed throughout the year by parthenogenesis and viviparous reproduction. Sexual reproduction takes place during severe winter season. A single female in her life produces about 40-50 nymphs. All young ones are apterous in viviparous females. The winged forms are produced when the population increase suddenly in overcrowding, partial starvation, high temperature and low relative humidity conditions. The nymphs moult three times and nymphal period lasts for 8-10 days. The adult period is for 3-4 days and total life cycle normally completes in 14-21 days during summer and 35-40 days in winter.

Management

1. Timely sowing of coriander plays an important role in occurrence of the aphid. If time of sowing is manipulated, the attack of this aphid is automatically reduced.

2. Lady bird beetles *Cocinella septumpunctata* and *Menochilus sexmaculatus* are very important predators of this aphid under field condition, so always conserve and protect them.

3. Aphid incidences can be controlled by two sprays with monocrotophos or quinalphos or endosulfan (0.1 per cent) at 10-15 days interval *i.e.,* at the time of floral initiation and 15 days thereafter.

Green Peach Aphid, *Myzus persicae* (Hemiptera: Aphididae)

Nature of Damage

This is a very serious pest of coriander. Damage of *M. persicae* is confined to the umbels only. Nymphs and adults suck the sap from the umbel. Due to honeydew secretion, sooty moulds are also developed on umbels which results in poor development of seeds.

Life Cycle

The female produces 5-15 ovipare. These are fertilized by the male and each ovipare then lays 4-14 eggs. The eggs undergo diapsuse during winter. Hatching of eggs coincides with vegetative growth of the plant. Newly hatched young once feed on soft portions of the plants and develops rapidly and start reproducing parthenogenetically.

Management

1. Yellow sticky traps attract the winged aphids and can be utilized in trapping the pest. For this purpose, number of materials like yellow polythylex, sticky traps and water traps have been used in minimizing the population of this pest.

2. Early sown crops upto the middle of November has low infestation then the crop sown after this period.

3. Natural enemies like Coccinellid predators *viz., Coccinella septumpunctata, Brumus suturalis, Menochilus sexmacalatus* and Adonis sp have been common predators of this aphid.

4. The pest is controlled by spraying endosulfan (0.07 per cent) twice just at the time of floral initiation and 15 days thereafter.

Cotton Whitefly, *Bemisia tabaci* (Hemiptera: Aleyrodidae)

Marks of Identification

The louse like nymphs, which suck the sap are sluggish creatures, clustered together on the under surface of the leaves and their pale-yellow bodies make them stand out against the green background. In adult stage, they are 1.0-1.5 mm long and their yellowish bodies are slightly dusted with a white waxy powder. They have two pairs of pure white wings and have prominent long hindwings.

Nature of Damage

Nymphs and adults both cause damage to the crop. They remain underside the leaves and suck the sap. Continuous sucking of sap from the leaves results in chlorotic spots and later on, the leaves coalesce and become brittle and finally drop-down from the plant prematurely. The honeydew secreted by the pest drop on the upper surface of the lower leaves and helps in development of hooty moulds which interferes with photosynthesis of leaves.

Life Cycle

The female generally lays eggs on the under surface of the leaves. The eggs are inserted in the leaf tissues. A single female lays 200-300 eggs either singly or in groups, which are yellowish in colour and turn brown to dark before hatching. Incubation period is 3-5 days. There are 4 nymphal instars, fourth nymphal instar is called pupa. Generally, nymphal period is 9-14 days. The pupal period lasts for 2-8 days. This pest generally completes its life cycle in about 13-20 days. There are 11-13 generations in a year.

Management

1. Judicious use of nitrogenous fertilizers and irrigation as well as proper spacing in plants can help in checking whitefly population build up.
2. Use of fish oil soap at 2 per cent, neem oil (0.5 per cent) or methyl-o-demeton (1 liter per ha) is advised for suppressing the pest population.

Coriander Cutworm, *Spodoptera exiguva* (Lepidoptera: Noctuidae)

Marks of Identification

The moths have dark spotted forewings and white hindwings. They are active at night but remain hidden under various shrubs in the day time. The colour of larvae is light green and they are 30 mm long.

Nature of Damage

The larvae cause damage to the crop. The damaged crop, on which larvae feed gives a webbed appearance. The older caterpillars, which feed in morning and evening, are voracious feeder. They feed large amount of leaves.

Life Cycle

A female lays upto 600 eggs in clusters. The eggs are spherical and resemble poppy seeds in shape and size, having lines radiating from the center. The egg clusters are covered with buff hairs. The eggs hatch in 1-3 days and the young caterpillars start feeding in groups. The larvae become full-fed in 15-20 days. Pupation occurs in soil. The pupal stage lasts for 5-7 days and the life cycle is completed in abut 30 days. There are 8-10 generations in a year.

Management

1. Activity of this pest can be suppressed by keeping field clean.
2. The pest can be suppressed by collecting and destroying the egg masses.
3. *Euplectus* sp. is a closely associated natural enemy of this pest.
4. Spray endosulfan (0.07 per cent) to suppress the pest.

Flower Stink Bug, *Agonoseelis nubile* (Hemiptera: Pentatomidae)

Marks of Identification

The pest is about 12 mm long and its body colour is green or dark green with black spots. The scutelum is quite long and conical in shape.

Nature of Damage

Both nymph and adults cause damage to plants. They suck sap from leaves, flowers and unmatured seeds. Due to heavy feeding chlorotic spots develope on leaves. In case of heavy infestation the flowers drop prematurely from the plants and the grains developed from these flowers become shriveled.

Life Cycle

Very less information is available regarding the life cycle of this pest. The female lays its eggs in stems of plant, and is active from February to October.

Management

1. Hand picking of insect in case of small field and kitchen gardens.
2. Spray malathion (0.04 per cent) or endosulfan (0.07 per cent) for its effective management.

Green Stink Bug, *Nezara viridula* (Hemiptera: Pentatomidae)

Marks of Identification

Green stink bug is bigger in shape and size in comparison to flower stink bug. Its body is 15 mm long and 8 mm broad. It is deep green coloured insect, whose scutelum is pointed.

Nature of Damage

Both nymphs and adults cause damage by sucking sap from stem, leaves, flower and green seeds. Due to infestation, the plant growth is badly affected, the flowers drop prematurely and the seed yield is reduced.

Life Cycle

The female lays about 300 eggs in cluster on the surface of leaves. Eggs are whitish or yellowish in colour when freshly laid but turn to pink before hatching. Eggs are barrel shaped and 1 mm long. The incubation period is of about a week. Nymphs are about 5-8 mm in length with beautiful colour patterns. They are brownish red in colour with multicolor spots. Freshly hatched nymphs remain in clusters around the egg raft and it is only after first moult disperse and start active feeding. The nymphal development takes about one month. There are four to five generations in a year.

Management

1. Hand picking of nymphs and adult bugs.
2. Spray dimethoate (0.03 per cent) or quinalphos (0.05 per cent) for effective control of this pest.

Cumin

Aphid, *Hyaloptenus arundini* (Hemiptera: Aphididae)

Marks of Identification

The adult and nymphs are lice like in shape and size and their body colour is deep green. Their colony is generally found on the soft portion of the plant.

Nature of Damage

Both nymphs and adults cause damage to the crop. They suck the sap from the soft portion of the plant. In early growth stage due to their attack the plant growth is badly affected, while in later stage the pest cause great reduction in the yield. Besides this, the pest extract honeydew on the leaves which attract the black mould, which again interferes in the photosynthetic activities of the plant and ultimately cause reduction in the crop yield.

Life Cycle

This aphid breeds throughout the year by parthenogenesis and viviparous reproduction. Sexual reproduction takes place during severe winter season. A single female in her life produces about 40-50 nymphs. All young once are apterous and viviparous females. The winged forms are produced when the population increases suddenly when over crowding, partial starvation, high temperature and low relative humidity conditions occur. The nymphs moult 3 times and nymphal period last for 8-10 days. The adult period is of 3-4 days and the total life cycle normally completes in 14-21 days during summers and as long as 35-40 days in winter.

Management

1. Always grow aphid resistant varieties of cumin *viz.*, VC-187, VC-154, VC-822 and VC- 33.
2. Spray dimethoate (0.03 per cent) for effective control.

Fenugreek

Aphid, *Aphis craccivora* (Hemiptera: Aphididae)

Marks of Identification

Both winged and wingless forms are present in this aphid. The mature adults measure 1.75–1.90 mm length. After first moult, nymph become yellowish green in colour and become 0.01- 0/04 mm long. As the nymph matures, its colour becomes darker. The mature adult becomes olive grey or dull grey in colour.

Nature of Damage

The nymph and adults suck the sap, usually from the underside of leaves. Infestation in early stages causes stunting of the plants as well as reduce their vigour. When the attack occurs at the time of flowering and pod formation, there is significant reduction in the yield. Due to honey dew secretion black mould developed on the leaves which interfere in photosynthetic activities.

Life Cycle

The offspring of the winged form may be wingless. Even without fertilization, the female may produce 8-20 young ones in a life span of 10-12 days. The young nymphs pass through four moults to become adult in 5-8 days. The apterous female starts producing broods within 24 hours of attaining that stage. Breeding occurs almost throughout the year, and there are many overlapping generations in a year.

Management

1. Natural enemies like Coccinellid predators *viz.*, *Coccinella septumpunctata*, *Brumus suturalis*, *Menochilus sexmacalatus* and *Adonis* sp. are common predators of this aphid.
2. Spray endosulfan (0.07 per cent) or malathion (0.04 per cent) is advised to protect the crop from this pest.

Flea Beetle, *Phyllotreta cruciferae* (Coleoptera: Chrysomelidae).

Marks of Identification

Adult beetles are elongate oval in shape and metallic bluish green in colour. Head has impunctate vertex and black antennae.

Nature of Damage

The adults mostly feed on the leaves by making innumerable round holes in the host plants. The affected leaves dry up. A special kind of decaying odour is emitted by the cabbage plants attacked by this pest.

Life Cycle

A single female lays 50-80 eggs in soil. Incubation period last for 5-10 days. The grubs feed on roots and do not cause much damage. Grub period lasts for 9-15 days. Pupal period is of 8-14 days. There are 7-8 generations of this pest in a year.

Management

1. The adults are parasitized by *Microctonus indicus*.
2. Spray quinalphos or chlorpyriphos (0.05 per cent) for effective control of this pest.

Fenugreek Leaf Miner, *Chromtomiya horticola* (Diptera: Agromyzidae)

Marks of Identification

The adults are two-winged flies having grayish black mesonotum and yellow frons. It is 2 to 3 mm long. The fully grown maggots are 3 mm long and 0.75 mm broad.

Nature of Damage

The larvae makes tunnels on the leaves interfere with photosynthesis and affect the growth of the plant making them look unattractive.

Life Cycle

The female lays eggs singly in leaf tissues. The eggs hatch in 2-3 days and larvae feed between lower and upper epidermis by making zig-zag tunnels. They are full grown in about 5 days and pupate within the galleries. The adult emerges from the pupae in 6 days. The life cycle is completed in 13-14 days and there are many generations in a year.

Management

1. The attack of this pest is much more in drought condition so timely irrigation to the crop is very important.
2. Destroy damaged leaves in its initial stage.
3. Chalcid, *Solenotus guptai* and *Rhopalotus thokerie* are the common parasitoid of maggots.
4. Spray dimethoate (0.03 per cent) or methyl-O-demetom (0.05 per cent) to protect the crop from this pest.

Leaf Webber, *Hymenia recurvalis* (Lepidoptera: Pyralidae)

Marks of Identification

Adult are very tiny moth, measuring 1.0 mm in length and 1.5–2.0 mm with wing expansion. The wings are dark green in colour with whitish spots. The larvae are dark green in colour with dark brown head.

Nature of Damage

The larvae bind leaves together with silken threads and then feed inside these leaves by scraping the green material. The infested leaves are not useful for consumption. In case of heavy infestation, the leaves become dry and flowers, pods and seed formation do not takes place.

Life Cycle

Female lays 150 eggs in veins of leaves. Incubation period range between 3-4 days and the newly emerged larvae are dark green in colour. They moult 5 times to become fully mature. The fully mature larvae pupate inside soil in a cocoon. Pupal period lasts for 8-11 days. Adult live for 3-5 days.

Management

1. Pluck the affected leaves to reduce the larval infestation.
2. The adult moths attract towards light so use light trap to destroy the adult moths.
3. Braconid, *Apantelis delhensis* and *Cardiachiles* are larval parasitoids of leaf webber.
4. Spray malathion (0.5 per cent) to control this pest.

Cinnamon

Cinnamon butterfly, *Chilasa clytia* (Lepidoptera: Papilioridae)

Marks of Identification

The adults are large, males have blackish brown wings with white spots on the outer markings, females have black wings with bluish white markings. Full grown larvae are pale yellow with dark stripes on the sides and are 25 mm in length.

Nature of Damage

This is the most destructive pest of cinnamon. The pest is generally seen in the field during December-June. The larvae feed on tender and slightly mature leaves with portions of veins.

Life Cycle

Eggs are laid on tender leaves and shoots and they hatch in 3-4 days. The larval stage comprising of 5 instars is completed in 11-17 days. The pupal period lasts for 11-13 days.

Management

1. The eggs are parasitized by *Telenomus remus*.
2. Spraying of quinalphos (0.05 per cent) for controlling the pest infestation is quite effective.

Polyphagous Pests

Locust

Locusts are members of the grass hopper family acrididae which included most of the short-horned grasshopper.

Types of Locust

There are some 5000 species of grasshoppers found in the world, but only 9 are recognized as locust of which 3 belong to the Indian subcontinent. These are as follows:

1. Migratory locust (*Locusta migratoria*)–Found in India, Europe, Africa, Sri Lanka, Pakistan, East and South Asia, Australia.
2. Bombay locust (*Patanga succinacta*)–Found in india, Sri Lanka, Malaysia.

3. Desert Locust (*Schistocerca gregaria*)–Found in India, Pakistan, Arabia, Middle East and North African countries.

Desert Locust

The desert locust is the most destructive of all the locusts. It invades in an area of 30 million square kilometer spreading over 60 countries from the west and north Africa to Assam in India and in 50 per cent of this area, breeding can occur.

Distribution of Desert Locust

The desert locust is an inhabitant of the dry grasslands of desert areas and is found in many countries of the world. Its distribution extends from Pakistan to Afghanistan, Iran, Iraq, Middle East and Northern Africa. In India, breeding grounds are located in Rajasthan, part of Gujarat and Hisar and Mohindergarh district of Haryana. These places are not the permanent home but are merely the outbreak areas where the locusts undergo change in their phase from gregarious to solitary.

Phases of Locusts

The desert locust is found in two phases:

1. The solitarious phase–When individuals are at low densities.
2. The gregarious phase–When they are at high densities

The transition from the solitarious phase to gregarious and vice versa is called the transient phase.

Difference Between Solitarious and Gregarious Phases

The characteristics of these phases are different from each other, particularly in the colour of their nymphs.

Solitarious Phase	Gregarious Phase
1. The nymphs (hoppers) are varied in colour according to the colour of the surrounding vegetation.	1. The nymphs are yellow or pink with distinct black markings.
2. The adult remain greenish grey throughout their life time.	2. The adults are pink on emergence, gradually turning grey and ultimately yellow, when sexually mature.
3. Numbers of molts are 5 to 6 (nymphs)	3. Number of moults are 5 times (nymphs)
4. Hoppers remain scattered on vegetation whereas adults during nights as isolated individuals.	4. Hoppers form groups or band and march long distances, whereas, adults generally fly during day time in swarms.

The Life Cycle

The desert locust like all other locusts and grasshoppers passes through three stages: egg, nymphs (hoppers) and adults.

Plate 13.96: Locust

Plate 13.97: Termite

Plate 13.98: Termite Damaged Groundnut Stem

Plate 13.99: Termite

Plate 13.100: White grub

Plate 13.101: Adult White Grub

Plate 13.102: Locust Plague

Plate 13.103: Termite Colony

Oviposition and Incubation

The female lays eggs in batches called egg-pods, they look like rice grains and are arranged like a miniature hands of bananas. The female bores into the ground with the valves at the rear of the abdomen and deposits a batch of eggs. She then fills up the hole above the eggs with a plug of froth. The pod is about 5 cm long and is laid with its top 5-10 cm below the surface. This is a surprising depth which requires a great extension of the abdomen. The desert locust lays pods containing less than 80 eggs in the gregarious phase and between 95 to 158 in the solitarious phase. The eggs are initially laid in bare ground and often, but not exclusively in sandy soil. As a rule, the female will not lay unless soil is moist at about 5-10 cm below the surface. In soft sandy soils, females have been known to lay when soil moisture is only found at depths below 12 cm. the number of egg pods a female lays depends on how long it takes for her to develop a pod and how long the female lives.

Hopper Development and Behaviour

After hatching, the emerging hoppers work their ways up the forth plug to the surface. They immediately moult to the first instar. The hoppers then pass through five instars shedding a skin (moulting) between each. At the first moult (called fledging) the young adult known as a fledging emerges.

Adult Development and Behaviour

It takes about ten days after fledging for the adult wings to become hard so that it is capable of sustained flight. The adult then remain immature until they encounter conditions which stimulate maturation. There is however, a maximum period. Adults in an area of lush vegetation with maximum day temperature of 35°C or more, and with rain to maintain the vegetation growth, can probably lay within three weeks of fledging. At the other extreme, immature adults can survive for six months or more under dry conditions. Adults cannot survive long under hot dry conditions with little to eat. Migrations to areas where rainfall is one way in which plagues collapse. Adults can also survive during winter in West Africa, South of the Sahara where it is relatively warm, but these adults do not breed.

Breeding Season and Migration of Swarms

The breeding of locust depends upon the rainfall and the subsequent vegetation. The eggs are laid in the sandy soil and adequate moisture is required before they can hatch. In India, there are two breeding seasons during the year, *viz.*, (1) The summer breeding season and (2) The monsoon breeding season.

Origin of New Locust Cycle

The following sequence of events generally mark the beginning of the locust cycle–(a) extensive breeding on coastal areas of Arabian countries as a result of heavy winter and spring rainfall and the formation of gregarious swarms. (b) migration of locusts from Arabian coast into the interior (Baluchistan, Afghanistan) in spring. (c) migration of these swarms in summer, to Sind and Rajasthan. (d) extensive breeding of these swarms in Rajasthan in July- August and September. (e) migration of the subsidiary gregarious swarms from Rajasthan to Punjab in late summer.

Development of Desert Locust in Various Broods in India

Sl.No.	Stage of Development	Summer Brood	Winter Brood
1.	Pairing and eggs laying	February–April	July–September
2.	Hoppers	March-May	July–October
3.	Adults	May–June	September–October

Management

1. The adults can be beaten to death with thorny stick, brooms or can be swept together and buried underground in heaps.
2. Methyl parathion (2 per cent) or endosulfan (2 per cent) @ 25 kg/ha if dusted on crops, trees and the ground is very effective.
3. If eggs are laid in well defined area, a trench may be dug around it, so that the young nymphs on emerging drop into it and can be buried alive. Fill the ditch with soil.
4. At night when the hoppers rest on bushes, they can be burnt with flame-throwers.
5. Poison bait such as the poisoned bran mash or sawdust, if scattered in the early morning or in the evening, are effective. Sodium fluosilicate or paris green is used as poison.
6. A number of birds attack locusts and of these, the common *myna* and the *tiliar* (starling) are the most important.

Termite or White Ants, *Odentotermes obesus, O. assumthi, Microtermes obesi, Coptotermes heimii* (Isoptera: Termitidae)

Host Range

It is a polyphagous pest and is one of the most destructive creatures of all cultivated crop plants throughout the world.

Marks of Identification

The termites are social insects and their colony organization is based on caste system.

(A) Reproductive Castes

(i) Winged Male and Female

Winged individuals of both the sexes are produced in large numbers during the rainy seasons. They are brownish in colour with two pairs of slender, dark brown, long narrow wings which are used for nuptial flight only and when they mate, the wings then drop off.

(ii) Queen

The queen is wingless, creamy white in colour, abdomen marked with transverse dark brown stripes, 60-80 mm in length and 10 mm in thickness. She is a mother of the colony and lives in a specially prepared royal chamber which is situated in the centre of the nest.

(iii) King

King is much smaller than queen and remains with queen in royal chamber.

(iv) Supplementary Reproductive

They are short winged or wingless creature of both the sexes. They may be present among the offspring of the primary reproductive and are not normally found in colonies and appear to replace the primary reproductives when the later die.

(B) Sterile Castes

(i) Workers

They are sterile, apterous individuals with non-functional reproductive organs, about 6-8 mm long, pale in colour, brown head and eyes are small or absent. Mandibles are well developed. They work for the whole colony by storing food, look after the young ones and build up the termitarium. In a colony large numbers of workers are present.

(ii) Soldiers

Bigger than workers, sterile, head and thorax well chitined, abdomen is delicate and dirty white in colour. They are 5-10 per cent in number in a colony. They are winged. The main function of soldiers is to protect the colony.

Nature of Damage

Termite feed on all the things which contain cellulose. In loamy or light soils and dry areas where proper facilities for irrigation are not usually available, the termite infestation is more serious. The infestation of termite is more on *rabi* crops as compared to *kharif* crops. It feeds on the roots of all the vegetable plants. As a result of infestation, the leaves of infested plants starts drying, such plant can easily be pulled out from the soil. In later stage, the whole plants are withered and dry.

Life Cycle

In rainy season, when the weather conditions are favourable, the colonizing forms emerge from their nest and fly in swarms for a little time. The swarm of colonizing individuals has members of both the sexes. Swarming usually takes place in the day time and most of the individuals of the swarms are destroyed by birds, etc. The survived individuals mate, shed their wings and burrow in the ground to form a new colony of which they become king and queen. Seven to ten days after swarming, the female lays the first batch of 100-130 eggs. The queen can lay upto 30,000 eggs per day. Eggs are kidney shaped and light yellow in colour. The incubation period is about one week in summer. Within 6 weeks, the larvae develop to form soldiers or workers as per the requirement of the colony. The reproductive castes mature in 1-2 years. There is only one queen in the colony and normally her longevity is for 5-10 years. The king's life is much shorter than that of queen and when he dies is replaced by new one (supplementary reproductives). Only one generation is completed in a year.

Management

1. Use well decomposed organic manure. Never use raw manure or cow dung. Remove dead or decaying matter or dry stubbles from the field to avoid termite infestation.
2. It is seen that irrigation is useful to protect the crop from termite.
3. Treat the soil with quinalphos 1.5 per cent or methyl parathion 2 per cent dust @ 25 kg per ha before sowing or planting the crop or seed treatment with chlorpyriphos 20 EC @ 6 ml per kg seed or endosulfan 35 EC @ 6 ml per kg seed has also been recommended for effective protection of various crops.
4. In standing crop apply chlorpyriphos 20 EC @ 4 liter per ha with irrigation water for effective protection of crop against termite infestation.

Floricultural Crops

Rose

Red Scale (*Aonidiella aurantii*) (Hemiptera: Diaspididae)

This pest appears in severe form before and after the rainy season. Both adults and young nymphs suck the sap from stem and mature shoots and cause harm to plant vigour. Infested shoots have reddish brown encrustation, bear few small flowers and in case of severe infestation, the entire plant can become dry.

Management

1. Selection of planting material free from scale infestation.
2. Cutting and burning or burying of heavily infested shoots is very effective.
3. Soil application of Carbofuran (1.0 kg a.i.) followed by application pongamia oil (10 per cent) to infested shoots, immediately after pruning provides effective control.
4. Spray of fenthion or chlorpyriphos (0.05 per cent) is very effective.

Thrips (*Rhipiphorothrips crucentatus*) (Thysanoptera: Thripidae)

The thrips attack starts with the new flush after pruning on open cultivated rose. Blackish brown adults and reddish nymphs of thrips found underside of tender leaves and beneath sepals and petals of buds and flowers. Female lays eggs in leaf tissues, newly emerged nymphs cause damage to foliage by rasping the tissue and sucking the cell sap that oozes out.

Management

1. Spraying of acephate (0.1 per cent) is very effective in controlling this pest.
2. Raking up of soil during winter and drenching with chlorpyriphos (0.1 per cent) kills inactive stages of thrips.

Aphid (*Macrosiphum roseae*) (Homeptera: Aphididae)

This pest is active during cool and cloudy weather *i.e.*, in October- February. The blackish green coloured aphids live and feed in cluster near growing shoots. They suck cell sap resulting in yellowing and deformation of leaves, retardation of growing shoots.

Management

1. Neem or pongamia oil (1.0 per cent) sprays followed by dimethoate (0.03 per cent) gives good control of this pest in open cultivated rose.
2. Imidacloprid (0.005 per cent) or cartap hydrochloride (0.05 per cent) controls aphid damage on rose inside polyhouse.

Two Spotted Spider Mite (*Tetranychus urticae*) (Arachnida: Trombidiformes: Tetranychidae)

The mite is very active in March to May when hot and dry conditions prevail. Cream coloured nymphs and red coloured adults with two black spots are found in colonies on lower side of mature leaves covered by webs. In case of severe infestation they spread to entire plant which includes buds and flowers. The mites scrape tissues and suck the sap resulting in discoloration, yellowing, drying and dropping of attacked leaves.

Management

1. Proper ventilation, clean cultivation and frequent watering to bring down temperature keep the pest under check.
2. Cutting and burning of heavily infested shoots should be done.
3. Spray of dicofol (0.05 per cent) or wettable sulpher (0.3 per cent) controls mite problem in roses.

Foliage Caterpillar (*Achaea janata, Spodoptera litura, Euproctis fraterna*) (Lepidoptera: Noctuidae) (Lepidoptera: Lymantriidae)

Mature larvae of these caterpillar feed voraciously in groups on leaves, buds and flowers resulting in severe defoliation and flower loss.

Management

1. Collection and destruction of leaves with egg masses and early instars reduces further infestation.
2. Spray of quinalphos or methyl parathion (0.05 per cent) control pests effectively.
3. Spraying of neem oil (1 per cent) or neem seed kernel extract (4 per cent) after insecticide sprays effectively check the pest.

Bud Borer (*Helicoverpa armigera*) (Lepidoptera: Noctuidae)

This pest attacks on rose on open cultivated field. Its attack is severe from January to March. Female lays cream coloured eggs singly on young buds. Hatched larva bore into buds by making holes and feed on growing petals. Caterpillars also damage flowers by eating petals and leaving excreta.

Management

1. Collection and killing of grown up larvae reduce further damage and population.
2. Spray endosulfan (0.07 per cent) gives good control.
3. Neem seed kernel extract (4 per cent) or neem oil (1 per cent) sprays offer protection to plant from damage.
4. Application of HaNPV 250 LE/ha causes considerable mortality of early instar larvae.

Chafer Beetle (*Adoretus*) (Coleoptera: Scarabeidae)

Black or brown or metallic green coloured adult beetles emerge from soil with the onset of monsoon *i.e.*, May- June after dusk. They feed on leaves, buds, flowers and tender shoots by making irregular holes and cuts resulting in severe defoliation.

Management

1. Deep digging after pruning exposes eggs, grubs and pupae to natural enemies.
2. Application of phorate or carbofuran (1.0 kg/ha) kills grubs in the soil.
3. Spray water extract of neem seed kernel (2.5 per cent) inhibits feeding by the beetles.
4. Fortnightly spraying of chlorpyriphos (0.05 per cent) during evening protects foliage from beetle damage.

Frog-hopper (*Cercopis vulnerata*) (Hemiptera: Cicadellidae)

The frog-hopper larvae sit at the corner of the twig in a whipped cream world, sucking sap from the stem. Later, they emerge as adults and become active hopping and leaping about and feeding on the leaves, twigs or buds, characteristic of the species. The adults are small in size rarely over half an inch long, with rather large heads.

Management

1. In case, if only few plants are affected, wipe out with finger and thumb which kills the larvae (nymphs) but at commercial level this is practically not possible.
2. At commercial level, the infestation of frog-hopper can be reduced by spraying Malathion 50 EC (0.2 per cent) which if necessary can be repeated at 15 days intervals.

Chrysanthemum

Aphid (*Macrosiphoniella sanborni*) (Hemiptera: Aphididae)

The greenish black nymphs and chocolate brown adults suck cell sap from tender shoots and lower side of leaves. Continuous sap feeding results in yellowing and premature leaf fall, loss of vigor and stunted growth of the plant. The pest also excrete honey dew which attract the black sooty mould. In severely infested plants flower buds reduce in size and remain unopened.

Management

1. The aphid can be controlled by spraying of dimethoate 0.03 per cent followed by spray of neem oil 1 per cent.
2. Spray of Imodacloprid (0.03 per cent) or cartap hydrochloride (0.05 per cent) also very effective.
3. *Verticillium lacani* (2 ml/1 litre of water) a pathogenic fungus can also reduce the aphid population effectively.
4. Release of one day old *Chrysoperla carn*ea larvae at 50:1 ratio (aphid: chrysopid) reduce pest population effectively in greenhouse.

Thrips (*Microcephalothrips abdominalis, Frankliniella* sp.) (Thysanoptera: Thripidae)

More than seven species of thrips attack on chrysanthemum of which above mentioned are very common. Slender shaped nymphs and adults feed on tender leaves causing mottling, distortion and bronzed appearance. Affected flowers look discolored and scorched.

Management

1. Spray of acephate 0.1 per cent is effective in reducing the thrips population in open field condition, while for polyhouse grown chrysanthemum, spray cartap hydrochloride (0.05 per cent).
2. Raking and dranching of soil with chlorpyriphos (0.1 per cent) kill pupal stage and avoid further infeststion.
3. Covering of young buds with polythene bags with holes prior to blooming protects flowers from thrips attack.

Bud Borer (*Helicoverpa armigera*) (Lepidoptera: Noctuidae)

This pest is active during September- October on chrysanthemum. Adult females lay eggs on bracts of buds and flowers. Larvae feed on petals and damage flowers resulting in considerable loss.

Management

1. Control measures are to be initiated when eggs are noticed on young buds.
2. Spray fenvalerate 0.01 per cent to control initial infestation.
3. Neem seed kernel extract 4 per cent as alternative spray followed by endosulfan 0.07 per cent offers protection to crop from the borer damage.

Leaf Folder (*Hedylepta indicata*) (Lepidoptera: Pyralidae)

This is a serious pest of chrysanthemum grown under field conditions. Pale white eggs are laid in groups on lower side of leaves which hatch in 10 days. Green colour larvae with white stripes and brown head fold the leaves and feed on chlorophyll resulting in skeletonization and drying. The larvae occasionally also damage flowers leading to heavy loss.

Management

1. Cutting and burning of rolled leaves with larvae reduces population.
2. Spray of fenvalerate 0.01 per cent followed by methyl parathion 0.05 per cent gives effective control against this pest.

Leaf Miner (*Liriomyza trifolii*) (Diptera: Agrozymidae)

This is again a serious pest of chrysanthemum in glasshouse or polyhouse condition. This pest is active during March-June. Eggs are deposited in leaf by puncturing the tissue. Larvae make tunnels in leaves and feed on chlorophyll leaving characteristic trails of mines. Damaged leaves bear extensive mining marks which turn brown and dry.

Management

1. Removal and destruction of weeds which act as alternate hosts is very important to protect the crop from leaf miner.
2. Cutting and burning of heavily mined leaves reduce further damage.
3. Installing of yellow sticky traps coated with chlorpyriphos 0.05 per cent traps adult population.
4. Dipping of chrysanthemum cuttings in margosan 3 per cent for 2 hrs and foliar spray of the same chemical @ 0.4 per cent reduce pest population.
5. Spraying of cypermethrin 0.003 per cent at initial stage of infestation followed by triazophos 0.06 per cent offers good control of leaf miner.

Spider Mite (*Tetranychus sp.*) (Acarina: Tetranychidae)

Mite colonies are usually seen on the upper surface of leaves. Mite damage starts in March and reaches to peak during May-June. Affected leaves become discoloured, dry and fall, in case of severe damage grouing shoots dry up.

Management

1. Removing and burning of heavily infested shoots checks build up of the population.
2. Spray dicofol (0.05 per cent), followed by pongamia oil (1 per cent) spray controls mite damage.

Marigold

Bud Caterpillar (*Heliothis armigera*) (Lepidoptera: Noctuidae)

Eggs are laid singly on young buds, larvae feed on developing buds and flowers by damaging florets and petals.

Management

1. Collecting and destroying of infested buds and flowers reduce further damage.
2. Spraying of endosulfan (0.07 per cent) gives effective control.

Aphid (*Aphis gossypii*) (Homeptera: Aphididae)

Aphids infest flowers on lower side and at the base of the petals and suck the sap causing discoloration, withering and deformation of flowers.

Management

1. Spray of dimethoate (0.03 per cent) at low infestation levels gives effective control.
2. Neem or pongamia oil (1.0 per cent) cause considerable reduction in aphid population.

Mite (*Tetranychus cinnabarinus*) (Acarina: Tetranychidae)

Mite attack on marigold in August-September. Colonies of mite infest leaves and cover entire plant with webs. Both nymphs and adults stages feed on the sap resulting in discoloration, drying of leaves and plants dry in case of severe attack.

Management

1. Spray dicofol (0.03 per cent) followed by pongamia oil (1.0 per cent) spray.

Carnation

Two Spotted Spider Mite (*Tetranychus urticae*) (Acarina: Tetranychidae)

This is one of the most devastated pest of carnation under polyhouse conditions. Its infestation is severe during summer. Mite colonies are found on upper surface of leaves, beneath leaf folds and are covered by dirty webs. Affected leaves and buds become discoloured and dry. Stunted growth and drying of plants are the symptoms in case of heavy damage.

Management

1. Cutting and burning of severely infested and dried plant parts to be done.
2. Washing of plants with water dislodges the webs.
3. Spraying of triazophos (0.06 per cent) or ethion (0.05 per cent) effectively control the mites.

Thrips (*Frankliniella schultzei*) (Thysanoptera: Thripidae)

The nymphs and adults mainly attack growing buds, flowers and feed on sepals and petals causing discolouration (silver streak). Affected young buds will not open leading to considerable flower loss.

Management

1. Spraying of acephate (0.1 per cent) or imidacloprid (0.005 per cent) or cartap hydrochloride (0.05 per cent) at the initial infestation stage.
2. Drenching with chlorpyriphos (0.01 per cent) reduces further incidence by killing thrips present in the soil.

Bud Borer (*Helicoverpa armigera*) (Lepidoptera: Noctuidae)

Initiation of bud borer is observed from May to September in polyhouse. Larvae bore into growing buds by making holes in it and damage flowers by feeding developing petals and reproductive parts resulting in petal drop.

Management

1. Collecting and killing of growing up larvae by putting them in 0.1 per cent chlorpyriphos solution reduces the pest population.
2. Spraying of endosulfan (0.07 per cent) or methyl parathion (0.05 per cent) when eggs are observed on buds.

Gerbera

Whitefly (*Bemisia tabaci*) (Hemiptera: Aleyrodidae)

This pest is active during moderately hot and humid condition (February-July). Nymphs and adults suck cell sap from leaves resulting in yellowing, wilting and drying of matured leaves and curling of young leaves. In case of severe infestation sooty mould develops on leaves.

Management

1. Excessive dampness and dark conditions should be avoided.
2. Remove and burning of heavily infested leaves check pest build up.
3. Clean cultivation and use of insect proof net are essential.
4. Spraying dichlorovos (0.05 per cent) or methomyl (0.05 per cent) reduces the pest population.

Leaf Miner (*Liriomyza trifolii*) (Diptera: Agromyzidae)

Its attack starts when plants are at 3 or 4 leaf stages. Larvae feed inside leaves by leaving characteristic mines. Damaged leaves turned brown and then dry. The affected plants become stunted in growth and produce week and small flowers.

Management

1. Plucking and burning of severely mined leaves.
2. Use of yellow sticky trap coated with chloriphos (0.05 per cent) gives considerable protection and encourage parasite *Diglyphys isaea*.
3. Spraying of deltamethrin (0.0015 per cent) followed by triazophos (0.06 per cent) when infestation begins give good control.

Thrips (*Thrips palmi*) (Thysanoptera: Thripidae)

Thrips are active during September–November and March–May. Adult and nymphs suck the sap from the upper surface of leaves and petals of growing flowers. Damaged young leaves curl up while matured ones become brittle and turn bronze. Affected buds get stunted and do not open properly. Affected flowers become brown and discolored.

Management

Spraying of dimethoate (0.03 per cent) in combination with pongamia oil (1 per cent) at early stage (during August and February) and drenching of beds with chlorpyriphos (0.1 per cent) checks the pest buildup.

Aphid (*Myzus persicae*) (Homoptera: Aphididae)

Aphids mainly attack growing flowers. They colonize on lower side of petals and spread to entire flowers. Aphids suck the sap from petals leading to discoloration, withering and deformation of flowers. Development of sooty mould on honey dew excreted by the pest makes flowers look sticky with black patches.

Management

Spraying of methomyl (0.1 per cent) or imidacloprid (0.005 per cent) gives good control of aphid.

Bud Caterpillar (*Spodoptera litura*) (Lepidoptera: Noctuidae)

Larvae cause severe damage to flower buds before they open as the larvae are protected by bracts on both open and greenhouse conditions. Damage results in petal drop, reduce quality of flowers due to excreta and severe flower loss.

Management

1. Spraying of methyl parathion (0.05 per cent) or methomyl (0.05 per cent) gives good protection from this pest.
2. Installing pheromone traps after first shower are quite helpful in bringing down pest population.

Gladiolus

Thrips (*Taeniothrips simplex*) (Thysanoptera: Thripidae)

This is a very serious pest of gladiolus. This appears during cloudy weather. The nymphs and adults rasp epidermal layers and suck the plant sap resulting in silvering browning and distortion of affected leaves and buds. They also attack corms in storage. Attacked corms look sticky, shriveled and produce week plants.

Management

Spraying of acephate (0.1 per cent) or dimethoate (0.03 per cent) when the initial symptoms are noticed at two week interval.

Cutworm (*Agrotis segetum*) (Lepidoptera: Noctuidae)

This pest attack is severe in hilly areas. The larvae feed on tender parts of the plant. Clay coloured fully grown larvae hide in soil during day time and cut the plants at ground level during night. Plants are vulnerable to attack up to third leaf stage.

Management

1. Deep ploughing of soil exposes pupae to natural enemies.
2. Use of poison bait made out of rice bran, jaggery and chlorpyriphos (0.1 per cent) (10:1:0.5 Kg) in the field kills the larvae.

Tuberose

Weevil (*Mylocerus* sp.) (Coleoptera: Curculionidae)

Ash coloured adult weevils feed leaves from edges and produce characteristic holes and also damages tender shoots. Grubs feed on roots and tunnel into bulbs causing wilting of attacked plants.

Management

1. Soil treatment with malathion dust (0.1 per cent) before planting for grub control.
2. Spraying of fenthion (0.1 per cent) protects the leaves from weevil damage.
3. Neem oil (1 per cent) is effective in checking adult feeding.

Bud Caterpillar (*Heliothis armigera*) (Lepidoptera: Noctuidae)

The larvae bore into buds and fed on flowers resulting in considerable flower loss.

Management

Spraying fenvalerate (0.05 per cent) followed by endosulfan (0.07 per cent) at 10 days interval.

Orchids

Armoured Scale (*Diaspis boisduvalli*) (Hemiptera: Diaspidae)

This is a serious and destructive pest on orchids. This pest colonize on lower side of leaves, stem and shoots. They suck the plant juice and creat toxic effect on plants resulting in burning and drying of damaged plants.

Management

1. Removal of heavily infested leaves avoids spread of scales.
2. Spraying of dimethoate (0.03 per cent) or ethion (0.05 per cent) followed by a sprey of pongamia oil (1 per cent) controls the scale.

Mealy Bug (*Pseudococcus maritimus* and *Ferris virgata*) (Aemiptera: Pseudococcidae)

These two scales cause considerable damage in orchids. They suck the plant sap and cause deformation of leaves and retardation in growth of the plant. They excrete honeydew which attracts ants and sooty mould develops in case of severe infestation.

Management

1. Collection and destruction of heavily infested plant parts reduce further spread of the pest.
2. Spraying methyl parathion (0.05 per cent) or methomyl (0.05 per cent) when crawling of ants on plant are noticed, provide effective control of pest.
3. Pongamia oil (1 per cent) sprays also check mealy bug infestation.

Thrips (*Anaphorathrips orchidii*) (Thysamoptera: Thripidae)

Both nymphs and adults suck the sap from tender leaves, growing buds and flowers. Affected leaves develop brown streaks while buds and flowers are distorted and discoloured causing flower loss.

Management

Spraying of imidacloprid (0.007 per cent) at 10 days interval, when early symptoms are noticed, effectively checks thrips damage.

Snail (*Achatina fullica*) (Mollusca: Achatinidae)

Snail attack is observed during monsoon season. Adult and young stages devour the plants by feeding on leaves at night. They scrap leaves leaving only veins and cut tender shoots.

Management

1. Hand picking of grown up stages and killing them by dipping in 5 per cent salt solution reduces the pest population.
2. Application of 5 per cent metaldehyde dust on plants gives effective control of the snail.
3. Neem oil (1 per cent) or soap nut extract (6 per cent) sprays protect foliage from damage.

Medicinal Crops

Solanum (*Solanum viarum*)

Fruit Borer, *Leucinodes orbonalis* (Lepidoptera: Pyralidae) and *Helicoverpa armigera* (Lepidoptera: Noctuidae)

Female moths lay eggs on tender leaves, buds and fruits. *H. armigera* larvae feed on foliage initially and later bore into developing fruits by making holes and feed on internal contents. Early instars of *L. orbonalis* initially bore into tender shoots, resulting in wilting and drying of damaged shoots. Grown up stages bore into the fruits and feed inside.

Management

1. Collection and destruction of bored fruits reduce larval population.
2. Spraying of cypermethrin (0.005 per cent) or deltamethrin 0.0015 per cent followed by a spray of endosulfan (0.07 per cent) beginning with flowering stage.

Fruit Fly, *Dacus latifrons* (Diptera: Tephritidae)

Adult flies lay eggs inside young fruits by puncturing. Hatched maggots enter into fruit and feed on the pulp. Damaged fruits rot, turn black and drop causing severe fruit loss. Loss is high in rainy season.

Management

1. Collection and destruction of fallen fruits reduce fruit fly population.
2. Spraying of deltamethrin (0.0015 per cent) or neem oil (1 per cent) followed by dimethoate (0.03 per cent) effectively controls fruit fly damage.

Foliage Caterpillar, *Spodoptera litura* (Lepidoptera: Noctuidae) and *Spilosoma oblique* (Lepidoptera: Arctidae)

Eggs are laid in clusters on lower surface of leaves. Hatched larvae are gregarious in nature and damage leaves by scraping resulting in skeletonization and drying. Later instars feed voraciously on leaves causing severe defoliation.

Management

1. Collection of egg masses and early instar larvae and killing them effectively reduce the pest population.
2. Use of poison baits, consisting of rice bran, jaggery and chlorpyriphos (10:1:0.0) kills grown up stage.
3. Spraying of methyl parathion 0.05 per cent or acephate 0.01 per cent is highly effective.

Isabgol (*Plantago ovate*)

Aphid, *Aphis gossypii* (Homoptera: Aphididae)

Yellowish green and brown coloured aphids colonize the plants on tender shoots and lower surface of leaves. Both adults and young stage suck the sap causing curling, withering and retardation in growth of the infested plant parts.

Management

1. Spraying neem or pongamia oil (1 per cent) effectively reduces aphid population.
2. Spray dimethoate (0.03 per cent) for its effective control.

Seed Beetle, *Lasiderma serricorne* (Coleoptera: Anobiidae)

This is a serious pest in storage. Both grubs and adults cause damage by feeding on the internal contents of the seed.

Management

1. Treatment of seed with neem or pongamia oil 1 per cent protects them from damage.
2. Seed treatment with chlorpyriphos 0.05 per cent or endosulfan 0.07 per cent protects seed from damage.

Aswagandha (*Withania somnifera*)

Epilachna Beetle, *Henospilachna vigintioctopunctata* (Coleoptera: Coccinellidae)

Clusters of egg are laid on lower surface of leaves. Grubs and adults feed on leaves by scraping resulting in skeletonization, drying of leaves and defoliation of attacked plants.

Management

1. Collection and destruction of eggs, grubs and adults.
2. Spraying of malathion 0.1 per cent or fenthion 0.1 per cent protect the foliage from the beetle damage.

Mealy Bug, *Coccidohystrix insolitus* (Hemiptera: Pseudococcidae)

White and wexy coated nymphs and adults infest lower side of leaves in colonies and suck the sap. Affected leaves get deformed, turn yellow and drop off in severe cases. Honey dew excreted by bugs favours development of sooty mould which affects the plant growth adversely.

Management

1. Removal and destruction of severely infested plant parts check further spread.
2. Spraying of acephate 0.1 per cent or methomyl 0.05 per cent followed by a spray of pongamia oil 1 per cent checks the pest infestation.

Ash Weevil, *Mylocerus viridanus* (Coleoptera: Curculionidae)

Adult weevils cause damage to leaves by eating from edges in a characteristic manner resulting in severe foliage loss if populations are very high. Grubs feed on roots leading to wilting and drying of the plant.

Management

1. Tapping of the adults into insecticide solution reduces their population.
2. Spraying of fenthion 0.05 per cent protects foliage from pest attack.
3. Soil application of malathion dust or drenching of soil with chlorpyriphos 0.1 per cent kill the grubs.

Opium (*Papaver somniferum*)

Cut Worm, *Agrotis ipsilon* (Lepidoptera: Noctuidae)

Female moth lays egg on lower surface of leaves or on the moist soil. Dark brown larvae cut young plants at ground level and damage shoots of grown up plants after dusk and hide during day time, attack by the pest leads to retardation in plant growth and yield loss.

Management

1. Deep ploughing during summer exposes pupae.
2. Flooding of infested field with water dissolved with 0.1 per cent chlorpyriphos.

Capsule Borer, *Helicoverpa armigera* (Lepidoptera: Noctuidae)

Larvae fed initially on tender leaves and later bore into developing capsules by making holes.

Management

1. Spraying of fenvalerate (0.05 per cent) or endosulfan (0.07 per cent) when the larvae are noticed.
2. Spraying of neem kernel extract 5 per cent or neem oil 1 per cent also checks feeding by the larvae.

Aphid, *Myzus persicae* (Homoptera: Aphididae)

Nymph and adults colonise on growing shoots and cause damage by sucking the sap. Affected shoots get deformed, wither and retard in growth.

Management

Spray dimethoate 0.03 per cent reduces aphid population.

Sarapagandha (*Rauvolfia serpentine*)

Scale, *Saissetia coffeae* (Hemiptera: Coccidae)

Brownish black scale infests lower surface of leaves along midrib and veins. They also infest petioles and tender shoots and deplete the sap from plant. Affected leaves turn yellow and drop off while plants loose vigour and become stunted in growth.

Management

1. Removal and burning of heavily infested plant parts ring down scale population.
2. Spraying of chlorpyriphos 0.05 per cent or acephate 0.1 per cent effectively controls the scale attack.

Grasshopper, *Orthacric simulans* (Orthoptera: Acrididae)

Both young and adult stages damage the crop by making irregular holes on leaves and cutting tender shoots. Attack on young plant results in plant loss.

Management

1. Spraying of neem oil 1 per cent or neem kernel extract 4 per cent prevents feeding by the hoppers.
2. Foliar application of malathion 0.1 per cent effectively checks damage.

Leaf Folder, *Glyphodes vertumalis* (Lepidoptera: Pyralidae)

Larvae fold leaves together and feed on green tissue causing drying of leaves and stunted growth of the plant in case of severe attack.

Management

1. Collection and destruction of dried and infested plant parts reduce the pest population.
2. Spraying of profenophos or ethofenprox 0.05 per cent controls the larval damage.

(14)
Stored Grain Pests and their Management

Safe storage of agricultural products/grains is a problem in almost all the tropical countries because of high humidity and warm climate. Storage of agricultural produces particularly cereals, pulses and oilseeds for a long duration is associated with the losses in quality, quantity, viability and monitory values. The main bio-deteriorating agents are rodents, insects, birds, mites and microorganisms. Insect occupies key position among them. The government of India had constituted a committee for the assessment of storage losses in 1966 under the chairmanship of Dr. V. C. Pance. According to the interim report of the committee, the post harvest losses have been estimated to the extent of 9.33 per cent. The break up is as under:

1.	Threshing yard loss	:	1.68 per cent
2.	Transport losses	:	0.15 per cent
3.	Processing	:	0.92 per cent
4.	Storage		
5.	Rodent	:	2.50 per cent
6.	Birds	:	0.85 per cent
7.	Insects	:	2.55 per cent
8.	Moisture	:	0.68 per cent
	Total		9.33 per cent

The effect of humidity on the development of insect pests of grains and other dry products are intimately associated with that of temperature. The insects feeding on dry materials need a certain minimum level of moisture in their food.

The moisture requirement varies from species to species but practically all of them need more than 10 per cent moisture and the optimum is around 14 per cent. Therefore, grain having less than 10 per cent moisture is considered safe for storage. Micro-organism also develop at high moisture content. Likewise, temperature is one of the most important factors of the environment, as the multiplication of insect pests of stored grains depends on it. The minimum temperature at which these insects are able to develop and multiply in between 15.5 and 18.3 °C. Most of the insects can live for long periods at lower temperature but their activity is reduced very much. The optimum temperature for most of the species lies between 28.0 to 32.0 °C.

Losses are manifested in several ways by above mentioned pests like weight loss, food loss, quality loss, loss of goodwill (reputation), monetary loss and seed loss, etc.

Loss of Weight

Weight losses result from evaporation of moisture, component parts of a product being eaten by insect, rodent and birds, and from allowing quantities to spill from container in which the produce is transported or stored.

Food Loss

Weight loss during storage due to a loss of moisture is a measure of food loss. This type of loss may result from exposure of the product to extremes of temperature and humidity. During drying, processing and storage from the development of fungi or from the attack of insects, rodents and birds.

Overexposure to sun destroys certain vitamins and causes oxidation of carotene. High temperature during artificial drying cause loss of thiamine content in rice. The physical conditions prevailing during the storage of brown rice cause a reduction in thiamine content reported as amounting to 75 per cent loss over four years of storage when in bags and about 8.6 per cent loss when stored in bulk in silos.

Loss in Quality

In general quality of produce is assessed and the products are graded on the basis of appearance (uniformity of size and colour, texture, and dirt content), but smell and flavour are included as quality criteria, particularly for products like spices. In some cases chemical data such as oil content, acidity, moisture content and the presence or absence of toxins are also included.

Loss of Goodwill

There is a loss of international reputation. If produce from a particular country is regularly found to be infested, then this information spreads internationally and other countries will tend to assume contamination and produce will either not accepted at all or prior fumigation will be required.

Monetary Loss

The losses occurred in the form of weight loss, food loss, quality loss and loss of goodwill ultimately lead to monetary loss of farmers. Because they have to sell their product immediately after harvest when prices are low. If stored for a season all these losses are bound to occur.

Seed Loss

The importance of the availability of good seed is recognized by all farming communities and, irrespective of the level of farming, seed is given special attention. Despite this, consideration losses of seed occur mainly due to lack of information about the factors causing poor germination and to poor facilities for safe storage. Seed loss is apparent through reduced germination, abnormal growth of rootlets and shoots and reduced vigor of the plant.

According to U.N. report published in 1975 for India stated that up to 10 million tons of grains are lost annually through faulty storage. This 10 mt. was 10 per cent of food grain production during 1974 and twice the amount, which India imported in 1974. This quantity of food grains was more than enough to meet the world food deficit during the year, which was 7.5 mt. India is having about 143 million hectares cultivable land and annual food grain production of 210 mt achieved in 2002-03. About 65-70 per cent of total food grains is stored by the farmers in traditional bulk and bag storages for food, feed and seed purposes, till the next harvest. Remaining 30-35 per cent is procured and stored by public agencies like Food Corporation of India (FCI), Central Ware Housing Corporation (CWC), State Ware Housing Corporation (SWCS), Co-operatives, rural godowns and other state Govt. agencies. At present, Food Corporation of India is having a buffer stock of 42 million tones of food grains. Looking to data of last years food grain production (210 mt.) nearly 10 per cent of this is lost every year because of damage caused by insect pests and faulty storages, which comes to 21 mt. Even if 50 per cent of food grains is saved by adopting safe storage practices than 10.5 mt. food grains could be saved annually which could either be used domestically or export outside country to earn foreign exchange which really gives a boon to Indian Economy.

Therefore, the first step is to save what has been grown and harvested for human beings by protection against pests and by improving local processing methods to retain inherent natural nutritional value of produce. What man has reaped must be protected at the same time, further yields must be improved. Managing stored grain and to check post harvest losses, wisely with minimum loss and maintaining its quality is a major concern at all stages. How much, where and when to exercise this control depends largely on socio-economic conditions of farmers in the country.

Insect Pests of Stored Foodgrains

Rice Weevil (*Sitophilus* species)

Family-Curclionidae

Marks of Identification

They have a characteristic snout or rostrum which projects from the front of the head. The beetles biting mouthparts are located at the tip of the rostrum, and a pair of elbowed, clubbed antennae are located at the base.

Nature of Damage

Damage is done by adult and grub. Grains are hollowed out. Kernels are reduced to mere powder. The pest infests grains in field too. Adults cut circular holes. Heating takes place during heavy infestation, which is known as dry heating.

The three major *Sitophilus* species that are significant pests of stored grain are *S. oryzae* the rice weevil, *S. zeamais* the maize weevil, and *S.granarius* the granary weevil. In appearance, *S. oryzae* and *S. zeamais* are identical with distinct yellow to orange coloured spots on their wing covers. *S. granarius* lacks spots on its elytra and is a rich chocolate brown colour. Adult beetles are 2.5 to 3 mm long. *S. granarius* tends to be the larger of the three species. These three beetles are major pests of stored grain while *S. oryzae* and *S. granarius* may be found attacking all major cereals–*S. zeamais* is most often recorded in maize but is capable of infesting and developing on other cereals.

Biology

Eggs

The adult female bores a hole in the grain surface using its biting mouthparts. One egg is laid in this hole which is then plugged with a jelly substance sealing the egg inside the grain. The number of eggs laid by a female during its lifespan may vary from 100 to 450.

Larvae

After hatching, the small legless larva feeds on the endosperm of the grain. Larvae are relatively immobile and pass through a number of instar stages gradually increasing in size. Larval stage lasts for about 20 days.

Pupae

When fully grown, the larva passes into an inactive pupal stage and gradually assumes the shape and form of the adult. Pupal period is of 3-6 days.

Adult

Adult lives for 3-5 months and active feeder and are strong fliers. Infestations can cause considerable grain heating. There are 6-8 generations in a year.

Lesser Grain Borer (*Rhyzopertha dominica*)

Family-Bostrychidae

Marks of Identification

The adult beetle is small, between 2.5 to 3 mm long, has a distinctive cylindrical shape, and is dark brown to black. The head, tucked underneath the thorax, is invisible when viewed from above.

Nature of Damage

Adult comes out from the grain leaving a irregular hole. In bagged storage, irregular messy waste flour spots indicates infestation of this pest. Heating is very common. Localized infestation is almost a rule. Both adult and larvae cause damage and are voracious feeders. As such grain kernels are reduced to mere shells. The damaged kernels remain engulfed in a film of waste flour.

Biology

Eggs

Laid in clusters as females actively bore through grains. Eggs are laid outside the kernel and young larvae bore into the grain to complete their development. Under optimum conditions the female lays up to 500 eggs during its lifespan. Eggs laid on stored commodities at moisture levels as low as 8 per cent can still hatch.

Larvae

The cream coloured larvae have biting mouthparts and three pairs of legs. The young larvae are mobile but become immobile as they complete their development concealed within grain or flour. The larvae normally pass through four instars during which their size increases. All larvae have usually bored into grain (or a suitable hard substrate) by the third instar.

Pupae

The mature fourth instar enters into an inactive pupal stage within the grain and gradually assumes the form of the adult.

Adult

When the pupal stage is completed, the newly formed adult emerges from the grain by chewing through the outer grain layers. Entire life cycle takes 25 days under optimum condition. The adult beetle is long lived and is a strong flier when conditions are warm. They are adept burrowers and produce large quantities of flour. They are also capable of chewing their way through many types of packaging materials including jute, waxed paper and some polyethylene films. There are 6-7 generations in a year.

Saw-toothed Grain Beetle (*Oryzaephilus surinamensis*)

Family–Silvanidae

Marks of Identification

The adult beetle is 2.5 to 3 mm long and can be readily identified from other beetle pests of stored products by the distinctive shape of its thorax–the section between the head and the body. It has six jagged, saw-toothed projections along the outer edges of the thorax. They are very active and their habit of infesting cracks and crevices makes them difficult to detect when inspecting buildings for their presence.

Nature of Damage

Adult and larvae cause roughening of grain surface and off odour in grain. Grains with higher percentage of brokens, dockage and foreign matter sustain heavy infestation, which leads to heating of grain.

Biology

Eggs

Laid at loosely in crevices in the commodity. An average 375 eggs are laid during the life of an adult female. Incubation period is of 3-17 days, limited to 5 days at higher temperature.

Larvae

The larvae pass through four instars during their development. They are tolerant to dry conditions and development has been recorded at relative humidity as low as 10 per cent. Larvae are mobile and not concealed. Larval period last for 14-20 days.

Pupa

The pupal stage is formed within a cocoon spun by the mature larvae, and broken grains or other food particles may be used in its construction. Pre-pupal and pupal period lasts for 7-21 days.

Adult

Long lived live upto 3 years, feeds, flies, and will rapidly walk long distances. They can easily enter packaged food, and prefer to live in cracks and crevices. Its multiplication is quick in rainy season and coastal areas.

Flour Beetles

Red Rust Flour Beetle (*Tribolium castaneum*)
Confused Flour Beetle (*T. confusum*)
Family–Tenebrionidae
Marks of Identification

The rust-red flour beetle is between 3 to 4 mm long, flattened, reddish brown in colour and parallel sided. Eyes are crescent-shaped. The confused flour beetle is slightly larger, 4 to 4.5 mm long. Although similar in appearance, the confused flour beetle does not have a distinct club formed by the last three segments of each antenna, it has a distinct ridge above each eye, and the eyes are set further apart when viewed from underneath. Larvae are elongate, light brown. The beetle is a major pest of stored products, especially grain and milled cereal products.

Nature of Damage

Both adults and larvae feed on milled products. Floor beetles are secondary pests of all grains and primary pests of flour and other milled products. In grains, embryo or germ portion is preferred.The confused flour beetle prefers more finely divided commodities than the rust-red flour beetle.

Biology

Eggs

Laid at random in the commodity and more than 1,000 eggs may be produced by a female during its lifespan. Incubation period is of 5-12 days depending upon temperature.

Plate 14.1: Rice Weevil–
Sitophillus **spp.**

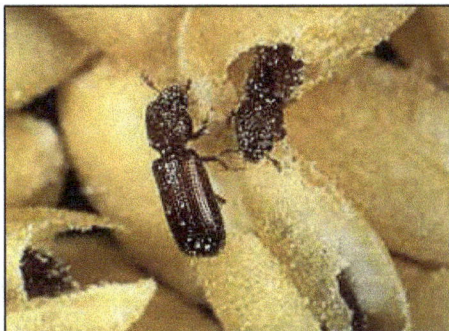

Plate 14.2: Lesser Grain Borer–
Rhyzopertha dominica

Plate 14.3: Long-headed Flour Beetle–
Latheticus oryzae

Plate 14.4: Pulse Beetle–
Callosobruchus **spp.**

Plate 14.5: Drug Store Beetle–
Stegobium paniceum

Plate 14.6: Cigarette Beetle–
Lasioderma serricorne

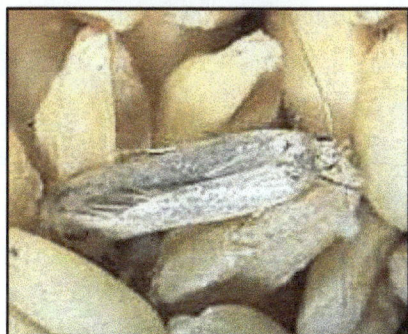

Plate 14.7: Angoumois Grain Moth–
Sitotroga cerealella

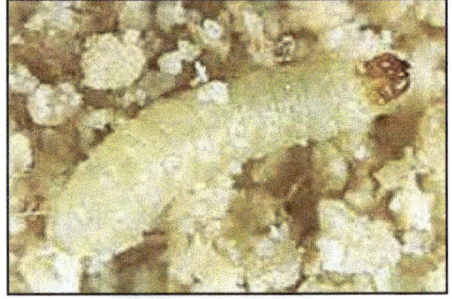

Plate 14.8: Indian Meal Moth–*Plodia interpunctella*

Plate 14.9: Adult Beetle and Larvae of the Rust-red Flour Beetle *Tribolium castaneum*

Plate 14.10: Saw-toothed Grain Beetle–*Oryzaephilus surinamensis*

Plate 14.11: Warehouse Moths–
Ephestia species

Larvae

Light brown colour, found free living and mobile in stored products. The larva has a pair of legs on each of the three segments immediately behind the head. Their size varies widely with age. The larva passes through several stages or instars during which it's size increases. The larval period is 3-12 weeks which is influenced by environmental conditions.

Pupae

When fully grown the larva enters into an active pupal stage and gradually changes into the form of the adult beetle. Pupal period lasts 5-9 days.

Adult

The newly emerged adults of both species are creamy white coloured but quickly darken to their typical red-brown to brown colour. Adults are long lived, actively feed, and heavy infestations often leave an unacceptable taint. Life cycle is completed in 4-5 weeeks. There are many generations in a year.

Khapra Beetle (*Trogoderma granarium*)

Family–Dermistidae

Marks of Identification

The beetle is small, oval, pale-red brown or black in colour. The male is almost half the size of the female. The antennae are small and clavate type.

Nature of Damage

It is a primary pest. It damages the grain starting with germ portion, surface scratching and devouring the grain. Actually it reduces grain into frass. Damage is confined to peripheral layer of bags or 30-45 cms. in bulk storage.

Biology

Eggs

The eggs are laid singly among the grains. The eggs are round, semi-transparent, white in colour. The eggs hatch in 6-16 days.

Larvae

The newly hatched larvae is yellowish white in colour later on turns brownish followed by reddish brown. The larval period is of 19-37 days. They moult 4 times.

Pupae

Pupation takes place in the last larval skin among the grains or in cocoon. Pupal period lasts for 4 to 27 days.

Adult

The adult beetles may survive for 25 to 35 days. The life cycle is completed under favourable condition in 34 to 76 days. There are 22-24 generations in a year.

Pulse Beetle (*Callosobruchus* spp.)

Family–Bruchidae

Marks of Identification

The adult beetles are small, squat with long conspicuous serrate antennae, colour is brownish grey with characteristic elevated ivory like spots near the middle of the dorsal side. Elytra do not cover the abdomen completely.

Nature of Damage

Larvae eat up the grain kernel and make a cavity. Adults are harmless and so not feed on stored produce at all.

Biology

Eggs

Female lays 80-100 egs laid singly, glued to the surface of the grain and pods (in field). Incubation period is of 5 days.

Larvae

On hatching, bore into seed, making round, translucent 'windows' in seed before pupation. Larval stage lasts for 15-50 days.

Adult

Emerges through 'window' leaving neat round hole. Short lived, does not feed on seed, runs quickly and flies very well. The life cycle completes in 23 days. There are 7-8 generations in a year.

Long-headed Flour Beetle (*Latheticus oryzae*)

Family–Tenebrionidae

Marks of Identification

The adult beetle is relatively small (2.5 to 3 mm), slender, oval shaped, and has a distinctive yellowish-brown colour. The head protrudes well past the front eyes, and the antennae are short and thickened, with a five segmented club.

Nature of Damage

Secondary pest of foodgrains and is important pest of milled products. Both adults and larvae feed on stored products.

Biology

Eggs

The female adult lays about 300 eggs singly during her life span. The eggs which are sticky when laid are usually covered in flour particles. Incubation period is of 7-12 days.

Larvae

The free living cylindrical larvae are white in colour and pass through 6 to 7 instars. Larval period is of 15-80 days depending upon temperature.

Pupae

The pupae are white and are found naked amongst the foodstuff. Pupal period is of 5-10 days.

Adults

The slow moving adult is long-lived surviving up to 6 months. Life cycle is completed in 25 days at 35 °C and 70 per cent R.H.

Cigarette Beetle (*Lasioderma serricorne*)

Family–Anobiidae

Marks of Identification

A small beetle 2 to 2.5 mm long, oval in shape and brown to dark brown in colour. It resembles the drug store beetle (*Stegobium paniceum*), however segments of each antenna are uniform in length and shape and the wing covers or elytra are not striated. The larvae is white in colour and short and stout in shape.

Nature of Damage

The tobacco or cigarette beetle attacks a wide range of stored products throughout tropical and sub-tropical climates and is particularly damaging to stored tobacco, cigars, cigarettes and cocoa beans.

Biology

Eggs

The eggs are laid singularly in crevices or folds in the substrate. The adult female produces about 100 eggs in its short life-span of 25 to 30 days.

Larvae

The white, fleshy larvae are distinctly hairy and pass through 4 to 6 instars. Newly hatched larvae cannot attack whole grain. The larvae are mobile and readily enter small opening of packaged foodstuffs in search of food.

Pupae

When mature, the larve pupate within a thin cocoon built amongst the substrate. The adult spends some days within the cocoon before emerging from and boring through packaged materials.

Adult

In daylight, the adult beetle shuns light hiding in cracks and crevices. In warm conditions, it flies readily at dusk and is attracted to artificial light. The species is not cold-hardy and can overwinter in temperate climates.

Drug Store Beetle (*Stegobium paniceum*)

Family–Anobiidae

Marks of Identification

A small beetle 2 to 3.5 mm long, cylindical and light brown in colour. It resembles the cigarette beetle (*Lasioderma serricorne*), but is readily identified by the distinctive elongated shape of the last three segments at the end of each antenna, and the wing covers or elytra are striated. Larvae is white in colour and short and stout in shape.

Nature of Damage

The drug store beetle is a cosmopolitan pest that can infest almost any dry animal or plant material. It is particularly noted as a pest of pharmacies where it infests drugs, and domestic premises attacking breakfast foods, biscuits and herbs, amongst its catholic diet.

Biology

Eggs
The female lays about 75 eggs randomly amongst stored foodstuffs during its life-span of 40 to 90 days.

Larvae
The larva progresses through four to six instars.

Pupae
The last instar larvae constructs a cocoon within which it pupates.

Adult
The adult does not feed and fly.

Warehouse Moths (*Ephestia* species)

Family–Phycitidae

Marks of Identification

The adults are short lived and do not feed. They fly well and are active at dawn and dusk. Wings grey with vague darker markings when alive. The larvae (caterpillars) are light pink with a small black spot at the base of each hair. The black pigment at the base of each hair is a key feature in identifying *Ephestia* species.

Nature of Damage

They are important pests of flour mills, food-processing plants and in dried fruit. Silk webbing produced by caterpillars can stop the product flowing freely and make it hard to handle.

Biology

Egg
Female lays 200 eggs on the jute bags. Incubation period is of 2 to 17 days.

Larva
The larvae is grey white in colour. They move freely near the produce, feeding until mature. One larva is observed to consume germ of as many as 64 wheat kernels in its life cycle. The larval period lasts for 20 to 35 days.

Pupa
A silken cocoon for pupation is spun by the mature larvae. The pupal period lasts for 10 days.

Adult

The adult lives for 12-14 days. The life cycle is completed in 3 to 6 weeks.

Indian Meal Moth (*Plodia interpunctella*)

Family–Phycitidae

Marks of Identification

The adult insect has cream and brown wings when the insect is alive. The larvae (caterpillars) are creamy white with no small black spot at the base of each hair.

Nature of Damage

They are important pests of flour mills, food processing plants and in dried fruit. Silk webbing produced by caterpillars can stop the product flowing freely and make it hard to handle.

Biology

Egg

Eggs are laid either singly or in group. A single female lays about 200 to 300 eggs.

Larva

The larvae are dirty white in colour. They fed on embryo portion. Larval period lasts for 20 to 35 days.

Pupa

The fully grown larvae make silken cocoon. Pupal period lasts for 10 days.

Adult

The adult moths are pale yellow with red brown wings. The adult live for 12-13 days. The life cycle is completed in 3 to 6 weeks.

Angoumois Grain Moth (*Sitotroga cerealella*)

Family–Gelechiidae

Marks of Identifications

Wings pale greyish brown with black spot towards tip (in fresh specimens) and are smaller in size than other storage moths.

Nature of Damage

This moth is a pest of whole cereal grain and will attack grain before harvest, particularly maize. Damage to grain in storage only occurs in the surface layer since adults are unable to penetrate deeply.

Biology

Eggs

Laid on grain surface. Incubation period being 4 to 30 days.

Larvae

Immobile, develop concealed within a single grain. Larva under goes 4 instars and larval period lasts 12-23 days.

Pupa

Larvae spins silken cocoon inside the grain. Pupae are brown in colour. Pupal period lasts for 7 to 10 days.

Adult

On hatching, pupal case is often left protruding from grain. Adult moth is short lived, does not feed, flies. There are 4-5 generations in a year.

Rice Moth (*Corcyra cephalonica*)

Family–Phycitidae

Marks of Identification

The adult moth is about 15 mm long and pale yellowish brown in colour, with faint dark lines along the length of the wings. The young larvae is creamy white in colour.

Nature of Damage

Larvae is only responsible for damage. It pollutes foodgrains with frass, moults and dense webbing. In case of whole grains, kernels are bound into lumps upto 2 kg.

Biology

Eggs

Laid on gunny bags, walls and among grains. A female lays 100 to 200 eggs. Incubation period is of 7 days.

Larvae

The young larvae feeds on broken grains. The larval period is of 21 to 60 days.

Pupa

Fully mature larvae pupate in silken cocoon in grain. The cocoon is white. Pupal period lasts for about 10 days.

Adult

Adult is biggest among foodgrain infesting moths. Antennae do not cross over wings while insect is at rest. Adult lives for 7-15 days.

Control of Pests of Stored Grain

Following are the important methods which can help in safe storage of foodgrains particularly at farmer's level:

1. Preventive measures
2. Curative methods

1. Preventive Measures

"Prevention is better than cure", hence the following preventive measures are recommended:

(*i*) Hygiene or Sanitation

(a) Threshing floor/yard should be clean, free from insect infestation and away from the vicinity of village/granaries.

(b) Clean the harvesting and threshing machines before their use.

(c) Trucks, trollies or bullock carts which are used for transportation of foodgrain should be made free from insect infestation.

(d) Clean the storage structures/godowns before storage of newly harvested crop.

(e) All dirt, dust, rubbish, sweepings and webbings, etc. should be removed from the stores and dumped or destroyed.

(f) All the cracks, crevices, holes existing in the floors, walls, ceiling should be plastered with mud or cement permanently.

(g) All the rat burrows should be closed with the mixture of broken glass pieces and mud and then plastered with mud or cement.

(h) White wash the stores before storage of foodgrains.

(i) Foodgrains should be kept in stores which are insect and moisture proof.

(j) Proper stacking of bags also help in grain protection.

(*ii*) Disinfestation of Stores/Receptacles

Before the use of receptacles or stores, they should be disinfested with appropriate residual insecticides preferably by spraying malathion 50 per cent EC with dilution 1:100 and applied at the rate of 3 lit/100 sq m.

(*iii*) Legal Method

Entry of an insect which is not found in the particular area can be prevented by imposing of plant quarantine regulations under Destructive Insect Pests Act 1914.

2. Curative Measures

The infestation of stored grain insect pest can be controlled by the following methods-

(*i*) Non chemical control measures

(*ii*) Chemical control measures

(i) Non-chemical Control Measure

The measures where chemicals are not used for the control of insect pests of stored grains are:

(a) Ecological Control Measures

The safety of stored grains from insect pests largely depends on proper management of three factors *viz.*, temperature, moisture and availability of oxygen. All these factors are required for rapid development and multiplication of insects. Hence, they have to be properly manipulated through design and construction of storage structure or godowns and storage practices so as to create ecological conditions unfavorable for attack by insects.

Temperature ranging from 20 °C to 40 °C accelerate the development of insect but above 42 °C and below 15 °C retards reproduction and development, while prolonged temperature above 45 °C and below 10 °C may kill the insects. Heating of grains to 50 °C will be lethal to insects but it is not advisable, because the grains are affected and lose their viability. Moisture is the critical factor in safe storage of foodgrains. Grains stored at around 10 per cent moisture content escape from the attack of insect (except khapra beetle). In storage oxygen is consumed by grains and insects during respiration and carbon di oxide is produced. Insect respires at the rate of 20,000 to 1,30,000 times than that of the same weight of the grains. Thus, oxygen level will reduce below 1 per cent and carbon di oxide level will automatically increases which will be lethal to all the stages of insects.

(b) Mechanical Control Measures

Amongst other methods mechanical methods are quite practicable, they are as under:

1. *Screening of grains*: Broken and cracked grains promotes the attack of stored grain pests. Hence, screening or sieving out of such grains reduce the insect infestation and gives a better appearance to the lot. Following steps make screening more effective:

   (*i*) Screening should be done regularly and away from the stores to avoid any reinfestation.

   (*ii*) Immediate destruction of screenings.

   (*iii*) Bags used for carrying the screenings should not be used again unless disinfested.

2. *Use of edible oils*: Coconut, groundnut or mustard oil at the rate of 0.25-0.5 per cent are effective for the protection of pulses against pulse beetle upto 6 moths.

(c) Cultural Control Measures

1. *Splitting of pulses*: Pulse beetle prefer to attack whole pulses and not split one. Split pulses escape from the attack and as such most suitable for storage whole pulses.

(d) Engineering Control Measures

1. *Airtight storage (Hermetric)*: Foodgrains kept in airtight sealed structure remain insect free, reason being that during storage the grain respire and liberates carbon di oxide. Excessive carbon di oxide accumulation and depletion of oxygen do not allow the insects to survive.

2. *Drying of grain*: In grain having moisture content below 10 per cent, most of the insect species do not survive or unable to multiply. In villages, the grains are dried by spreading a thin layer of grains under the sun or by dryers. However, it should be kept in mind that too rapid drying or over drying may damage seeds, develop wrinkles, cause cracking of seed coat and destroy vitamins and seed viability.

(*ii*) Chemical Control Measures

Amongst the present methods of insect control, chemical control is the most popular and perhaps most effective one. The chemical control measures include following methods:

(*a*) Prophylactic treatment

(*b*) Curative treatment

(a) Prophylactic Treatment

To prevent the insect infestation and cross infestation, it is done by spraying the insecticides. These insecticides should not, however, be sprayed on foodgrains directly. Some of the insecticides and their concentrations recommended for prophylactic treatments of stored grains are given.

(b) Curative Treatment

In spite of regular prophylactic treatment infestation develops due to hidden infestation or cross infestation, etc. The following insecticides and fungicides are used.

(*i*) *Knockdown chemicals:* Those insecticides which are capable of immobilizing or killing insects quickly. They are usually aimed against flying insects but may kill the insects on surface as well as in cracks and crevices. Such chemicals are pyrethrum spray, or fumigants.

(*ii*) *Grain protectants*: Mixing of chemical dusts with the foodgrains meant for human consumption is not recommended. However, they can be mixed with the grains meant for seed purposes only. Pyrethrum dust, DDT or BHC dusts are commonly used as seed protectants.

(*iii*) *Fumigants*: A fumigant is a chemical which at required temperature and pressure can exist in the gaseous state in sufficient concentrations to be lethal to a given pest organism. Some of the effective fumigants are as:

(*a*) *Ethylene di bromide (EDB)*: EDB is one of the most toxic chemicals to insect damaging stored foodgrains. The use of EDB, filled in ampoules has been found suitable for fumigation of food grains stored in household structures, holding small quantities, which can be made reasonably airtight. Ampoules are available in size of 3, 6 and 10 ml. EDB can be used on all foodgrains but not on milled products, oilseeds and moist grains.

(b) *Aluminium phosphide*: This is a solid fumigant available in the market in the form of tablet of 3 gm each packed in sealed tubes. The active ingredient is phosphine gas which is an acute toxic poison. When the tablet comes in contact with the moisture of the atmosphere, it liberates ammonia, carbon di oxide and phosphine. Phosphine gas being highly toxic to the pests of foodgrains.

(15)
Beneficial Insects

Beekeeping or Apiculture

Apiculture is derived from the honeybee's Latin name *Apis mellifera*, meaning 'honey gatherer'. Since bees do not collect honey but nectar from which honey is made, the scientific name should actually be *Apis mellifica* meaning 'honey maker'. Some 90 million years ago, flowering plants first appeared on earth. The wasp-like ancestors of bees took advantage of the food made available by flowers and began to modify their diet and physical characteristics. Since then, flowering plants and bees co-evolved. This eventually led to a complete interdependence, meaning that flowering plants and bees cannot live and reproduce without each other.

Bee keeping has been practiced in India since time immemorial. The earliest references date back to Vedas and Ramayana. But, scientific bee keeping with the help of simple machine is a comparatively new venture. The western method of frame-hive was first introduced in Bengal in 1882 and in the Punjab in 1883-84.

Types of Honey Bee

Honey bee belongs to the class Insecta, order hymenoptera and family apidae. These are the honey bee species:

1. Rock bee–*Apis dorsata*
2. Little bee–*Apis florea*
3. Indian bee–*Apiscerana indica*
4. European bee–*Apis mellifera*

Out of these five types, three are common in India. They are *Apis dorsata, Apis florea* and *Apis indica*.

Rock Bee (*Apis dorsata*)

It is commonly called as rock giant honey. This is the largest Indian species with an average size of about 20 m.m. It builds large comb (0.90 x 15 metres) on tree branches, under caves, or under roofs of high buildings. They are migratory species as during June and July they swarm to the hills, but in winter come back to the plains. They have been found up to the height of 3,500 feet above sea level. This species is yet to be successfully hived. This species is the highest honey gather (average 15 kg. per colony per year). Sometimes, the yield exceeds 30 kg. per colony per year. This bee is notorious for its ferocity and tendency to make unprovoked, sometimes fatal, mass attack on persons who approach its hive.

Little Bee (*Apis florea*)

It is commonly called as little bee. It is a miniature of the rock bee. It is a plain species and rarely occurs above 1000 feet of sea level. It builds small comb (about 15.24 cms) on the branches of trees, or in bushes, or under the wall of the buildings. The yield of honey from this type is very little (few ounces per colony per year), and the production does not compensate the labour undergone on it.

Indian Bee (*Apis cerana indica*)

Popularly known as Indian bee, it is commonly found in the plains and forests of India. There are several regional strains of it, of which plain, transitional and hill varieties are the three recognized types. *Picea* strain is found in hills at an altitude up to 7,000 ft, *Pironi* is distributed along the transition between an altitude of 3,000 to 4,000 ft, whereas, *Lighter indica* is a plain strain found up to an altitude of 1,500 ft. It builds several parallel combs (about 30 cms across) in protected places like hollow of trees, caves, in rocks and in other such cavities. Due to their mild nature and average output of honey between 3 kg. to 5 kg. per colony per year, they are amongst the best of the Indian variety to be hived in artificial conditions.

European Bee (*Apis mellifera*)

This is very common all over the Europe. This bee is similar to *Apis indica* in its habitats. There are several varieties and strains of this bee, amongst them the Italian variety is the best. It yields an average of 35 to 40 kg. of honey per year per colony. Attempts to domesticate this be in India or large scale has yet not been proved to be a success.

Castes of Honey Bee

Honey bee is a social insect. The nest of the honey bee is known as the bee-hive. A hive in summer consists of 32 to 50 thousand individuals, depending on the locality and food availability. There are three types of individuals in a colony, *viz.*, the Queen, worker and drone. Due to the existence of several morphological forms, bees are said to be a polymorphic species. All these three castes depend on each other for their existence.

Queen

It is a diploid, fertile female. The presence of queen in a colony is a must. The size of the body of queen is much larger than other castes of bees of the colony. Her legs are strong for she is always walking about on the comb. The queen has a sting curved like a scimitar at the tip of the abdomen, which in fact is a modification of ovipositor. The sting serves as an organ of defence but never uses it against anybody except her own caste. The queen is responsible for laying eggs for a colony. She lays about 1000 to 1500 eggs everyday and lives a life of two three years. However, the number of eggs laid per day may vary from individual to individual, and it has been found that a queen may produce as much as 6,000 eggs per day. Queen lays both fertilized eggs (from which females develop) and unfertilized eggs (from which males develop).

Worker

It is a diploid, sterile female. The size of a worker is the smallest among the castes but they constitute the majority of the bees in a colony. Their function is to collect honey, to look after young ones, to clean the comb, to defend the hive and to maintain the temperature of the hive. Numerous adaptations have occurred in the worker for performing various duties. The body is covered with branched hairs so that when a bee visits a flower, pollen grains adhere to the hairs and other parts of the body. The worker cleans off pollen grains with special structures the antenna cleaners on each foreleg, pollen brushes on all legs and pollen combs on hind legs. All pollen are stored in the pollen basket present on the outer surface of tibiae on hind legs. Water and nectar are gathered by sucking type mouthparts which are modifications of the maxillae and labium. Workers have a well developed sting at the tip of the abdomen which is a modified ovipositor.

Drone

It is haploid, fertile male. The males are larger than workers and are quite noisy. They are unable to gather food, but eat voraciously. They are stingless and their sole function is to fertilize the female (queen). The number of drones in a colony varies from 200-300, (1 per cent of the colony). The drone develops parthenogenetically from unfertilized eggs.

Life Cycle of Honey Bee

Each hive contains one queen, several drones and innumerable workers. During nuptial flight, many drones follow the queen. The drone which successfully copulates with queen looses its copulatory apparatus and ultimately dies. The sperms get stored in the spermathecae of the queen which after returning to the hive start laying eggs. The number of eggs laid may exceed 2,000 per day but it depends largely upon temperature and availability of food. A queen can lay both fertilized and unfertilized eggs. The entire process of egg laying is believed to be under voluntary control of the queen.

Eggs are small pearly white and spindle shaped. Unfertilized eggs are laid in much spacious drone cells where they develop into drones Larvae hatchout from the eggs after about three days of egg laying. The unfertilized eggs develop parthenogenetically into male and fertilized eggs develop into females. During first 2-3 days all larvae are fed on a special food the 'Royal Jelly' secreted by the hypopharyngeal glands of the young workers. After that coarser food, the 'Bee Bread', which is a mixture of honey and pollen grain is given. However, the queen forming larvae are fed on royal jelly for full larval life and they are taken for further development into a special chamber called the queen's chamber. Differential food types results into difference in size of the developing larvae. The larvae moults several times and on the seventh day enter the quiescent period of pupation. The worker bees cover the cell with a thin layer of wax. Inside the cell larva spins a delicate silken cocoon around itself and turns into a pupa. The adult comes out by cutting the wall of cocoon first and secondly by breaking the wax covering.

Table: Honey Bee Cycle

	Egg	Larva	Pupa	Total
Queen	3 days	5-1/2 days	7-1/2 days	16 days
Worker	3 days	6 days	12 days	21 days
Drone	3 days	6-1/2 days	14-1/2 days	24 days

Worker bees after coming out of the pupal case starts working and their duties changes with the advancing age. Collection of nectar and pollen from flowers is done in the last phase of their life, *i.e.* after 25th day of their birth.

Architecture of a Hive

The highest degree of nest construction among insects is found in bees. The structure of the hive is unsurpassed and unparalleled in the animal kingdom. The hive and comb of the bees are formed mainly by workers. A comb is a vertical sheet of wax, composed of a double layer of hexagonal cells projecting in both directions from central wax-sheet.

Comb hangs vertically downward, while cells are horizontal in position. The hexagonal shape of cells accumulates maximum space in minimum use of wax and labour. The wax used in building of a comb is secreted from the wax glands present in the abdomen of worker bees. This wax has the highest melting point *i.e.*, 140°F. Before use, the wax is masticated and mixed with secretions of the cephalic glands to convert into a plastic substance. The resinous substance called "propolis", prepared from pollen, is used in making the comb water-proof, and it also helps in filling the cracks and crevices in the hive.

The cells of the comb are of various types. The 'Storage cells', which contains honey and pollen are generally built on the margin and at the top of the comb. The 'brood cells', which contains the young stages are built in the center and the lower part of the comb. Brood chamber is further divided into three types, namely Worker-chamber, where developing workers are reared; Drone-chamber, where developing drones are reared and the Queen-chamber, which is larger than other and where the larvae developing into queens are reared. There is no special chamber for adults. They move on the surface of the comb.

Essentials for Starting a Beekeeping

    1. A beginner should learn in detail about the habit and behaviour of the bees.

2. To procure the hive and other tools related to bee-keeping. The khadi and village industries commission is providing all sorts of assistance including monetary grant and technical knowhow.

3. The hive should be placed in a locality rich in vegetation especially the flowering plant. If several hives are kept in a plot, the distance between two hives should be atleast six feet.

4. The hive should face East. It should receive sunlight during morning and evening and some shade during mid-day.

5. Water should be available nearby and an open space infront of hive entrance is necessary.

6. It is advisable and economical to collect bees from a particular locality.

7. Spring is the best season for starting beekeeping as during this season swarming occurs and bees can be easily procured to be lived.

Beekeeping Equipments

Bee Hive

The modern bee hive is based on certain principles and is called "movable frame hive". Hive is made up of wooden box. The box may be single walled or double walled. Single walled hive is light and cheap so it is more common, whereas the double walled hive is costly and heavy, it is however, more durable and provides better protection to the bees. The insulation provided by a single walled hive is not enough to tolerate the fluctuations of atmospheric changes outside. Double walled hive with sufficient insulation keeps the hive warm in winter and cool in summer.

Modern hive has a basal plate or bottom board on which is placed a wood box called the "Brood chamber" A bottom pore in brood chamber act as bees entrance. Inside the brood chamber, several frames hang vertically from the top. These fames can be removed independently, that's why the modern hive is called as movable frame hive. The distance between two frames is known as "bee space". This space serves as a passage for the movement of the bees and at the sametime, it is very important because it is the space in which bee should not form a comb. If the bee space is kept less than the correct size, the bees join up the two combs and if the space is more than the correct size, bee forms an independent comb in between two frames. In both cases, it is impossible to remove the frames independently. There is another similar chamber above the brood chamber called as "Super". This chamber is meant for the storage of honey only and here queen is never allowed to enter. To prevent the entrance of queen into super, queen excluder is used between brood and super chambers. On the top of the super there is an inner covering and then a roof.

Different types of hive are being used in different parts of the world. In India generally, three types of bee-hive namely Langstroth, Newton and Jeolikote are in practice. The detail of the size and number of the frames used in different types of hive are given in tabular form below:

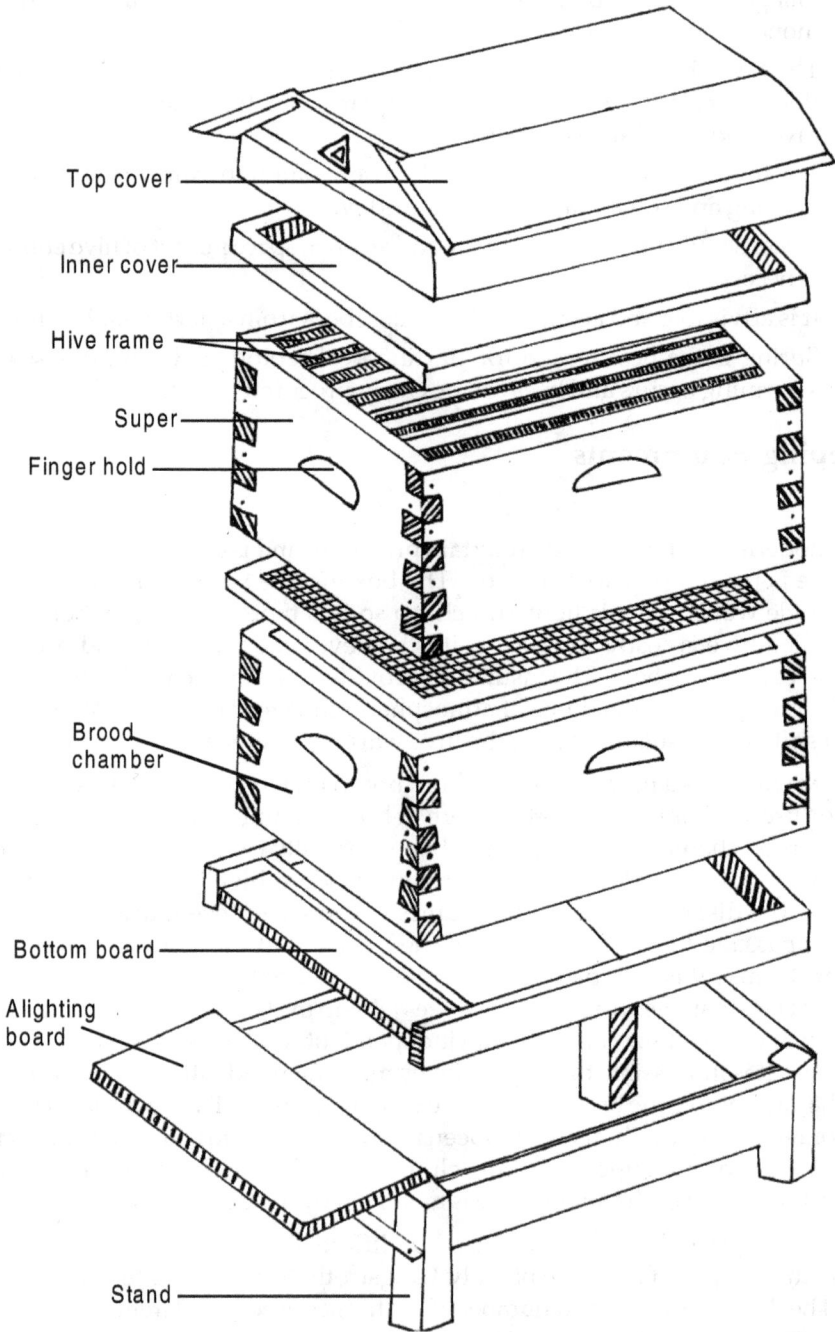

Top cover

Inner cover

Hive frame

Super

Finger hold

Brood chamber

Bottom board

Alighting board

Stand

Figure 15.1: Langstoth Ten-frame Hive

Plate 15.1: Bee Hive

Plate 15.2: Bees at Hive

Type of Hive	Number of Frames		Size of Frames		Recommendation
	Brood	Super	Brood	Super	
Newton	7	7	8"x5$\frac{1}{2}$"	8"x2$\frac{1}{2}$"	It should preferably be used in the plains
Jeolikote	8	8	12"x7"	12"x3$\frac{1}{2}$"	-do-
Langstroth	11	11	17$\frac{5}{8}$"x9$\frac{1}{8}$"	17$\frac{5}{8}$"x9$\frac{1}{8}$"	It should preferably be used in hill region

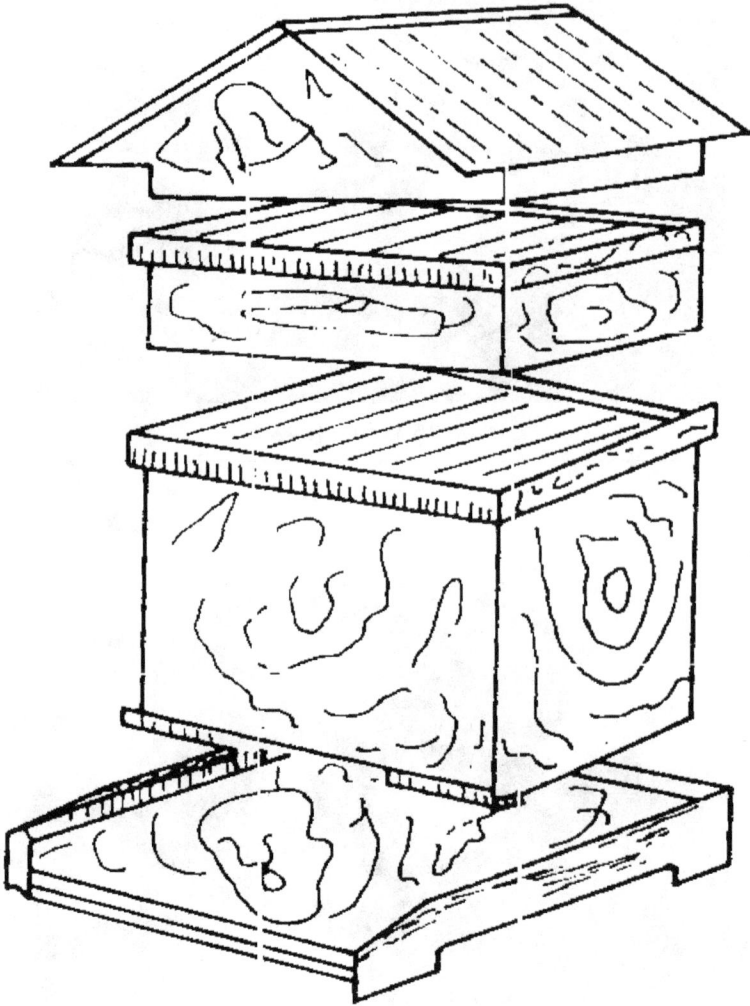

Figure 15.2: A Newtons Beehive

The hive is painted with two coats of white colour (sometime green or yellow also), which not only protects the hive from weather conditions but also help the bees to easily recognize their hive.

Queen Excluder

This consists of a frame fitted with metallic wire net assembled together 0.150 inches apart. It is utilized for preventing the queen's entrance from the brood chamber to the super

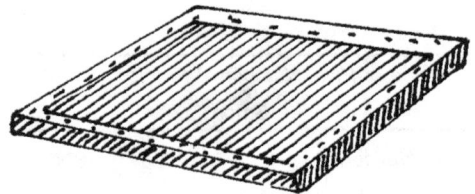

Figure 15.3: Queen Excluder

chamber. The holes in the net do not cause any inconvenience to the workers to pass through it.

Comb Foundation

It is a sheet of bee wax on both sides of which exact shape of different cells of the comb is made in advance. The sheet is cut to the size of hanging frames and are fitted inside it. It can be used for several years.

Bee Gloves

They are leather gloves used by bee keepers to protect heir hands from the sting of the bees.

Bee Veil

A bee veil is a covering to protect the bee-keeper's face from the sting of the bees. It is made up of fine net, usually silken through which bees cannot pass.

Smoker

It is device used by bee-keepers to subdue the bees if irritated during hive inspection. The materials used for producing smoke are anything like rotten wood, chips, wood latches, waste papers etc. In many parts of the world carbolic acid is used as bee quiter in place of smokers. It is powerful antiseptic, a good repellent and is used in diluted form.

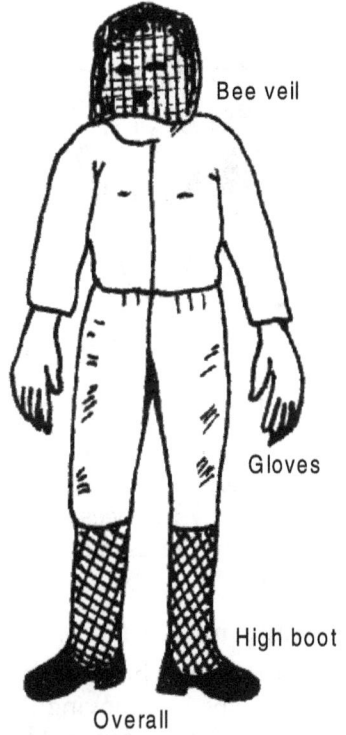

Figure 15.4: Bee Gloves

Figure 15.5: Comb Foundation

Figure 15.6: Bee Veil

Figure 15.7: Smoker (Different Sizes & Shapes)

Hive Tool

It is a flat, narrow and long piece of iron which help in scraping the dirty materials deposited by the bees especially bee glue and superfluous pieces of comb on the inner walls of the hive.

Figure 15.8: Hive Tool

Uncapping Knife

It is a long, broad iron piece which helps in removing the cap of the hive for inspection at regular intervals.

Bee Brush

It is a large brush often employed to brush off bees from honey combs particularly at the time of honey extraction.

Queen Introducing Cage

It is a pipe made up of wire nets through which the queen cannot pass. Both the ends of the pipe are opened. This tool is used for keeping the queen arrested in the hive for about twenty four hours, so that she gets acquainted with the hive as well as the worker bees. After putting the queen in the cage both the ends of the pipe are closed with "queen candy" (made by kneading fine powdered sugar with a little honey).The cage is placed inside the hive. In about twenty four hours time a hole is

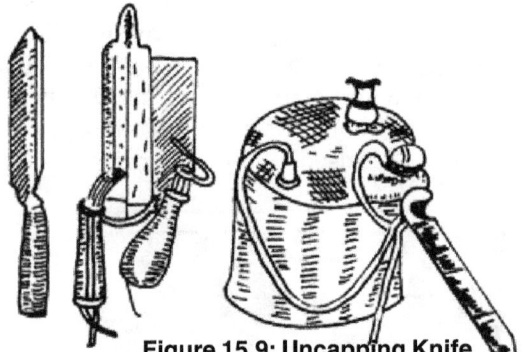

Figure 15.9: Uncapping Knife (Left Simple Knife, Right Electric Knife)

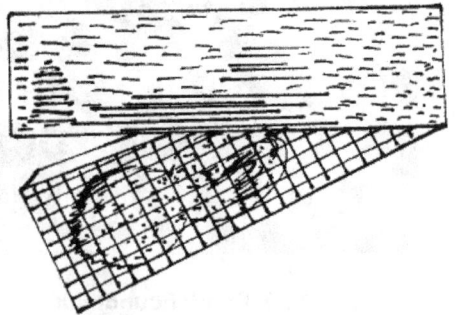

Figure 15.10: Queen Introducing Cage

made in candy due to eating of candy by workers as well as queen, thereby releasing the queen. If the bees fail to eat the candy within 48 hours, the queen should be released directly.

Feeder

During drought and lack of natural food, bees are fed with artificial food. Sugar syrup taken in a basin is placed over the frames of the brood chamber. For preventing the bees from sinking in the syrup, few green grass blades are placed in the basin along with the sugar syrup.

Honey Extractor

This instrument is used for extracting honey from the frames without destruction to the comb. It is a drum made up of metal, having several pockets around a rotating wheel. The frames are made to hang from these pockets and the pockets are made to rotate round a central axis. This rotation creates a centrifugal force which separates the honey from the comb. The collected honey is taken out from the drum through a hole at the bottom. After extraction of the honey, the combs and frames are utilized again.

Figure 15.11: Honey Extractor

Hive Entrance Guard

During swarming season a device similar to queen excluder is placed in front of the hive entrance which prevents the escape of queen.

Important Points Regarding Handling of Bees

1. The hive should not be opened too often. It can however be opened twice or thrice a month. The time for examining the hive in summer is morning and evening, whereas it is the warmer part of the day in winter.
2. There should be a constant watch to place additional chamber at the time when bees want to expand their nest. Such expansion may occur during the peak flowering season.
3. During swarming season the bee keeper should be on the alert and should keep an eye on the behaviour of the bees.

Swarming

It is a natural phenomenon whereby mass movement of bees from one place to another takes place. Bees also undergo swarming especially during spring season. All the bees comes out of the hive, they get divided into two or three colonies and move in different directions. Queen and drones also follow them. In case the queen does not leave the hive, there are chances that bees may return back to their hive.

During swarming season, following precautions should be taken in advance.

1. Ample space for expansion of nest should be provided.
2. There should be sufficient ventilation in hive.
3. Hive should always have a young and vigorous queen.
4. During swarming season queen's wing should be clipped.

If all these measures fail then bees should be transferred to a different hive which give them a chance to survive.

Beekeeping Calendar

Spring Season

Bee-keeping calendar starts with the increased activity of honeybee colonies in spring due to congenial conditions and availability of spring bee flora in abundance. Early spring is the best time for first detailed examination and through inspection of the honeybee colonies for brood rearing, food reserves and presence of queen bee, removal of winter packing, clearing the bottom board, provision of more space to the colony, multiplication (division) of the colonies, control of ectoparasitic mite, replacement if worn out hive parts. General cleanliness and maintenance of the apiary site are the important operations to be undertaken during spring season.

Summer Season

The management operation should be aimed at lowering down the hive temperature, exploring the main honey flow, provision of shade, water, space and ventilation. Preparing colonies for maximum honey collection and honey extraction are some of the major operations to be given due consideration, during summer.

Monsoon Season

Monsoon season being dearth period is the harshest for honeybees. Cleanliness of bottom board, increasing ventilation, safe storage of extra drawn combs, artificial feeding to the colonies, control of robbing, curbing drone population, uniting weak or queenless colonies with average strength queen right colonies and control of bee enemies and diseases are the important management operations aimed at helping the honeybee colonies during this harsh period.

Autumn Season

Autumn or post monsoon is the second best season for honeybees. During this season, honeybee colonies do not issue swarms. Hence except swarm control, all the management operations discussed for spring also hold good during autumn season. In this season surplus honey should be extracted well in time before it gets granulated in the combs.

Winter Season

Winter season is the second harsh period for bees especially in the hilly areas. Examination of the colonies on a calm and sunny day, feeding the colonies and gradual shifting of the colonies to sun should be taken up before the winter sets in.

Protection from chilly winds and providing the winter packing are the other important operations to be taken care of during winter months.

Major Beekeeping Operations

Division of Colonies

☆ For dividing the colonies, there should be new drones or drone brood in the colony.

☆ Put an empty box and locate the queen bee and place this on the corner frame.

☆ Now remove a frame, which is full of eggs and put this into new box including bees.

☆ Now put frames on both sides of this frame having closed and open brood as well as honey and pollen.

☆ Care should be taken so that queen bee must remain in old box.

☆ Displace the new box 3 feet away from the old box.

☆ Then do the same with old box in opposite direction. After 15-16 days of division new queen will be ready and will start laying eggs, 10-1 days after mating.

Control of Swarming

In March–April due to lack of space, queen bee exit from colony or box with half of workers, known as swarming. The remaining colony becomes queen less. For stopping swarming, never allow congestion in colony.

☆ Out wax sheet frames and super chamber, time to time.

☆ While joining super chamber, put one or two frames into upper chamber from lower chamber and put new empty frames in lower chamber.

☆ If colony strength is high then divide it.

☆ Crash the queen cells, which are made prior to swarming.

☆ Swarming can also be controlled by cutting the wings of queen bee also by putting entrance guard.

Honey Extraction

☆ Honey is extracted in the months of May-June and then in end of November.

☆ Honey extraction should be done in wire mesh screened room called honey house.

☆ If provision of screened room is not there then remove frames in day time and extract honey at evening when bee stop working.

☆ For extracting honey, select the fully filled and sealed frame.

☆ Then dislodge the bees by giving a sharp jerk to the frame or brush off with a suitable bee brush or tree twig.

☆ Such honeycombs are fitted in empty chambers and taken to the honey house.

☆ To draw honey out of combs, waxen seals of honeycombs are uncapped with uncapping knife by placing the honeycombs in the drip tray to collect waxen capping and dropping honey.

☆ Uncapped honey frames are put into the honey extractor in such a way that they are not imbalanced.

☆ Honey extractor should be rotated gently.

☆ Gradually speed should be increased to about 300 rotation per minute.

☆ After extracting honey from one side, the extraction should be stopped and sides of the frames should be changed.

☆ Again rotate the extractor to extract honey from other side.

☆ When sufficient honey gets accumulated, it should be poured out of the extractor and strained through specific sieves/double fold muslin cloth before bottling.

☆ After honey extraction, empty frames are given back to the colonies as quickly as possible.

☆ Entrances of the hives should be narrowed down.

☆ Crevices and cracks in the hive parts are plugged. Honey spilt over the hive parts or in the apiary should also be washed to prevent robbing by other colonies or wild bees.

Feeding the Colonies

☆ In winter and rainy season if there is a need of feeding then provide sugar solution and water (1 part sugar, 1 part water) to the colonies.

☆ Put this solution in feeder or 1 liter container and keep it inside the boxes.

☆ If there is lack of space inside the box then space can be provided by providing super chamber.

☆ While feeding put some thin sticks or tiles in the feeder so that these can fed by sitting over these sticks so that they survive from drowning in the solution.

☆ Some times in winter bees do not take feed from feeder.

☆ In winter conditions sugar syrup should be used in concentrated form and should be filled in the frames.

☆ Feeding should be done in the evening. Before feeding, plug all the cracks and crevices of the box.

Uniting the Colonies

☆ Unite the weak or queen less colonies in the evening.

☆ Colonies with higher strength and having queen should be placed at the bottom.

☆ Remove the cover, inner cover and Hessian cloth so that frames become naked from above.

☆ Put newspaper on this and cover the chamber.

☆ Now create small cuts on newspaper between the frames with hive tool so that air can pass, but not the bees immediately. Now place the queen less or weak colony over it without bottom board as we place super chamber.

☆ Fill the joint between these boxes with mud so that bees of upper chamber couldn't escape. Bees will move through the gaps of paper. On third day, remove the paper.

☆ If queen of colony gets killed on account of any reason, then workers start laying eggs. Workers lay unfertilized eggs, so only drones are produced which are smaller in size.

Winter Packing

Winter packing is done by two ways:

(a) Inner Packing

In this operation, first of all empty down combs are removed from the hive. Then push remaining frames, having pollen and brood to one side with the help of hive in order to reduce the space between them. Then wrap dry straw within polythene or newspaper and place it within the hive between empty space.

(b) Outer Packing

Outer packing is done by covering the top of box with thick gunny bag.

Control of Robbing

☆ In monsoon and winter season when there is scarcity of feed, mightly colonies start stealing honey from weak colonies. This is called robbing. Around the box where robbing occurs, robber bees trying to enter. There is also fighting of bees. Some times robber bees kill queen bee. In order to stop robbing, plug all the gaps of box with mud. Make the entrance of box narrow, so that only one bee can enter at one time. Around the box where robbing is taking place, creat smoke or spread kerosene oil so that robber bees remain away from box due to smell. Interchanging the place of robber bees with robbed bees will also check robbing. If problem persist then keep the robber colony 3 km outward for few days.

Diseases of Honeybee

American Foulbrood (AFB)

AFB is the most serious bacterial disease of honey bee brood caused by the bacterium *Paenibacillus larvae*. The disease is transferred and initiated only by the spore stage of the bacterium. The reason this disease is so serious is that the spores can remain viable and last indefinitely on beekeeping equipment. It is extremely contagious and spreads easily via contaminated equipment, hive tools, and beekeeper's hands.

Symptoms

1. Hive may show less than normal bee flight with dead bees on the bottom board. The colony may appear weak after opening the hive.
2. Capped brood is uneven with puncture holes in the caps of brood cells.
3. Colonies with heavy infestation often display irritable behavior.
4. AFB has a distinct "foul" odor that can help in alerting the beekeeper to a disease problem.
5. With a toothpick, lift punctured cap and remove content of brood cell. The larval remnant may be a light brown mass sunk onto the bottom side of the brood cell. If the mass is ropy when withdrawing the toothpick from the cell, there is a strong indication of American Foulbrood disease (AFB).
6. Over time, the larval remains in the cell will dry and harden into a dark brown leathery scale on the bottom side of the brood cell. A single scale contains millions of spores that remain viable for decades. Bees can not remove scales from cells.

Control

1. Antibiotic-resistant AFB (r-AFB) has become established in BC. Antibiotics must be used for treatment purposes only. Do not use antibiotics as a prophylactic (=preventive) measure.
2. Inspect regularly, especially when disease has been reported in the area or after the colony has been placed in crop pollination.
3. When AFB has been confirmed, kill the bees and burn all the equipment. or shake bees onto foundation and burn all the old equipments. Feed the bees with medicated sugar syrup at two week intervals until foundation has been drawn out.
4. Reduce the exchange of hive equipment between hives and apiaries.
5. Replace 20 per cent of all brood frames each year so that after a few years, no brood frame is older than five years.
6. Apply hygienic management practices, including clean clothing, hive tools, and gloves.

Chalkbrood

Chalkbrood is a disease of bee brood caused by a fungus, *Ascosphaera apis*. The larvae must ingest the spores of the fungus for infection to occur. It only infects larvae that are three to four days old.

Symptoms

1. The infected larvae are quickly covered by white cotton-like mycelium of the fungus which eventually fills the entire cell. The white/grey mass soon

hardens, forming a hard, shrunken mummy which is easily removed from the cell The larva in the cell will resemble a chunk of chalk, hence, the name of the disease is chalkbrood.

2. Mummified larvae in front of the hive and on the bottom board are easily detected. Mummies on the bottom board may not necessarily indicate a serious problem, but confirm hygienic bee behaviour.

3. High incidence of Chalkbrood mostly indicates poor hygienic behaviour and stress due to weather, poor management or diseases.

Sacbrood

Sacbrood is caused by a virus. The infected larva dies and the tissue disintegrates into a brown watery solution held by the larval outer skin. The skin sac can be removed intact from the cell. The cell is often uncapped but may also be closed and the cap punctured similarly to AFB cells.

Sacbrood occurs when colonies have been under stress. Bees generally clear up the disease and remove affected brood themselves.

Control

No control measures available.

European Foulbrood (EFB)

EFB is a bacterial disease of honey bee brood. It is generally considered less virulent than American foulbrood and colonies sometimes recover from infection. Its field symptoms are easily confused with those of AFB, but there are important differences.

Symptoms

1. EFB shows up when the colonies have been under stress due to other diseases, colonies nearby, poor management and weather.

2. EFB is easily controlled with standard antibiotic treatments.

3. EFB affects bee brood much the same as AFB except that the disease affects open brood *i.e.* the larvae are affected before they are capped.

4. Affected cells show discoloured larvae often in twisted positions with visible tracheal tubes.

5. The brood has a "sour" odour, distinctly different from AFB.

6. EFB scales are easily removed from the cell (compared to AFB scales).

Control

1. Inspect brood frames regularly.

2. Remove all frames with significant numbers of affected cells.

3. Spray or sprinkle antibiotics (oxytetracycline) dissolved in 250 ml of sugar syrup over the colony every 3-4 days for 10 days

4. Requeening provides a distinct break in the brood cycle of the colony, allowing the bees to clean up existing disease. It may also provide new bees with better cleaning behaviour, *i.e.* less susceptible to disease.

5. Minimize robbing by preventing sugar spillage. Do not barrel feed.

6. Apply hygienic management practices. Clean hive tools, smoker and gloves after inspection of each apiary. Clean clothes regularly.

7. Replace brood frames after five years.

Nosema

Nosema is caused by the microsporidian (protozoa) *Nosema apis*, a small, unicellular organism that is unique to honey bees. It is the most widespread of the adult honey bee diseases.

Symptoms

1. The disease only affects adult bees by parasitizing their midgut. Adult bees have difficulty with controlling their faecal discharge. In heavy infestation, hive bodies are often smeared with faecal deposits.

2. The disease is often not detected because affected bees are either inside the colony (in winter) or in the field, where they die.

3. Nosema impairs the digestive process and causes bee starvation.

4. Nosema is often confused with dysentery which produces similar symptoms.

Control

1. Nosema disease occurs when colonies are under stress and poorly managed. The condition is often exacerbated when there is moisture build up and poor air circulation in the hive.

2. The antibiotic fumagillin is highly effective. The best natural defense is a strong healthy colony with a prolific queen and sufficient food stores. Rapid replacement of adult bees will minimize or eliminate the disease.

Tracheal Mites *Acarapis woodi*

Tracheal mites infest the tracheal system of the adult honey bee. Levels are highest during the winter and spring. Mites prefer adult bees less than four days old. Once they are on the bee, mites are attracted to carbon dioxide emissions and enter the spiracles located on the thorax which lead to the tracheal system. They puncture the wall of the trachea and suck the hemolymph of the bee. Tracheal mites live, breed and lay eggs in the tracheal system. The adults and eggs plug the tubes of the trachea which impairs oxygen exchange. They also spread secondary diseases and pathogens since they puncture the trachea in order to feed. Individual bees die due to the disruption to respiration, damage to the tracheae, microorganisms entering the hemolymph, and from the loss of hemolymph. Honey production may be reduced when over 30 per cent of the population is infested. Infested bees will be seen leaving the colony and crawling on the grass just outside the hive. They will crawl up the blades of grass or the hive, fall back down and try again. The wings may be disjointed and the bees unable to fly. The abdomens may be swollen. In late stages of infestation, bees will abscond from the hive.

Control

One method for controlling tracheal mites is use of menthol. The temperature must be above 60° F in order for menthol to work. The bees breath in the vapor which, is believed to desiccate the mites. Menthol must be removed during a nectar flow in order not to contaminate honey.

Varroa Mites (*Varroa destructor*)

Varroa mites are external, obligate parasites of worker and drone honey bees. Varroa mites are visible with the naked eye and look somewhat like a tick. They feed on the hemolymph of adult bees and the developing brood. The reproduction cycle of the mite takes place inside the cells. Female mites (foundresses) enter the brood cells of last stage worker or drone larvae just prior to the cells being capped. There she will deposit five to six eggs over a period of time while feeding on the brood. The first egg laid will be unfertilized and develop into a male. The subsequent eggs will be fertilized and develop into females. The eggs hatch and the young mites begin to feed on the developing pupa. It is normal for mating to occur between siblings. The adult female mites along with the original female mite(s) leave the cell when the bee emerges. The female mites will enter another cell or attach themselves to an adult bee to feed. Varroa mites are transported from colony to colony by drifting or robbing bees.

Symptoms

Visible symptoms of Varroa mite damage can be evident on newly emerged bees which is due to mite feeding on the immatures within the cell. The newly emerged bees may be smaller than normal, have crumpled or disjointed wings and shortened abdomens. The lifespan of the newly emerged bee is also reduced. Severe infestations of Varroa mites within the cell (5 or more foundresses) can cause death to the pupa. Other symptoms of mite infestation are rapid colony decline, reduced adult bee population, evacuation of the hive by crawling bees, queen supersedure, spotty brood, and abnormal brood with symptoms resembling European foulbrood and sacbrood disease.

Control

Cultural methods for Varroa mite control include using drone comb or bottom screens to trap Varroa mites. Varroa mites prefer drone brood. Using a few frames of drone comb per colony draws mites into the cells which are then capped by the worker bees, trapping the mites within. The frames are removed and put into the freezer 24-48 hours to kill the mites. Bottom screens are also an effective control method. They are basically a wood-bound screen that is placed underneath the brood chamber. The mites fall through the screen onto the bottom board or ground. The mites are thus separated from the bees and eventually die.

Natural Enemies of Honeybee

Wax Moths

The greater wax moth (*Galleria mellonella*) is the most serious and destructive insect pest of unprotected honey bee comb in warmer regions. Wax moths primarily infest stored equipment but will invade colonies whose worker bee population has been weakened by disease, queenlessness, failing queens, pesticide kills or starvation. Newly hatched larvae are white but successive instars are medium to dark grey on the top with creamy white undersides. The larval head capsule is brown. Wax moth larvae prefer dark combs because they contain a variety of nutrients such as entrapped pollen and larval skins. The larvae grow rapidly and migrate toward the edges of the frames or corners of the supers to spin a cocoon and pupate.

Damage occurs as the larvae burrow into the comb feeding on the wax, larval skins, pollen and honey. As the larvae chew through the comb they spin a silk lined tunnel through the cell walls and over the face of the comb. These silk threads can tether emerging bees by their abdomens to their cells and they die of starvation because they are unable to escape from their cell. This phenomenon is termed galleriasis. In severe infestations, the wax comb, wooden frames, and sides of the hive bodies can be heavily damaged.

Control

The most effective method for preventing wax moth damage in hives occupied by bees is to maintain strong colonies. The bees will remove the moth larvae and repair the damage as it occurs. Stored equipment can be protected against wax moths by fumigating it with paradichlorobenzene crystals or by stacking honey supers in a criss-cross fashion in open sheds. The penetrating air and daylight discourage colonization by moths.

Usefulness of Honey Bee

The most important part played by bees is production of honey, which is a nourishment of high food value.

Honey

It is a sweet, viscous, edible fluid obtained by honey bees from nectar and pollen secreted by plants. The flight radius of the Indian honey bees is between ½ to one mile, which is much less than their Western counterparts who cover a radius of 3 to 5 miles. The collection of nectar is very hard and strenuous work, as for collecting 500 gms. of nectar, a bee has to make about 10,000 flights. Out of this 500 grams, about half gets evaporated. For collecting one pound of honey, bees have to make more than fourty thousand trips.

Chemical Composition of Honey

An average sample of honey is composed of water, sugar (levulose, dextrose, sucrose, dexrin), ash (minerals like calcium, iron, phosphate and manganese), about

8 components of vitamin B complex (Pantothenic acid, Biotin, Pyridoxin, Choline, Ascorbic acid, Thiamine, Riboflavin and Miacin). Besides, this honey is an antiseptic and contains formic acid as the preservative. The colour, flavour and odour of honey usually depends on flowers from which nectar is gathered. One kilogram of honey contains 3200 calories and is a energy rich food. It is easily digestable, hence given to the infants. The optimum temperature for storage of honey is 70°F, below which it looses its colour and the glucose molecule cyrstallises.

Bee Wax

It is a wax of high melting point (about 140°F) secreted by wax glands of worker bees. It is utilized in the construction of hive. This wax is used by human beings for several purposes like manufacturing of cosmetics, cold creams, shaving creams, polishes, candies, ointments, lipsticks, lubricants, in modeling works etc. It is also used by apiculturists in the formation of comb-foundation bases for modern bee live.

Propolis and Balms are other collections of bee from the plants. These substances are utilized in repairing and fastening the comb. Bees are good pollinators and responsible for cross pollination in several varieties of leguminous plants. Thus, they play a good role in agriculture and horticulture. Lastly, the sting of bee which is a source of annoyance to man is supposed to be the cure of few diseases. It is used in manufacturing of Ayurvedic medicines. Honey is supposed to be blood purifier, a cure against cough and cold, sore throat, ulcers of the tongue, ulcer of stomach and intestine, etc. It is prescribed for heart and diabetic patients and is useful for kidney and lung disorders.

Economics of 10 Honey Bee Colonies

Sl.No.	Item heads	Item Cost (Rs.)	
		Apis mellifera (Italian Bee)	Apis cerana (Indian Bee)
A. Expenditure			
1.	Bees	4000	2000
2.	Hives	8000	6000
3.	Stands	1500	1000
4.	Bee veil	35	35
5.	Gloves	50	50
6.	Smoker	200	200
7.	Hive tool, knife	50	50
8.	Comb foundation	300	200
9.	Sugar	350	250
10.	Medicine	50	50
	Total	**14,485**	**9,785**

Contd...

Contd...

Sl.No.	Item heads	Item Cost (Rs.)	
		Apis mellifera (Italian Bee)	Apis cerana (Indian Bee)
B. Income			
11.	Honey (@ Rs. 50/kg)	7500	3000
12.	Bees wax (@ Rs. 120/kg)	300	160
13.	Multiplied bee colonies (Min 5)	2000	1000
14.	Rental colonies for pollination (5) @ 300/-	1500	1500
	Total	**11,300**	**5,660**
C. Capital Invested (Sum of 1-8)		14,085	9,485
D. Recurring Expenses (Sum of 9 and 10)		400	300
Profit (Income-recurring expenses; B-D)		**10,900**	**5,360**

Sericulture

History

There is no authentic information regarding the origin and use of silk. The ancient literature is based on two views. According to one view, silk industry originated for the first time in India at the foot of the Himalayas, and from there it spread to other countries of the world. Second view, which has a greater acceptance, is that this industry originated in China about 3000 B.C. According to this, a Chinese Princess Siling Chi was the first to discover the art of reeling an unbroken filament from a cocoon. This art was kept a close secret for nearly 3000 years. This art later on spread to the rest of the world through several agencies like civil war refugees, war prisioners, marriage of royal families, etc.

Silk

Silk is the result of secretion of silk glands. They are a pair of long tubular and coiled glands lying one on each side of alimentary canal of the Caterpillar. *Fibroin*, a sort of fibrous protein is secreted by each gland which at first is in the fluid condition. These glands are connected with a very narrow tube like structure known as spinneret which is a part of the hypophyarnx. The liquid secretions of two glands passes through the spinneret which transform them into a single thread. *Sericin* which causes the two fibres of fibroin to unite is secreted by a pair of accessory gland situated at the anterior region of silkgland. Two streams of fibroin along with Sericin are expelled through the spinneret due to contraction and expansion of the body of Caterpillar. This sticky secretion after coming in contact with the air is converted into a fine, long and solid thread of silk.

Properties of Silk

Silk threads are very fine, soft and light in weight. They are very thin but strong having high elastic property. When a cross section of the silk thread is observed under microscope, it is roughly in the figure of 8. Main inner portion (70-80 per cent) is made up of fibroin (true fibre) which is rurrounded by a thin covering of Sericin (gum covering). There is also little quantity of waxy and colouring material. Both fibroin and sericin are proteinous in nature. Fibroin is insoluble in water and is made up of glycine, alanine and tyrosine. Sericin is easily soluble in water and is composed of sericin, alanine and leucine.

Types of Silk

Insects belonging to families Bombycidae and Saturnidae of order lepidoptera and class Insecta produces silk of commercial value. There are many species of silk-moth which can produce the silk. But only few have been useful. Mainly four types of silk have been recognized which are secreted by different species of silk worms.

(*i*) Mulbery Silk

It is secreted by the caterpillar of *Bombyx mori* which feeds on mulberry leaves.This silk is supposed to be superior in quality to the other types due to its shining and creamy white colour.

(*ii*) Tasar Silk

It is secreted by caterpillars of *Antheraea mylitta, A. paphia, A. oyeli, A pernyi, A. proyeli* etc. This silk is coppery colour. They feed on the leaves of Arjun, Asan, Sal, Oak and various other secondary food plants.

(*iii*) Eri Silk

It is produced by caterpillars of *Attacus ricini* which feed on castor leaves. Its colour is also creamy white like mulberry silk, but is less shining than the latter.

(*iv*) Muga Silk

It is obtained from caterpillars of *Antheraea assama* which feeds on Som, Champa and Moyankuri.

Different Types of Silkworm and their Hosts

Type of Silk	Type of Silk Insects	Food Plants
Mulbery	*Bombyx mori*	*Moras alba* (Mulberry)
Tasar	*Antheraea mylitta*	*Terminalia arjuna* (Arjun)
	Antheraea paphia	*Terminalia tomentosa* (Asan)
	Antheraea royeli	*Sorea robusta* (Sal)
	Antheraea pernyi	*Zizyphus jujuba* (Plum) etc.
Eri	*Attacus ricini*	*Ricinus communis* (castor)
Munga	*Antheraea assama*	*Tetraanthera monopetala* (Som),
		Michalia oblonga (champa),
		Listea citrata (Moyankuri)

Life History of Silkworm

Sericulture (Mulbery Type)

This is multivoltine. Mulberry silkworms are of domesticated type because they can be reared indoors. Large and healthy cocoons are selected during harvesting season for the next crop. These cocoons are kept in well ventilated cages. They emerge after few days. Males and females are easily distinguishable (female larger in size and abdomen broader than the males). The matured moth measures about 1½"-2" in wing span and is pale creamy in colour. These moths are kept in pairs (one male and one female) in coupling Jars for about 24 hrs. They copulate during this period and after that the females are transferred to egg-laying boxes made up of card-board or earthen-ware. Each female lays about 300-400 eggs in the next 24 hrs. These eggs are very small in size, white in colour, seed like in appearance and commercially they are

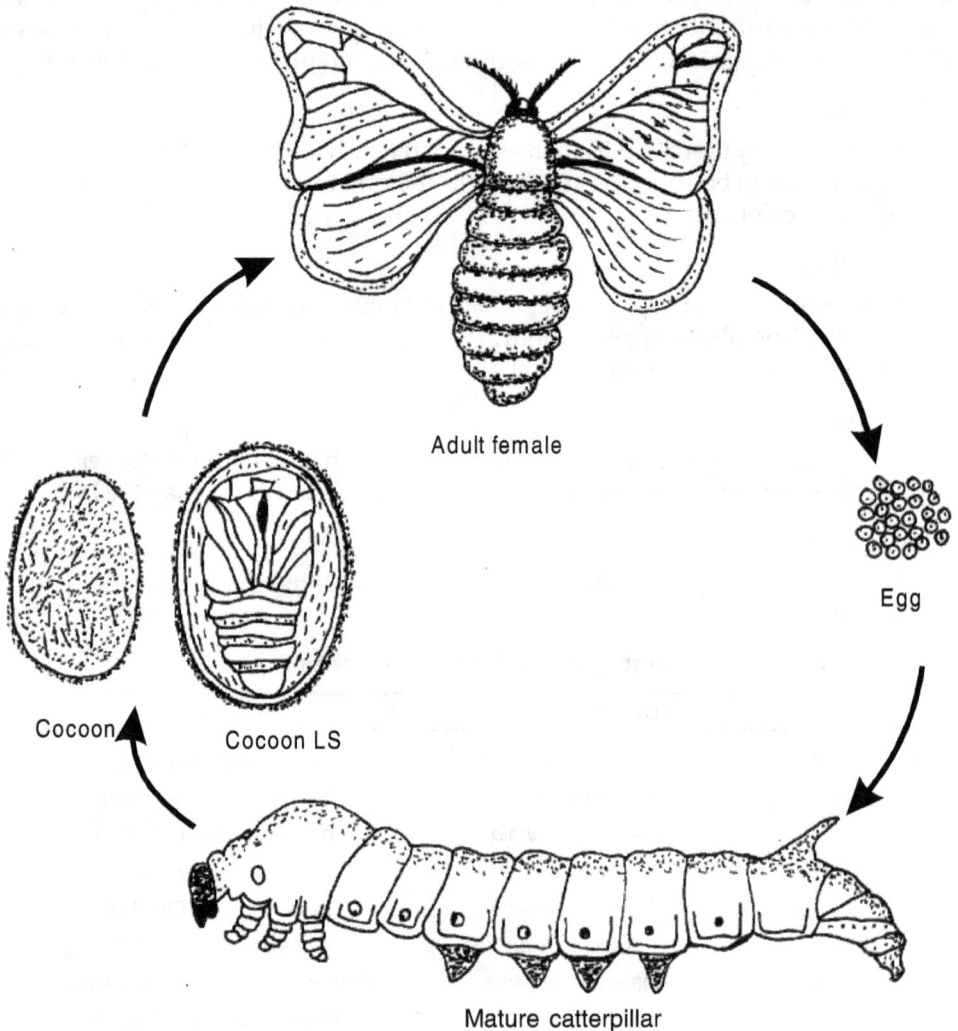

Adult female

Egg

Cocoon

Cocoon LS

Mature catterpillar

Figure 15.12: Life Cycle of Mulberry Silkworm

known as "seeds". A great care is required in selection of these seeds because the whole industry is based on these seeds. They should be healthy and free from any sort of disease. For this, tissue fluid of egg-laid moths are examined under microscope. If an evidence of any type of disease is found, the eggs laid by that particular moth are destroyed. These disease free seeds are also supplied to villagers and tribal rearers. The eggs are kept in incubators (at 75°F) in small boxes where hatch 7-10 days. The small larvae are known as caterpillars which are subjected to the process of rearing.

1. Cultivation of Food Plants

The larvae of *Bombyx mori* feed on leaves of mulberry, so for undisrupted and regular rearing, a continuous supply of healthy mulberry leaves is essential. For this,

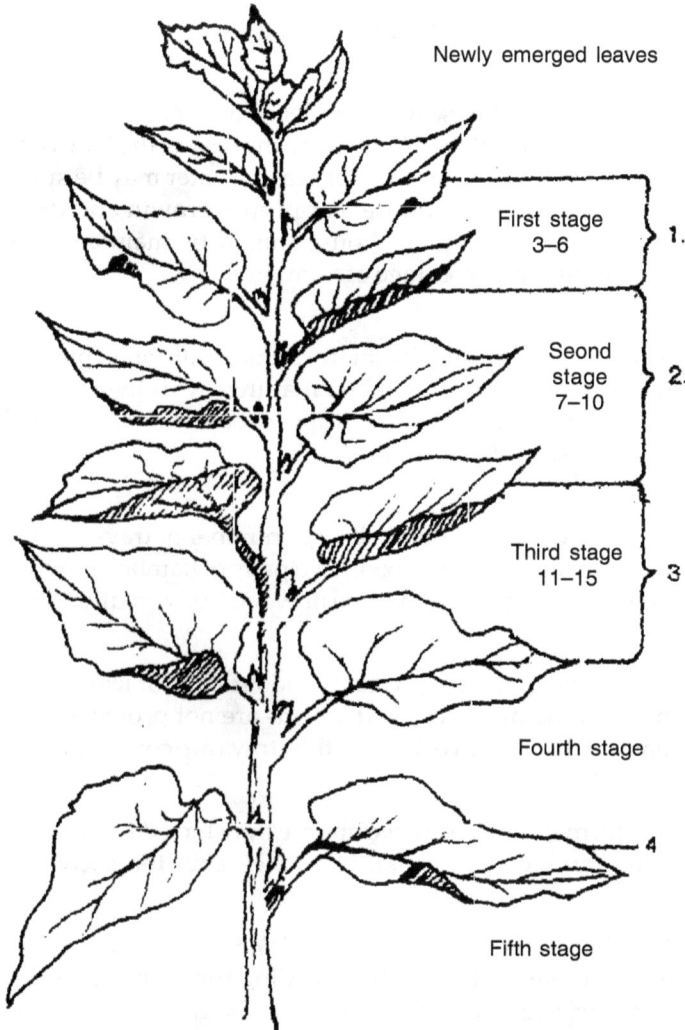

Figure 15.13: Selection of Leaf for Silkworm Rearing

a systematic and regular cultivation of food plants is maintained. Although mulberry leaves have high food value, their cultivation and maintenance is a costly affair. Cost of silk production is directly proportionate to the cost of mulberry leaves. Therefore, for large scale production of silk at a cheaper rate, it is essential to reduce the cost of mulberry leaves. One of the techniques for the purpose is to raise the mulberry trees from "grafting", rather than from cuttings or seeds. They are economical because they require little attention and maintenance once they are established. At state grafts nursery center at Khawa in Mysore improved mulberry grafts have been produced by grafting Japanese variety with local varieties. This has improved both the quality as well as the quantity of the yield. These leaves are also supplied to cultivators at reasonable rate.

2. Rearing of Silkworms

Tools and materials necessary for rearing silkworms:

House

Any building or thatched house which is well ventilated may be used for rearing the worms, but mud-walled thatched houses are the best as they are cool in summer and warm in winter season. During summer season water may be sprinkled inside the thatch to lower the high temperature. For proper growth and development of silkworms, the temperature inside the house should be maintained more or less between 70-75°F with similar percentage of humidity.

Feeding Trays

Freshly hatched worms are kept in flat trays alongwith small pieces of mulberry leaves. These trays are made up of bamboo mattings with their edges turned up, which afford a raised border made by stout stripes of bamboos. On the back of the tray two stripes are firmly fastened longitudinally.

Machan

Machans are needed to accommodate large number of trays in a limited space. Machans are easily and best made by fixing two pairs of bamboo or wooden poles in the ground and tying across bars of bamboo or wood horizontally.

Nets

Large amount of excreta, dirty products and remains of leaves may fall on the trays from the holes of the upper trays. If worms are not protected from these by-products they may get diseased. To prevent this, trays are covered with the nets.

Spinning Trays

Before cocoon formation, mature worms are transferred to special type of trays known as spinning trays or chandraki. Here they spin the cocoon without any disturbance.

Copper Sulphate, Sulphur and Some Other Germicides

Before the start of rearing the machan, feeding trays, spinning trays, nets and everything used in rearing except the leaves are washed with $CuSo_4$ solution or other antiseptic chemicals. The remaining germs are further killed by fumigation of sulphur.

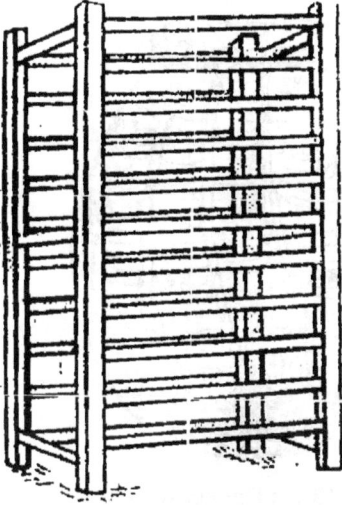
Figure 15.14: Insect Rearing Stand

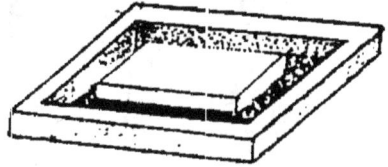
Figure 15.15: Cement Tank

Figure 15.16: Wooden Tray

Figure 15.17: Bamboo Basket

Figure 15.18: Knife

Figure 15.19: Plank

Tiny caterpillars which hatche from the egg measures 5-7 mm. in length. They are transferred to feeding trays already having chopped tender leaves of mulberry. These caterpillars move on the leaves in a characteristic looping manner. Their body is rough, wrinkled grayish in colour. They are made up of 12 segments which is distinct into three parts *i.e.,* head, thorax and abdomen. The head bears mandibulate mouth-parts with which they feed upon the leaves. The thorax is 3 segmented and all the segments bear a pair of true jointed legs. The abdomen which has 10 segments is provided with five pairs of unjointed, stumpy *prolegs* or *pseudolegs*. (One pair each in segment 3rd, 4th, 5th, 6th and 10th) a short dorsal and horn (on the 8th segment) and a series of *spiracles* on lateral sides. These larvae feed voraciously on mulberry leaves and grow very quickly. They stop feeding, become inactive after four to five days, and

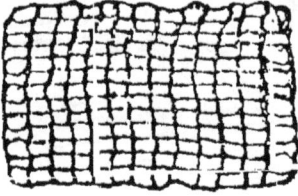

Figure 15.20: Cleaning Net

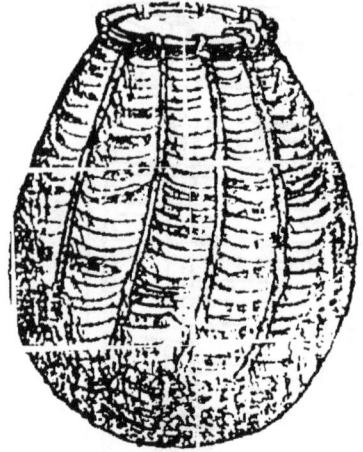

Figure 15.21: Basket for Leaves Collection

Figure 15.22: Chandrika

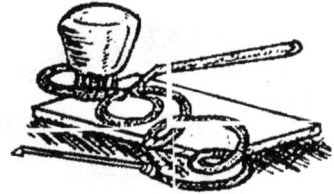

Figure 15.24: Sprayer

Figure 15.23: Foam Sheet

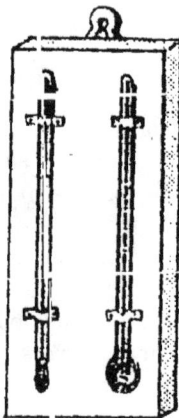

Figure 15.26: Parafine Paper

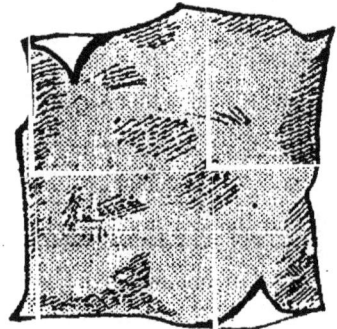

Figure 15.25: Dry-wet Thermometer

Figure 15.27: Feather

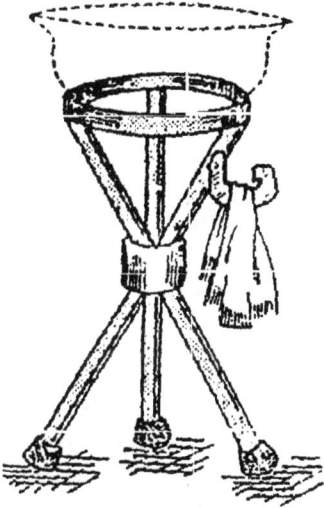

Figure 15.28: Basin Stand

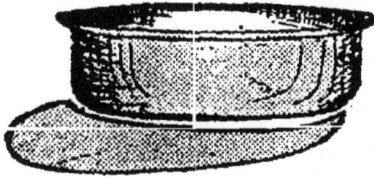

Figure 15.29: Leaf Collection Chamber

Figure 15.30: Basin

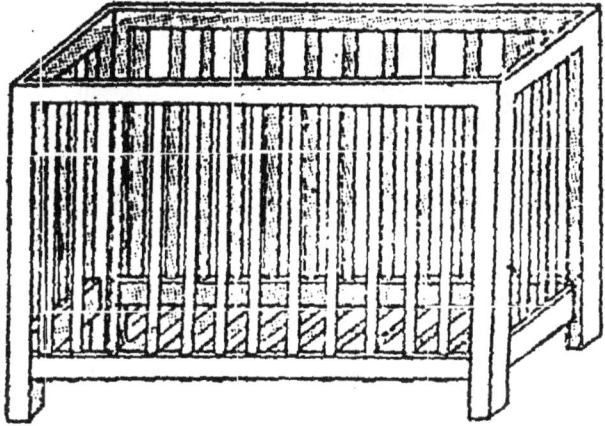

Figure 15.31: Feeding Stand

then Ist moulting takes place. The 2nd stage larvae resemble the Ist stage larvae except that they are slightly bigger in size. They also eat voraciously for 7 days, then 2nd moulting takes place and 3rd stage larvae are formed. The larvae repeat this process for four times. The maturity is achieved in about 45 days since the time of hatching and matured caterpillar now measures 7-10 cms. in length. By this time the formation of a pair of salivary glands is completed. Since these *salivary glands* secrete silk, they are also called as *silk-gland*.

When the matured caterpillars stop feeding they are transferred to *spinning trays*. They excrete their last excreta and begin to secrete the sticky secretions from the silk-gland through a very narrow pore situated on the hypopharynx. The secretion is continuous and after coming in contact with the air sticky secretion is converted into a fine, long and solid thread of silk. The thread becomes wrapped around the body of larva forming a *pupal case or cocoon*. This process continues for 3-4 days, at the end of which the caterpillar is enclosed within a thick, somewhat hard, oval, whitish or yellowish cocoon. Within 15 days, the caterpillar is transformed into a brownish *pupa or chrysalis*. Active metamorphic changes takes place during pupation in which abdominal prolegs disappear, while the thorax develops two pairs of wings.

The pupa is finally metamorphosed into young adult moth in about 12-15 days. This young moth or imago secretes an alkaline fluid to soften one end of the cocoon and then escapes by forcing its way out of the softened silk. Soon after the emergence, the silk moths mate, lay eggs and die. Just after the formation of cocoons, healthy cocoons are selected and are kept in cages for the next crop.

3. Reeling of Raw Silk from Cocoon

For the production of standard variety of raw silk latest technology is employed for speedy and economical reeling of raw silk. This constitutes an important aspect of sericulture because cocoon production is directly related to reeling industries.

Before reeling the thread, the cocoons are dipped in a container of hot water for more than 10 minutes. During this period, they are continuously stirred with a rod. Due to this, their outer portion is loosened and removed in the form of long tapes and end of the continuous filament is found. The filaments of several cocoons are picked up and passed through the 'glass eye' on the reel. The thread thus reeled forms the 'raw silk' of commerce. About 1 kg. of raw silk is obtained from nearly 55,000 cocoons.

The following precautions should be taken during the rearing of mulberry silkworms:

(*i*) The worms should never be kept overcrowed in a tray.

(*ii*) Dried or dusty wet leaves should never be fed to the worms.

Plate 15.3: Silk Moth Rearing Tray and Caterpillars

(iii) No doubt free ventilation is a must but wind should not be allowed directly over the worms.

(iv) There should be equal distribution of leaves among the worms.

(v) Worms which are under process of moulting should not be disturbed otherwise they may die or mouting may be delayed.

(vi) There should be no dust at the floor of the house. For this, it should be well plastered with cowdung or mud at regular intervals.

(vii) Smoking should be strictly prohibited in the rearing room.

(viii) Worms should not be handled with dirty hands otherwise they may get diseased. They should be handled only after washing the hands thoroughly with antiseptic solutions and drying the hands.

(ix) One should enter the rearing house only after putting off the shoes, chappals, etc.

(x) Very hot or drinking water may be sprinkled over the feeding trays.

Sericulture (Tasar Type)

Although Tasar and Munga silkworms are wild in nature, but attempts have been made and are in progress to domesticate them too. Success has been achieved in this regard to some extent. A technique of "controlled rearing" has been evolved by the Central Tasar Research Station, Ranchi.

Cultivation of Food-plants

Since they are wild in nature, cultivation of food plants is not at all necessary. The worms are mounted on the food plants in the nearby forests. But still for convenience, plantation of food plants can be done. Their primary food plants are Asan, Arjun, Sal, Oak, etc., and there are large number of secondary food plants.

For the cultivation of the food plants first of all particular piece of land selected for the purpose is prepared (ploughing, leveling, manuring etc.) and then the saplings are planted after sufficient rain fall. The distance between two saplings should be 20-25 ft. Watering, manuring, and ploughing of the soil around the saplings are done at regular intervals according to the need. They are protected from cattles, other animals and villagers. Proper care is taken till they have attained a considerable height. Bushes are pruned 3-5 weeks before the start of the rearing reason. It is not advisable to rear the worms on a plant every year because in that case sufficient foliages will not be available for the developing larvae. To overcome this problem, the land in which rearing is to be done is divided into two plots. In a particular plot rearing should be done every alternate year. For speedy and healthy growth of the offshoots, it is necessary to give proper care and attention to host plants which includes ploughing, manuring, watering and pruning at regular intervals.

Rearing of Silkworms

They are bivoltine i.e. two crops in a year, one from August-October and other from October-December. It is from August to December that the tasar insects are active and for the rest of the year they are inactive *i.e.;* under dipause. The active and inactive phases of life of tasar insects is controlled by environmental and hormonal factors. The moulting hormone *ecdysone* plays a vital role. Ecdysone is secreted by prothoracic gland. When their secretion stops then moulting stops, worms become inactive, and when secretion starts worms become active again.

Healthy cocoons are selected during the harvesting season for the next rearing. These cocoons are kept in well ventilated cages. In favourable season emergence of male and female moths take place. The size of the tasar silk-moths is larger than the other silk-moths. The female which is yellowish or deep brown in colour is larger than the brick-red coloured male. Wings generally have an eye-spot. For successful mating monias made up of palm leaves are used. In each monia, a pair of moth is kept. Within 24 hrs., they are expected to complete their mating process. After completion of this period, monias are opened and males are allowed to fly away. Females are transferred to earthenware pots or card-board boxes in which they deposit their eggs in next 24 hrs. These egg laid females moths are subjected to pathological test. If they suffer from any disease, the eggs laid by the particular moth is destroyed. Only eggs of disease-free moths are kept for hatching. The eggs are washed in 5 per cent formaline solution and then kept in incubators in small hatching boxes. In about 7-10 days the egg hatches.

The freshly emerged larvae are yellowish in colour, covered with bristles and measures less than ½" in length. They are mounted on the bushes of the host-plants already pruned for the purpose. With the help of their mandibulate type of mouth-parts they feed on the tender leaves of host-plants. They grow in size and after 3-4 days they become inactive and Ist moulting takes place. Structural details of caterpillars of tasar-worms are similar to that of the mulberry-worms with slight variations. With successive moults, the size increases and colour changes. There occurs four moultings in the life-cycle of a tasarworm and thus five larval stages exists. The fifth larval stage is of the longest duration (15-20 days) and it measures approximately 4"-5" in length and 50 gms. in weight. It takes about 40-50 days for a freshly hatched larvae to attain the size of full grown and healthy caterpillar which is capable of spinning a cocoon.

After passing out the last excreta, the larva takes rest for a while, then it becomes active in search of a suitable place for spinning the cocoon. After selecting a suitable position for formation oring, which is generally above a node, the caterpillar crawls down to form a *hammock* by tying a few leaves with Silk-threads. The hammock is generally in the shape of a cone or a cup with an opening on the top. After hammock formation, larvae comes out from the hammock to form a *ring* around the twig. First of all, peeling of the bark in a circular pattern takes place with the help of a pair of powerful mandibles. Around this scare the silk is thrown in a semicircular manner and within few minutes, a strong ring of silken thread is formed. Ring formation is followed by the formation of the *peduncle.* Soon after ring and peduncle formation larvae enters the hammock and start spinning the cocoon. The spinning of cocoon is completed after 4 to 6 days of its commencement and the larvae inside pupate after 4 to 6 days.

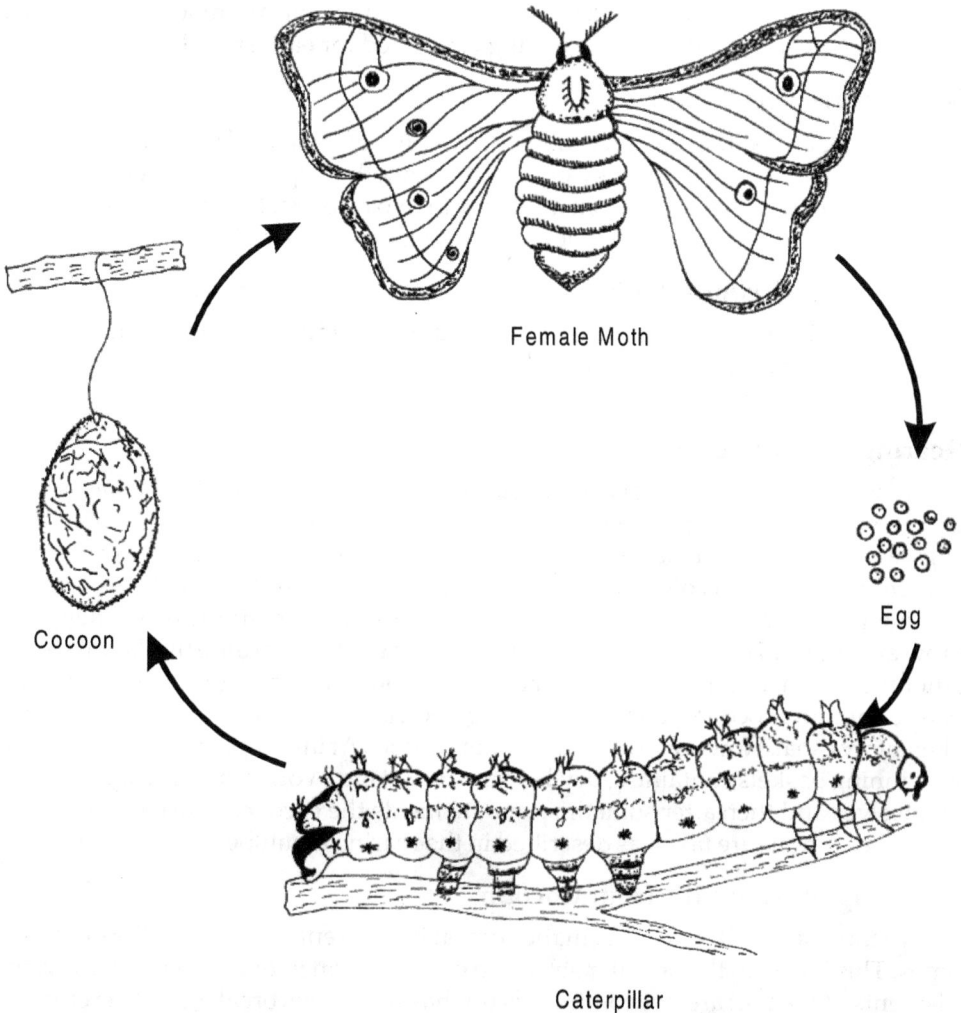

Female Moth

Egg

Cocoon

Caterpillar

Figure 15.32: Life Cycle of Tasar Silkworm

Cocoon is a hard protective covering secreted by the silk-glands of the larva. Tasar cocoons have three parts; rings, peduncle and main body of the cocoon. These cocoons are very hard. During rearing season, a continuous watch is maintained to prevent the tasar worms to be eaten up or destroyed by insectivorous birds, bats, rats, squirrels, lizards, predators, parasites, etc.

Reeling of Raw Silk from the Cocoon

The mechanism is slightly different than the mulberry ones due to hard nature of the cocoons, which is due to presence of a gummy substance. Reeling of the silk filament cannot be done unless this dried gummy substance is softened. For this special cooking technique is employed. The cocoons are dipped in 0.5 per cent Na_2Co_3 solution for 18 hours. Thereafter, the cocoons are subjected to steam cooking under 15 lb./inch-2 pressure for 2-1/2 hrs. For better tensile strength of the filament, these cocoons after 24 hours are treated with 0.5 per cent formaldehyde solution for 15 minutes. Excess of water from the cocoons is removed by squeezing. The reeling of the silk filaments from these can be done with improved reeling machines. Four cocoons are used constantly as long as the reeling continued for each spindle.

Sericulture (Eri Type)

This type of silk is obtained from the worms of *Attacus ricini* which feeds on castor leaves. After mulberry silkworms, it is the another type which can be artificially domesticated. But they differ from mulberry in two respects. Firstly, the silk filament is not continuous and secondly, the cocoon is made up of two layers.

Cultivation of Food-plants

Cultivation of castor plants is done like the other crops of the year and this can be done on a larger scale. A fixed distance is maintained between two plants. Plants are not allowed to bear fruits and seeds, with the help of regular pruning.

Rearing of Silkworms

Not only is the life-history of *Attacus* more or less similar to that of *Bombyx* but also the rearing technique, tools and materials are the same. On an average 75-85 eggs laid by a single female. They hatchout in about 8-12 days. Young caterpillars are of greenish-yellow in colour with black spots. These caterpillars are transferred to feeding trays and are supplied with small pieces of fresh castor leaves. They feed, grow and moult. The larvae undergo four moults and thus there are five larval stages. After the fourth moult, the caterpillar eats voraciously and becomes full grown. Just before spinning cocoons, they stop feeding and eject out the whole excreta from the alimentary canal. Now they are ready to spin cocoons At this stage they are transferred to spinning baskets containing dry leaves, straws, etc. Worms spin cocoons around themselves and then after sometime pupate inside the cocoons. During rearing the same precautions are taken as described in the rearing of mulberry Silkworms.

Spinning of Raw Silk from Cocoon

Extraction of silk thread from the cocoon is different from that of the other Silk types. This is due to the peculiar structure of the cocoon in two respects. Firstly, the filaments of the Silk are not continuous but has numerous breakages. Therefore, in this case, moths are allowed to emerge from the cocoons in the natural way. Secondly, the cocoons are double layered, inner contains cast off skin of pupa and excretory remains of the larvae and thick and hard outer layer contains the silk threads. Due to the discontinuous nature of the silk filaments, it is not reeled but spun from the outer layer of the cocoon. For removing the dirty inner portion of the cocoons, a reversing machine is used which turns the cocoon inside out. These reversed cocoons are washed several times alternately with cold water, hot water containing caustic soda, and cold water. After this, spinning is done with the help of various machines like, Pusa continuous machine, Takli or Charkha.

Eri Silk, no doubt, is also produced in Bihar but Assam is the largest producer of Eri and Munga Silk.

Munga Silk

Munga silk is only produced in Assam. This is also a wild variety like tasar. It's life history and rearing technique is almost like that of tasar-worms except the food-plants, which are in this case-Som, Champa and Moyankuri.

Diseases and Enemies of the Silkworms

This profitable industry is often threatened not only by various diseases resulted from the viral, fungal, bacterial and protozoan infections but also by insect predators, birds and other higher animals. Diseases are of the following types.

(*i*) Pebrine

It is the most serious disease of the Silkworms and if not checked will wipe out the entire industry. It is caused by an internal protozoan parasite (*Nosema bombycis*) which is contagious as well as hereditary. To cope with this serious disease, body tissue fluid of the female moths should be examined under microscope. If pebrine spores are found, the eggs laid by those females should be destroyed. If the infections is of a mild nature then treatment of eggs with warm water (47°C) will serve the purpose.

(*ii*) Flacherie

It is a bacterial infection which generally occurs due to unhealthy and dirty conditions of silk house. Prevention of this disease can be done by improving the conditions of the Silk houses like better ventilation, keeping it neat and clean and provision for sufficient light.

(*iii*) Grassarie

It is a sort of viral infection which is also caused by unhealthy conditions. General cleanliness and separation of infected worms serves the purpose. Good food enables the worms to resist these diseases to a greater extent.

(*iv*) Muscardine

It is due to fungal infection caused by *Beuveria bassiana* which proves fatal. The infection mainly takes place through the skin due to body contamination. It can be controlled using disinfectants with 3-5 per cent formalin solution.

Predators

1. Uzi fly, *Tricholiga bombycis*. This parasite fly causes great loss and is a serious menace to Silk industry. They lay their eggs near or inside the body of the caterpillar. Larvae upon hatching feed on the tissue of the caterpillar which ultimately dies.
2. *Canthecona fincellata* and tachnid fly. They possess long and stout rostrum, with the help of which, they penetrate the body wall of silk worms. A good amount of haemolymph oozes out and worms ultimately die.
3. Ants, hornets, crows, kites, bats, rates, squirrels and lizards feed upon silk worms thereby causing great loss to the silk industry.

Use of Silk

Bulk of silk fibres produced is utilized in preparing silk clothes. Uses of pure silk is decreasing gradually due to its high cost value and costly maintenance. Production of systematic fibres has posed a serious threat to the silk industry. Clothes in which silk fibres are combined with other natural and synthetic fibres are in great demand not only in India but also in foreign countries. Seeing this demand many textile industries are manufacturing clothes like Teri-silk, contsilk etc. Besides silk being used as garments it is also used in other industries and for military purposes. It is used in the manufacture of fishing fibres, parachutes, catridge bags, insulation coils for telephones and wireless receivers, tyres of racing cars, filter clothes for flour mills, and in medical dressing and suture materials.

Lac Culture

Lac is a natural resin of animal origin. It is secreted by an insect, known as lac-insect. In order to obtain lac, these insects are cultured and the technique is called lac-culture. It involves proper care and regular pruning of host plants and propagation, collection and processing of lac.

History

Lac has been used in India from time immemorial for several purposes. From the epic of Mahabharat it has been recorded that Kauravas built a palace of lac for the destruction of Pandavas (*Lakshagrah*). We come across references of lac in the Atharvaveda and Mahabharata, so it is presumed that ancient Hindus were quite familiar with lac and its uses.

Systematic Position

A number of species of lac insects are known, of this *Lacifer lacca* is by far the most important and produces the bulk of the lac of commercial value. It belongs to family lacciferidae.

Food Plants

The insects live as a parasite, feeding on the sap of certain trees and shrubs. The important trees on which the lac insects breed and thrives well are Kusum (*Schleichera oleasa or trijuga*), Palas (*Butea monosperma*), Ber (*Zizyphus jujuba*), Babul (*Acacia arabica*), Khair (*Acacia catcchu*), Arhar (*Cajanus indicus*). Before coming to the actual mechanism of lac secretion and its processing, it is advisable for a lac-culturist to have detailed knowledge of lac insect and its life cycle. The adult lac insect shows a marked phenomenon of sexual dimorphism. The male and female insect vary in shape, size and also in presence or absence of certain body parts.

Structure of Male Lac Insect

It is larger in size and red in colour. The body is typically divided into head, thorax and abdomen. The head bears a pair of antennae and a pair of eyes. Mouth parts are absent so a male adult insect is unable to feed. Thorax bears three pairs of legs. Wings may or may not be found. Abdomen is the largest part of the body bearing a pair of caudal setae and sheath containing pennies at the posterior end.

Structure of Female Lac Insect

It is smaller in size. Head bears a pair of antennae and a single proboscis. Eyes are absent. Thorax is devoid of wings and legs. The loss of eyes, wings, and legs are due to the fact that the female larvae after settling down once, never move again and thus these parts become useless and ultimately suffer from atrophy. Abdomen bears a pair of caudal setae. It is the female lac insect which secretes the bulk of lac for commerce.

Fertilization

After attaining the maturity, males emerge out from their cells, walk over the lac incrustations. The male enters the female cell through anal tubular opening and inside female cell it fertilizes the female. After copulation, the male dies. One male is capable of fertilizing several females. Females develops very rapidly after fertilization. They suck more sap from plants and exude more resin and wax.

Life Cycle

The females after fertilization are capable of producing eggs. But it has been noticed in case of lac insects that the post fertilization development start when the eggs are still inside the ovary. These developing eggs are oviposited into the incubating chambers (formed inside the female cell by the body contraction of females). A female is capable of producing about one thousand eggs (average 200-500). Inside incubating chamber, the eggs hatch into larvae.

The larvae are minute, boat shaped, red coloured and measure little over half milimetre in length. Larva consists of head, thorax and abdomen. Head bears a pair of antennae, a pair of simple eyes and a single proboscis. All three thoracic segments are provided with a pair of walking legs. Thorax also bears two pairs of spiracles for respiration. Abdomen is provided with a pair of caudal setae.

These larvae begin to wander in search of suitable center to fix themselves. This mass movement of larvae from female cell to the new off-shoots of host plant, is termed as "swarming". The emergence of larvae from female cell occurs through anal tubular opening of the cell and this emergence may continue for three weeks. The larvae of lac are very sluggish and feed continuously once they get fixed with the twig. In the meantime, the larvae start secreting resinous substance around their body through certain glands present in the body. After some-time the larvae gets fully covered by the lac encasement, also known as lac cell. Once they are fully covered, they moult and begin to feed actively. The cell produced by male and female differ in shape, and can be easily distinguished sometimes later. Male cells are elongated and cigar shaped. There is a pair of branchial pores in the anterior side and a single large circular opening covered by the flap in the posterior side. It is through the posterior circular opening that the natured male lac insect emerges out of its cell. Female cell is oval, having a pair of small branchial pores in anterior side and a single round anal tubular opening in posterior side. Through the anal tubular opening are protruding waxy white filaments, secreted by the glands in the insects body, which is an indication the insect inside the cells is alive and is in healthy condition. These filaments, also prevent the blocking of the pore during excess secretion of lac.

Larvae moult in their respective cells. It is the second stage larva which undergoes predopupation for a brief time, whereby it changes into adult stage. Now the male emerges out from its cell, moves on lac incrustation and enters the female cell for fertilization. In this way the life cycle is completed.

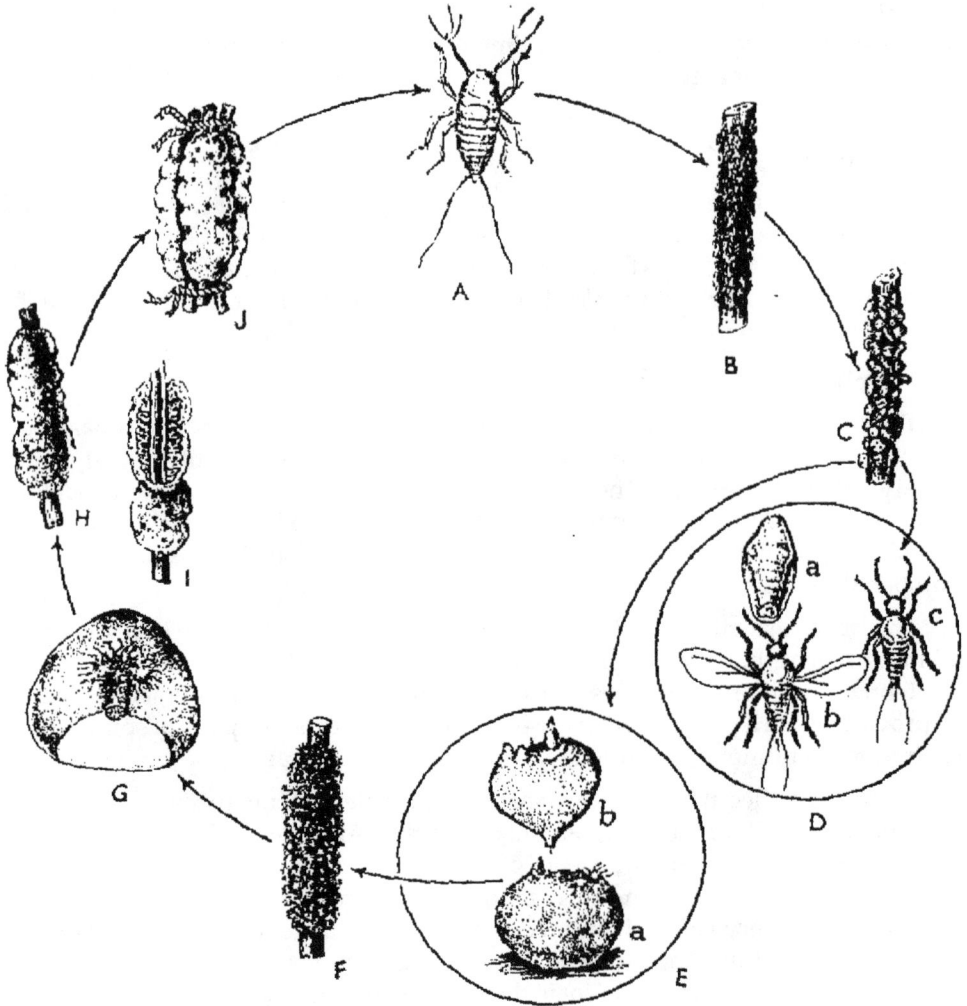

Figure 15.33: Life Cycle of Lac Insect

Lac Secretion

Lac is a resinous substance secreted by certain glands present in the abdomen of the lac insects. The secretion of lac begins immediately after the larval settlement on new and tender shoots. This secretion appears first as a shining layer which soon gets hardened after coming in contact with air. This makes a coating around the insect and the twig on which it is residing. As the secretion continues the coating around one insect meet and fuses completely with the coating of another insect. In this way, a continuous or semi-continuous encrustation of lac is formed on the tender shoots.

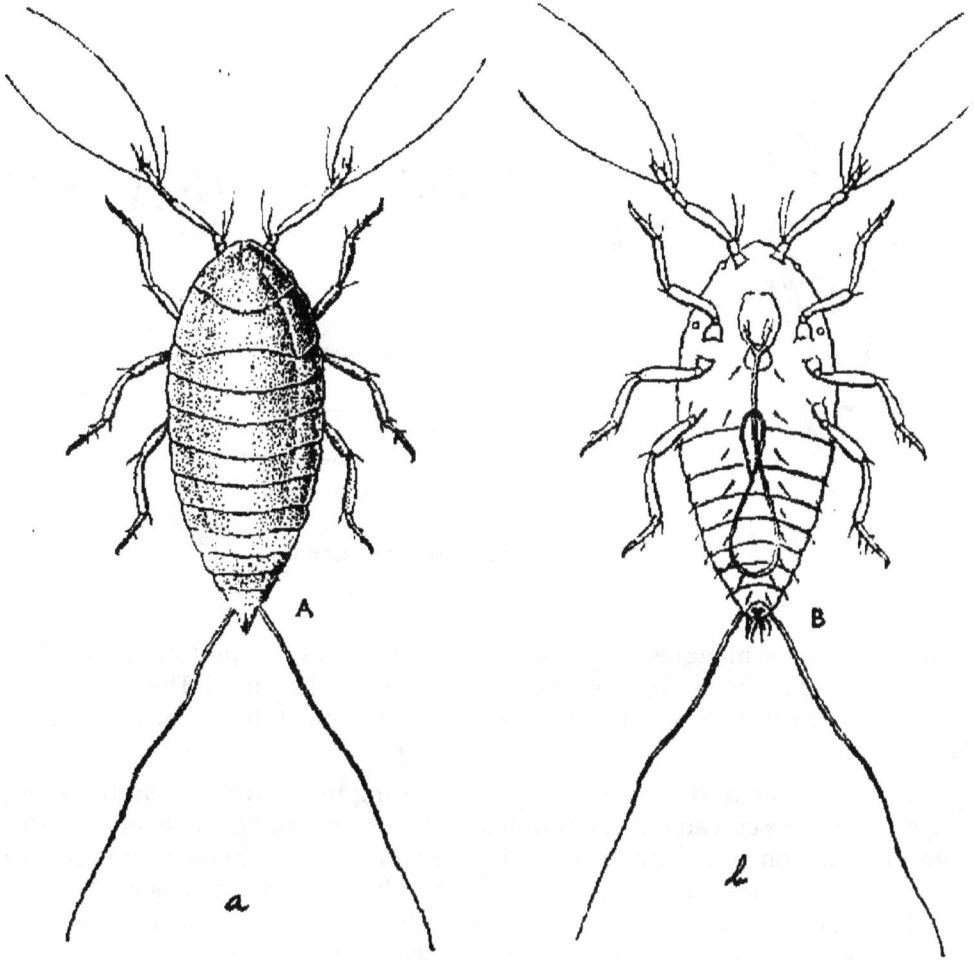

**Figure 15.34: Newly Emerged Nymph of *Laccifer lacca* (Kerr);
A: Dorsal view; B: Vetnral view (Clear with KOH) (After Imms and Chatterji, 1915)**

Cultivation of Lac

Cultivation of lac involves proper care of host plants, regular pruning of host plants, infection or inoculation, crop-reaping, control of insect pests, forecast of swarming, collection and processing of lac.

The first and perhaps the most important pre-requisite for cultivation of lac is proper care of the host plants. It is the host plants on which lac insects depend for their food, shelter and completion of their life cycle. There are two types of cultivation of host plant. One is that plants should be allowed to grow in their natural way and the function of lac-culturist is only to protect and care for the proper growth of plants. Another way is that a particular piece of land is taken for the purpose and systematic plantation of host plant is made there. Regular watch is necessary in this case by

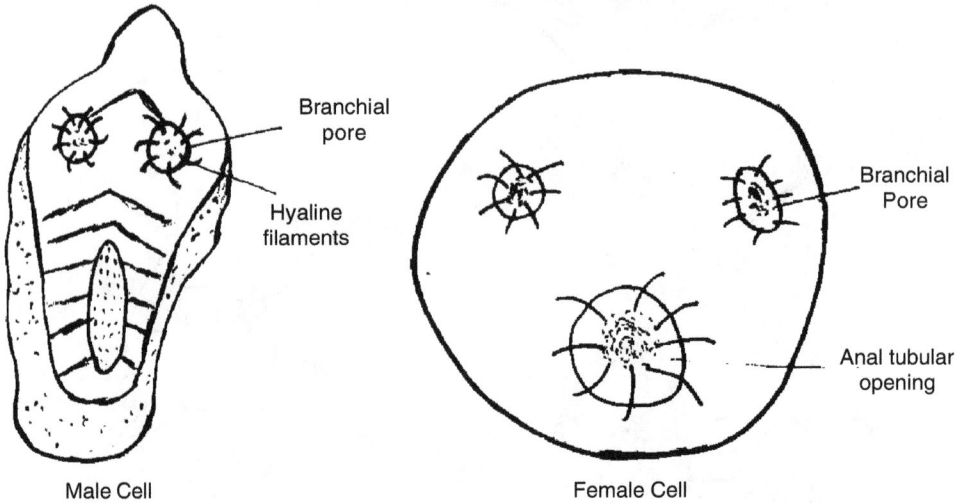

Figure 15.35: Different Forms of Lac Cell

providing artificial manures, irrigation facilities, ploughing and protecting the plants from cattle and human beings for which the land should be fenced. The larvae of lac insects are inoculated on host plants only after the host plants have reached a proper height.

The lac larvae feed on the cell sap by inserting their proboscis in the tender twigs. The proboscis can only be inserted in the tender young off-shoots. For this before inoculation, pruning of lac host plants is necessary. The branches less than an inch in diameter are selected for pruning. Branches half inch or less in diameter should be cut from the very base of their origin. But the branches more than half inch diameter should be cut at a distance of 1-1/2 inch from the base.

Inoculation

The method by which the lac insects are introduced to the new lac host plant is known as inoculation. This may be of two types, namely "Natural infection" and Artificial infection".

When infection from one plant to another occurs by natural movements of insect, it is called natural infection. This may be due to overcrowding of insect population and nonavailability of tender shoots on a particular tree.

Artificial infection takes place through the agencies other than those of nature. Prior to about two weeks of hatching, lac bearing sticks are cut to the size of six inches. They are called "Brood lac". Brood lac are then kept for about two weeks in some cool place. When the larvae start emerging from these brood lac, they are supposed to be ready for inoculation. Strings could be used for tying the brood lac with the host plant. In longitudinal infection the brood lac is tied in close contact with host branches. In lateral infection the brood lac is tied across the gaps between

Lac incrustation

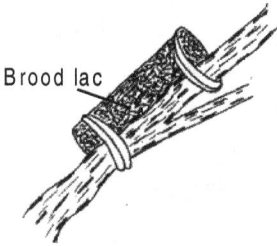

Brood lac

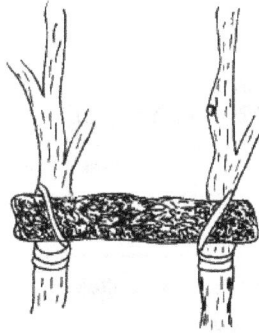

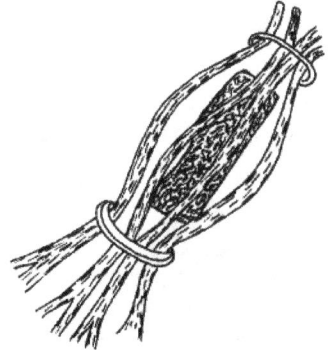

Longitudinal infection   Lateral infection   Interfaceal infection

Figure 15.36: Three Different Ways of Artificial Inoculation of Lac

two branches. In interlaced method, brood lac is tied among the branches of several new shoots.

Lac Crops

The lac insects repeat its life cycle twice in a year. There are actually four lac crops since the lac insects behave in two ways, either they develop in Kusum plants or develop on plants other than Kusum. The lac which grows on Non-Kusum plants are called as "Ranjeeni lac", and which grows on Kusum plant is called as "Kusumi lac".

Four lac crops have been named after four Hindi months in which they are cut from the tree. They are as follows:

Ranjeeni Crop

(*i*) Katki

Lac larvae are inoculated in June-July. Male insect emerges in August-September. Female gives rise to swarming larvae in October-November and the crop is reaped in Kartik (October and November).

(*ii*) Baisakhi

Larvae produced by Katki crop are inoculated in October-November, male insects emerges in February-March, females give rise to swarming larvae in June-July, the crop is reaped in Baisakh (April-May).

Kusumi Crop

(*i*) Aghani

Lac larvae are inoculated in June-July, male insect emerges in September, female gives rise to swarming larvae in January-February and crop is reaped in Aghan (December-January).

(*ii*) Jethwi

The larvae produced by Aghani crop is inoculated in the month of January-February, male emerges in March-April, female gives rise to swarming larvae in June-July and the crop is reaped in the month of Jeath (June-July).

Different Crops of Lac

Infection with Swarming Larvae	Emergence of Male Insect	Crop reaped	Female gives Rise to Swarming Larvae
Rajneeni or Nankusumi Crop			
Katki	August	Oct.-Nov.	Oct.-Nov.
Baisakhi (Oct.-Nov.)	Feb.-March	April-May	June-July
Kusumi Crop			
Aghani (June-July)	September	Dec.-Jan.	Jan.-Feb.
Jethwi (January)	March-April	June-July	June-July

Scraping and Processing of Lac

Lac cut from the host plant is called as "stick lac". Lac can be scraped from the twigs before or after the emergence of larvae. If it is used for manufacturing before the emergence of larvae, the type of lac produced is called as "Ari lac", and if it is used for manufacturing purpose after swarming of larvae has occurred, the lac is said to be "Phunki-lac".

The scraping of lac from twig is done by knife, after which they should not be exposed to sun. The scraped lac is grinded in hard stone mills. The unnecessary materials are sorted out. In order to remove the finer particles of dirt and colour, this lac is washed repeatedly with cold water. Now at this stage it is called as "Seed lac" and is exposed to sun for drying. Seed lac is now subjected to the melting process. The melted lac is sieved through cloth and is given the final shape by molding. The final form of lac is called "Shellac". Colour or different chemicals may be mixed during melting process for particular need.

Natural Enemies of Lac and their Control

Lac enemies impose a challenge to the lac culturist, as they not only decreases the population of lac insects, but also retard the production and quality of lac. Damage caused to lac insects may be grouped under two heads, (*a*) damage caused by insects (*b*) damage caused by animals other than insects.

Insect enemies of lac crop may be predators and parasites. The common parasites of lac insect are known as "Chalcid". They are small, winged insects which lay their eggs inside the lac coat either on the body of the lac insect or inside the body of the lac insect. The larvae which hatches from these eggs feed upon the lac insects, thereby causing mortality of their host. Damage done by this parasite constitute about 5-10 per cent of the total destruction of the lac crop.

Damage done by the predators are of greater intensity (35 per cent of the total destruction). The major predators of lac insects are *Eublemna anabilis* (the white moth) and *Holococera pulverea* (the blackish grey moth). They not only feed on lac insects but also destroy the lac produced by them. Squirrels, monkey, rat, bat, birds (wood peckers), etc., are the enemies other than insects which destroy the lac crop in different ways. Damage is also done by climatic factors such as excess heat, excess cold, heavy rain, storm and partly by the faulty cultivation methods.

Control

Damage caused by the above mentioned animals can be reduced to certain extent by use of the following methods.

Cultural Method

The amount of damage by infection can be reduced to a greater extent by taking care during the culture of lac insects, especially at the time of inoculation. The brood lac showing minimum enemy attack should be selected for inoculation and should be cut from the host plant very near to the time of emergence of larvae (about one week before the emergence). This will reduce the chances of parasite attack on the emerging larvae at new place (host). The brood lac used for inoculation should be removed from the new hosts branches as soon as the emergence of larvae stops (Approx. 3 weeks after inoculation). It reduces the chance of transfer of enemies to the new host plant from the brood lac. The infected brood lac, not fit for inoculation or the used up brood lac should not be retained for long. The lac should be scraped at once and the rest may be crushed or dropped into fire in order to destroy the predators and parasites. The delay in processing also gives chances to the enemy insects to escape into field. So, the manufacturers should try to convert stick lac into seed lac as soon as possible. By these cultural methods, the future production can be saved from infection to some extent.

Artificial Method

During the crop reaping, it is not always possible for the manufacturers to convert the huge amount of stick lac to seed lac at a time. To avoid the spreading of enemies at this time from stocked stick lac, simple artificial method can be used. Bundles of stick lac should be tied with stones and immersed in fenced water (river or ponds) for about a week. This will kill all the parasitic and predator insects as they can not survive in water.

Biological Method

It is an indirect method for killing the parasitic and predator insects. For this purpose, hyper-parasitic insects are used which attack the parasitic insects of lac and kill them. These hyper-parasitic insects are however not harmful to the lac crop.

Use of Lac

Lac has been used for the welfare of human beings from the great olden days. No doubt the development of many synthetic produces have made its importance to a little lesser degree, but still it can be included in the list of necessary articles.

Lac is used in making toys, bracelets, sealing wax, gramophone records etc. It is also used in making grinding stones, for filling ornaments, manufacturing of varnishes and paints, silvering the back of mirror, encasing cable wires, etc. Waste materials produced during the process of stick lac is used for dying purpose.

Appendices

Appendices

Appendix I
List of Pesticides/Pesticides Formulations Banned in India

A. Pesticides Banned for Manufacture, Import and Use (24 Nos.)

1. Aldrin
2. Benzene Hexachloride
3. Calcium Cyanide
4. Chlordane
5. Copper Acetorsenite
6. Cibromoclhloropropane
7. Endrin
8. Ethyl Mercury Chloride
9. Ethyl Parathion
10. Heptachlor
11. Menazone
12. Nitrofen
13. Paraquat Dimethyl Sulphate
14. Pentachloro Nitrobenzene
15. Pentachlorophenol
16. Sodium Methane Arsenate
17. Tetradifon
18. Toxafen
19. Aldicarb

20. Chlorobenzilate	Use banned w.e.f. 17-7-2003
21. Dieldrin	–do–
22. Maleic Hydrazide	–do–
23. Ethylene Dibromide	–do–
24. TCA (Trichloro acetic acid)	–do–

B. Pesticide/Pesticide Formulations Banned for Use but their Manufacture is Allowed for Export (3 Nos.)

25. Nicotin Sulfate
26. Phenyl Mercury Acelate
27. Captafol 80% Powder–(Use banned w.e.f. 17-7-2003)

C. Pesticide Formulations Banned for Import, Manufacture and Use (4 Nos.)

1. Methomyl 24% L
2. Methomyl 12.5% L
3. Phosphamidon 85% SL
4. Carbofuran 50% SP

Appendix II
List of Pesticides Refused Registration

Sl.No.	Name of Pesticides
1.	Calcium arsenate
2.	EPM
3.	Azinphos methyl
4.	Lead arsenate
5.	Mevinphos (Phosdrin)
6.	2,4,5-T
7.	Carbophenothion
8.	Vamidothion
9.	Mephosfolan
10.	Azinphos ethyl
11.	Binapacryl
12.	Dicrotophos
13.	Thiodemeton/Disulfoton
14.	Fentin acetate
15.	Fentin hydroxide
16.	Chinomethionate (Morestan)
17.	Ammonium sulphamate
18.	Leptophos (Phosvel)

Appendix III
Pesticides Restricted for Use in India

Sl.No.	Name of Pesticides
1.	Aluminium phosphide
2.	DDT
3.	Lindane
4.	Methyl bromide
5.	Methyl parathion
6.	Sodium cyanide
7.	Methoxy Ethyl Mercuric Chloride (MEMC)

Appendix IV

Insecticides Registered Under Section 9(3) of the Insecticide Act, 1968 as on 1st May 2004

Sl.No.	Name of Pesticides
1.	2,4-Dichlorophenoxy Acetic Acid
2.	Acephate
3.	Acetamiprid
4.	Alachlor
5.	Allethrin
6.	Alphacypermethrin
7.	Alphanaphthyl Acetic Acid
8.	Aluminium Phosphide
9.	Anilophos
10.	Atrazine
11.	Aureofungin
12.	Azadirachtin (Neem Products)
13.	*Bacillus thuringiensis* (B.t. & Bs.)
14.	Barium carbonate
15.	Benomyl

Sl.No.	Name of Pesticides
16.	Beta cyfluthrin
17.	Bifenthrin
18.	Bitertanol
19.	Bromadiolone
20.	Butachlor
21.	Captan
22.	Carbaryl
23.	Carbendazim
24.	Carbofuran
25.	Carbosulfan
26.	Carboxin
27.	Cartap Hydrochloride
28.	Chlorofenvinphos
29.	Chlorimuron ethyl
30.	Chlonnequat Chloride (CCC)
31.	Chlorothalonil
32.	Chlorpyriphos
33.	Cinmethylene
34.	Clodinafop-propinyl (Pyroxofopropinyl)
35.	Clomazone
36.	Copper hydroxide
37.	Copper oxychloride
38.	Copper sulphate
39.	Coumachlor
40.	Coumatetralyl
41.	Cuprous oxide
42.	Cyfluthrin
43.	Cyhalofop-butyl
44.	Cymoxanil
45.	Cypennethrin
46.	Cyphenothrin
47.	Dalapon
48.	Dazomet
49.	Deltamethrin (Decamethrin)
50.	Diazinon

Sl.No.	Name of Pesticides

51. Dichloro Diphenyl Trichloroethane (DDT)
52. Dichloropropene and Dichloropropane mixure (DD mixture)
53. Diclorvos (DDVP)
54. Diclofop-Methyl
55. Dicofol
56. Difenocenazole
57. Difenthiuron
58. Diflubenzuron
59. Dimethoate
60. Dinocap
61. Dithianon
62. Diuron
63. Dodine
64. D-trans Allethrin
65. Edifenphos
66. Endosulfan
67. Ethephon
68. Ethion
69. Ethofenprox (Etofenprox)
70. Ethoxysulfuron
71. Ethylene Dibromide and Carbon Tetrachloride mixture (EDCT Mixture 3 : 1)
72. Fenarimol
73. Fenazaquin
74. Fenitrothion
75. Fenobucarb (BPMC)
76. Fenoxaprop-o-Ethyl
77. Fenpropathrin
78. Fenthion
79. Fenvalerate
80. Ferbam
81. Fipronil
82. Fluchloralin
83. Flufenoxyuron
84. Fluvalinate

Sl.No.	Name of Pesticides
85.	Formothion
86.	Fosetyl-Al
87.	Gibberellic acid
88.	Glufosinate ammonium
89.	Glyphosate
90.	Hexaconazole
91.	Hydrogen Cyanide
92.	Imiprothrin
93.	Imazethapyr
94.	Imidacloprid
95.	Iprofenfos (Kitazin)
96.	Indoxacarb
97.	Iprodione
98.	Isoprothiolane
99.	Isoproturon
100.	Kasugamycin
101.	Lambdacyhalothrin
102.	Lime sulphur
103.	Lindane
104.	Linuron
105.	Mancozeb
106.	Malathion
107.	Mepiquate chloride
108.	Metalaxyl
109.	Metaldehyde
110.	Metasulfuron methyl
111.	Methabenzthiazuron
112.	Methomyl
113.	Methoxy Ethyl Mercury Chloride (MEMC)
114.	Methyl Bromide
115.	Methyl Chlorophenoxy Acetic Acid (MCPA)
116.	Methyl parathion
117.	Metolachlor
118.	Metoxuron
119.	Metribuzin

Sl.No.	Name of Pesticides
120.	Monocrotophos
121.	Myclobutanil
122.	Nickel Chloride
123.	Oxadiargyl
124.	Oxadiazon
125.	Oxycarboxin
126.	Oxydemeton-Methyl
127.	Oxyfluorfen
128.	Paclobutrazol
129.	Paradichlorobenzene (PDCB)
130.	Paraquat dichloride
131.	Penconazole
132.	Pendimethalin
133.	Permefurin
134.	Phenthoate
135.	Phorate
136.	Phosalone
137.	Phosphamidon
138.	Primiphos-methyl
139.	Prallethrin
140.	Pretilachlor
141.	Profenophos
142.	Propanil
143.	Propergite
144.	Propetamphos
145.	Propiconazole
146.	Propineb
147.	Propoxur
148.	Pyrethrins (pyrethrum)
149.	Quinalphos
150.	S-bioallethrin
151.	Simazine
152.	Sirmate
153.	Sodium Cyanide
154.	Spinosad

Sl.No.	Name of Pesticides
155.	Streptomycin + Tetracycline
156.	Sulfosulfuron
157.	Sulphur
158.	Tebuconazole
159.	Temephos
160.	Thiobencarb (Benthiocarb)
161.	Thiodicarb
162.	Thiamethoxam
163.	Thiometon
164.	Thiram
165.	Thiophanate methyl
166.	Transfluthrin
167.	Triadimefon
168.	Triallate
169.	Triazophos
170.	Trichlorofon
171.	*Trichoderma viride*
172.	Tricyclazole
173.	Tridemorph
174.	Trifluralin
175.	Validamycin
176.	Warfarin
177.	Zinc Phosphide
178.	Zineb
179.	Ziram

Pesticides and their formulation registered under the Insecticides Act.

Insecticides, registered under the Insecticides Act for use in Household/Public Health/ Rodent etc.

List of Manufacturer Registered for Production and Sale of Bio-Pesticides and Chemical Pesticides, their formulation, source of supply and indigenous manufacturers u/s 9(3) of the I, Act.

Note: The list is subject to change from time to time.

*Source of Appendices I & II: DPPQ & S Website as on 30-08-2004;

http: www.cibrc.nic.in/list_pest_bann.htm

www.cibrc.nic.in/reg_products.htm.

Glossary

Abdomen: The last of the three major body divisions of an insect.

Action threshold: The pest density at which a control tactic must be implemented to avoid an economic loss.

Active ingredient (AI): The component of a pesticide formulation responsible for the toxic effect.

Agroecosystem: A relatively artificial ecosystem in an agricultural field, pasture, or orchard.

Antenna, Antennae (pl.): A pair of sensory organs located on the head of an insect, above the mouthparts.

Aorta: The front-most, non-pulsating portion of the dorsal blood vessel of an insect.

Arthropod: Any of the invertebrate animal (such as insects, spiders, or crustaceans) having an exoskeleton, a segmented body and jointed limbs.

Augmentation: Biological control practices intended to increase the number or effectiveness of existing natural enemies.

Bacterium: A single-celled microscopic plant-like organism that does not produce chlorophyll.

Beak: Colloquial expression for the protruding mouthpart structures of a sucking insect (= proboscis).

Biological control: The use of living organisms, such as predators, parasitoids, and pathogens, to control pests, weeds, or diseases. Typically, it involves some human activity.

Biorational: Having a minimal disruptive influence upon the environment and its inhabitants (*e.g.*, a biorational insecticide). Broad-spectrum (insecticide): Active against a wide range of insects.

Bt: The bacterium, *Bacillus thuringiensis*.

Caterpillar: The immature stage (larva) of a butterfly, moth, or sawfly.

Chemical control: Pest management practices which rely on the application of synthetic or naturally-derived pesticides.

Class: A category of the classification scheme of living organisms ranking below a phylum and above an order (*e.g.*, Insecta).

Classical biological control: The importation of foreign natural enemies to control previously introduced, or native, pests.

Cocoon: A silken case formed by an insect larva for pupation.

Cole crops: Crops such as cabbage, broccoli, Brussels sprouts and other crucifers.

Complete metamorphosis: Type of insect development characterized by four distinct stages: egg, larva, pupa, and adult.

Conservation: Any biological control practice designed to protect and maintain populations of existing natural enemies.

Contact poison: A pesticide that is absorbed through the body wall as opposed to one that must be ingested.

Cucurbits: Vine crops such as cucumbers, melons, squash, and pumpkins.

Cultural control: Pest management practices that rely upon manipulation of the cropping environment (*e.g.*, cultivation of weeds harboring insect pests).

Density (insect populations): The number of insects per unit of measure (*e.g.*, beetles per square meter).

Diapause: A physiological state of arrested metabolism, growth, and development that occurs at a particular stage in the life cycle of an organism.

Dormancy: A recurring period in the life cycle of an organism when growth, development, and reproduction are suppressed.

Ecology: The study of an organisms interrelationship with its environment.

Economic threshold: *See* Action threshold.

Elytra: The thickened or leathery front or forewings of insects such as beetles.

Encapsulation: The surrounding of an invading body, such as the egg of a parasite, by insect hemocytes (blood cells) and the formation of a protective capsule.

Entomopathogenic: Insect-attacking organism.

Environmental impact quotient (EIQ): A relative value that estimates the environmental impact of a pesticide, by taking into account toxicity to natural enemies, wildlife, and humans, degree of exposure, aquatic and terrestrial effects, soil chemistry, etc.

Epizootic: A disease outbreak within an insect population.

Exoskeleton: A skeleton or supportive structure on the outside of an insect body.

Exotic: Introduced from another country or continent (*e.g.*, introduced insect pest).

Family: A taxonomic subdivision of an order, containing a group of related genera. Family names.

Forewing: The first or anterior pair of insect wings.

Fumigant: A substance which produces a gas, vapor, fume, or smoke intended to kill a pest.

Fungicide: Any substance that kills or inhibits the growth of a fungus.

Funicle: The portion of the flagellum of the antenna closest to the club.

Fungus, Fungi (pl.): Any of numerous plants lacking chlorophyll, ranging in form from a single cell to a body of branched filaments. Includes the yeasts, molds, smuts, and mushrooms.

Gene: A biochemical unit of hereditary, often coding for an entire protein.

Generalist: A pest or natural enemy that can utilize a wide range of species as host or prey.

Generation: Period from any given stage in the life cycle to the same life stage in the offspring. Typically from egg to egg.

Genetic engineering: The manipulation of the genetic material of an organism in order to achieve desirable characteristics.

Genus, Genera (pl.): A group of evolutionarily related species, sharing one or a number of characteristics.

Gradual metamorphosis: A type of insect development in which there is no prolonged resting stage (pupa). The three stages are: egg, nymph, and adult.

GV: Granulosis virus.

Habitat manipulation: Manipulation of agricultural areas and surrounding environment with the aim of conserving or augmenting populations of natural enemies (*e.g.,* the planting of a refuge for natural enemies).

Head: The anterior region of an insect, which bears the mouthparts, eyes, antennae and houses the brain.

Herbicide: A substance used to kill or control weeds.

Hermaphroditie: Having both male and female sex organs in one individual.

Hindwings: The second pair of wings of an insect.

Honeydew: The sugary liquid discharge from the anus of certain insects (Homoptera) such as aphids and scales.

Host: The organism in or on which a parasitoid lives; a plant on which an insect feeds.

Host plant resistance: The relative amount of heritable qualities possessed by a plant that reduces the degree of damage to the plant by a pest or pests.

Hyperparasite: A parasite whose host is another parasite.

Indigenous: Native to an area.

Inoculative release: The release of relatively small numbers of natural enemies that are expected to colonize, reproduce, and spread naturally throughout an area.

Insect growth regulator (IGR): A substance, natural or synthetic that controls or modifies insect growth processes.

Insecticide resistance: Genetically inherited ability to withstand doses of pesticide which would kill individuals from strains whose ancestors had not been exposed to the pesticide.

Insect resistant (plants): Tolerant or resistant to insect attack (as in plants). Individuals from strains whose ancestors had not been exposed to the pesticide.

Instar: The stage of an insects life between successive molts, for example the first instar is between hatching from the egg and the first molt.

Integrated pest management (IPM): An approach to the management of pests in which all available control options, including physical, chemical, and biological controls, are evaluated and integrated into a unified program.

Integument: The outer covering of the insect body that includes the cuticle and the epidermis.

Introduction (classical biological control): The importation of a natural enemy from a foreign country or continent, usually to control a pest also of foreign origin.

Inundative release: The release of relatively large numbers of natural enemies to suppress pest populations without the expectation that the natural enemies will colonize and spread throughout the area.

Larva, Larvae (pl.): The immature stage between the egg and pupa of insects having complete metamorphosis where the immature differs radically from the adult (*e.g.*, caterpillars, grubs).

Leafy greens: Lettuces and other leaf vegetables.

Least toxic: Having a minimal toxic effect upon non-target organisms.

Life Cycle: The sequence of events that occurs during the lifetime of an individual organism.

Maggot: The immature form (larva) of a fly or wasp, lacking legs and a well-developed head.

Mass-reared: Produced in large numbers as in natural enemies produced for release programs.

Mechanical control: Control of pests by physical means such as the use of screens or row covers.

Metabolism, Metabolic: Chemical changes that occur in living cells to provide energy for vital activities and to assimilate new material.

Metamorphosis: A change in body form during development of an insect.

Microbial: A microscopic organism; a germ.

Microbial insecticide: A preparation of microorganisms (*e.g.*, viruses or bacteria) or their products used to suppress insect pest populations.

Microsporidia: Single-celled life forms, related to Protozoa.

Mite: Any of several minute invertebrates belonging to the phylum Arthropoda, class Arachnida.

Morphology: Form or structure of an organism.

Multivoltine: Having more than one brood or generation per season.

Mycelium, Mycelia (pl.): A mass of interwoven filamentous 'threads' that make up the vegetative part of a fungus.

Native (insect or plant): Of local origin, not intentionally or accidentally introduced.

Natural control: The suppression of pest populations by naturally occurring biological and environmental agents.

Natural enemies: Living organisms found in nature that kill, weaken or reduce the reproductive potential of other organisms.

Nectar: The sugary liquid secreted by many flowers.

Nematode: An elongated, cylindrical worm parasitic in animal, insect, or plant, or free-living in soil or water.

NPV: Nuclear polyhedrosis virus.

Nymph: The immature stage, following hatching from the egg, of an insect that does not have a pupal stage and resembles its adult stage in all forms except size.

Ocelli: Simple eyes on some adult and larval insects. Typically there are three, which form an inverted triangle dorsally, the function of which is obscure.

Order: A taxonomic subdivision that contains groups of related families or superfamilies; usually ending in–ptera in insects.

Overwinter: A period of rest or hibernation by which insects survive the winter.

Oviposition: The laying or depositing of eggs.

Ovipositor: The egg-laying apparatus of a female insect.

Palp (palpus): Sensory organ attached to insect mouthparts that is used to test the quality of food.

Parasite: An organism that lives in or on another organism (the host) during some portion of its life cycle.

Parasitoid: An animal that feeds in or on another living animal, consuming all or most of its tissues and eventually killing it.

Parthenogenesis: Development of an insect, from egg to adult, without fertilization.

Pathogen: A disease-causing organism.

Pest: An organism that interferes with human activities, property, or health, or is objectionable.

Pest management: *See* Integrated pest management.

Pesticide: A substance that is used to kill, debilitate or repel a pest.

Pest-resistant crops: Crops that possess attributes which minimize damage by pests.

Phenology: The seasonal life history of an insect population.

Pheromone: A substance, such as a sex attractant, that is given off by one individual and causes a specific reaction in other individuals of the same species.

Phylum, Phyla (pl.): One of the major divisions of the animal kingdom.

Physical Control: Control of pests by physical means such as heat, cold, sound waves, etc.

Polyembryonic (eggs): A single egg that divides to form two or more (often hundreds) identical embryos.

Polyembryony: Having several embryos.

Population: A group of individuals of the same species within a given space and time.

Predaceous: Preying upon other organisms, predatory.

Predator: An animal that attacks and feeds on other animals, normally killing several individuals during its life cycle.

Pronotum: The upper, often shield-like, hardened body-wall plate, located just behind the head of an insect.

Protozoan: A microscopic, single-celled organism that is largely aquatic and includes many parasitic forms.

Pupa, Pupae (pl.): The nonfeeding stage between the larva and adult in insects with complete metamorphosis.

Puparium, Puparia (pl.): A case formed by the hardening of the last larval skin, in which the pupa is formed; usually of flies.

Pupate: To transform to a pupa.

Resistance (insecticide or pesticide): *See* Insecticide resistance.

Resistance (plant): *See* Host plant resistance.

Resurgence (pest): The development of large populations of pests that had previously been suppressed.

Sampling: Estimating the density of organisms (pests or natural enemies) or damage by examining a defined portion of the crop.

Scouting, Scout: *See* Sampling.

Septicaemia: Blood poisoning caused by pathogenic organisms.

Specialist: A pest or natural enemy that utilizes a narrow range of species for its host or prey.

Species: A group of individuals similar in structure and capable of interbreeding and producing fertile offspring. They are different in structure from other such groups and do not interbreed with them.

Spiracles: The external openings of the insect breathing (tracheal) system, found along the abdomen.

Spore: A reproductive structure developing in certain bacteria and fungi which is strongly resistant to environmental influences but becomes active under suitable conditions.

Stage (life stage): A distinct period in the development of an organism (*e.g.*, for some insects, egg, larval, pupal, and adult stages).

Stomach poison: An insecticide that is lethal only after it has been ingested by an insect, entering the insect body through the gut.

Systemic insecticide: An insecticide that is absorbed into plant sap and is lethal to insects feeding on or within the treated plant.

Thorax: The insect body region behind the head which bears the legs and wings.

Tolerance (host-plant resistance): The ability of a plant to withstand injury by pests.

Transformed (Bt-transformed): Transfer and expression of a gene (*e.g.*, for Bt toxin) into another organism.

Trap crop: A small area of a crop used to divert pests from a larger area of the same or another crop. The pests, once diverted to the trap crop, may be treated with an insecticide.

Univoltine: Having only one brood or generation per season.

Vector: An organism capable of carrying and transmitting a disease-causing agent from one host to another.

Virus: Any of various submicroscopic pathogens which can only replicate inside a living cell.

References

Ahamad, Shahid, 2009. *Plant Diseases Management for Sustainable Agriculture*. Daya Publishing House, New Delhi, xvi+373p..

Carter, George A., 2004. *Beekeeping: A Guide to the Better Understanding of Bees Their Diseases and the Chemistry of Beekeeping*. Biotech Books, New Delhi, 113p.

Chandola, R.P., 2010. *Dictionary of Entomology*. Daya Publishing House, New Delhi, vi+184p.

Chhillar, B S *et al.*, 2007. *Agricultural Acarology*. Daya Publishing House, New Delhi, xii+355p.

Cotton, Richard T., 2011. *Insect Pests of Stored Grain and Grain Products*. Biotech Books, New Delhi, v+241p.

Dhaliwal, G. S. and Arora, R., 2001. *Integrated Pest Management: Concepts and Approaches*. Kalyani Publishers, Ludhiana. 427p.

Fryer, J.C.F., 2010. *Insect Pests of Fruit Crops*. Biotech Books, New Delhi, 116p.

Gangawane, L.V. & Khilare, V.C., 2008. *Crop Diseases: Identification and Management: A Colour Handbook*. Daya Publishing House, New Delhi, xviii+244p.

Hiware, C.J., 2001. *Agro Cottage Industry Sericulture*. Daya Publishing House, New Delhi, xii+116p.

Huffaker, C.B. (Ed.) 1980. *New Technology of Pest Control*. John Wiley and Sons, New York, 788 p.

Kavane, R.K. & Sathe, T.V., 2011. *Wild Silk Technology*. Daya Publishing House, New Delhi, ix+224p.

Khan, Mohammad Ashraf *et al.*, 2011. *Silkworm Crop Protection: Concepts and Approaches*. Daya Publishing House, New Delhi, x+224p.

Kogan, M. 1998. Integrated Pest Management. Historical Perspectives and contemporary developments. *Ann. Rev. Entomol.* 43: 243–270.

Kumar, Arvind & Harbhajan Kaur, 2008. *Ecobiology of Aquatic Insects.* Daya Publishing House, New Delhi, xi+247p.

Maredia, K. M., Dakouo, D. and Mota-Sanchez, D., 2003. *Integrated Pest Management in the Global Arena.* CABI Publishing, CAB International, Wallingford, U.K. 512 p.

Mayee, C.D. *et al.,* 2004. *Biotechnological Approaches for the Integrated Management of Crop Diseases: Professor L V Gangawane Festschrift Volume.* Daya Publishing House, New Delhi, xxvi+315p.

Metcalf, R.L. and Luckman, W.H. (Eds.) 1982. *Introduction to Insect Pest Management.* John Wiley and Sons, New York, 577 p.

Mohanty, Prafullla Kumar, 2003. *Tropical Wild Silk Cocoons of India.* Daya Publishing House, New Delhi, v+197p.

Norton, G.A. and Munford, J.D., 1993. *Decision Tools for Pest Management.* CAB International, Wallingford, United Kingdom.279 p.

Patil, Vaishali J. & Sathe, T.V., 2003. *Insect Predators and Pest Management.* Daya Publishing House, New Delhi, vii+216p.

Pedigo, L.P. 2002. *Entomology and Pest Management.* Printice Hall of India Pvt Ltd., New Delhi 742p.

Pimentel, D. (Ed.) 1981. *CRC Handbook of Pest Management in Agriculture,* Vols. I, II, III CRC Press Inc. Boca Raton, Florida.

Prasad, D., 2005. *Crop Protection: Management Strategies.* Daya Publishing House, New Delhi, x+676p.

Prasad, D., 2008. *Insect Pest and Disease Management.* Daya Publishing House, New Delhi, xii+585p.

Roy, D.N. & Brown, A.W.A., 2010. *Entomology.* Biotech Books, New Delhi, ix+413p.

Sahayaraj, K., 2004. *Indian Insect Predators in Biological Control.* Daya Publishing House, New Delhi, xvi+336p.

Sathe, T.V. & Awate, Manisha R., 2009. *Crickets and Household Pests.* Daya Publishing House, New Delhi, viii+171p.

Sathe, T.V. & Bhoje, P.M., 2000. *Biological Pest Control.* Daya Publishing House, New Delhi, v+122p.

Sathe, T.V. & Bhosale, Y.A., 2001. *Insect Pest Predators.* Daya Publishing House, New Delhi, vi+169p.Sathe, T.V. & Girhe, B.E., 2002. *Mosquitoes and Diseases.* Daya Publishing House, New Delhi, vii+96p.

Sathe, T.V. & Jadhav, A.D., 2001. *Sericulture and Pest Management.* Daya Publishing House, New Delhi, vii+197p.

Sathe, T.V. & Jadhav, B.V., 2008. *Indian Pest Aphids.* Daya Publishing House, New Delhi, vii+211p.

Sathe, T.V. & Margaj, G.S., 2001. *Cotton Pests and Biocontrol Agents*. Daya Publishing House, New Delhi, vii+166p.

Sathe, T.V. & Oulkar, Jyoti M., 2010. *Insect Pest Management: Ecological Concepts*. Daya Publishing House, New Delhi, vii+235p.

Sathe, T.V. & Shinde, K.P., 2008. *Dragonflies and Pest Management*. Daya Publishing House, New Delhi, vii+179p.

Sathe, T.V. & Thite, S.H., 2004. *Shoot Feeding and Sericultural Trends*. Daya Publishing House, New Delhi, vii+157p.

Sathe, T.V. *et al.*, 2003. *Indian Pest Parasitoids*. Daya Publishing House, New Delhi, viii+145p.

Sathe, T.V. *et al.*, 2005. *Practicals in Basic Entomology*. Daya Publishing House, New Delhi, viii+84p.

Sathe, T.V., 2004. *Vermiculture and Organic Farming*. Daya Publishing House, New Delhi, ix+122p., figs., tabls., bib., ind., 23 cm

Sathe, T.V., 2005. *Basic Entomology: A Practical Manual*. Daya Publishing House, New Delhi, x+118p.

Sathe, T.V., 2006. *Fundamentals of Beekeeping*. Daya Publishing House, New Delhi, ix+100p.

Sathe, T.V., 2009. *A Textbook of Forest Entomology*. Daya Publishing House, New Delhi, ix+234p., 8 col. plts., figs., tabls., ind., 23 cm

Shamsuddin, Mohammed, 2009. *Silkworm Physiology: A Concise Textbook*. Daya Publishing House, New Delhi, xiv+212p.

Shukla, Abhishek, 2005. *Masalon mein Keet Prabandhan*. Daya Publishing House, New Delhi, viii+120p.

Shukla, Abhishek, 2008. *Krishi Keet Vigyan: Ek Parichaya*. Daya Publishing House, New Delhi, vii+310p.

Shukla, Abhishek, 2009. *A Handbook on Economic Entomology*. Daya Publishing House, New Delhi, vii+247p.

Singh, Tribhuwan *et al.*, 2010. *Silkworm Egg Science: Principles and Protocols*. Daya Publishing House, New Delhi, ix+278p.

Smith, E.H. and D. Pimentel (Eds.). 1978. *Pest Control Strategies*. Academic Press, New York, 334 pp.

Upadhyay, R.K.,Mukerji, K.G. and Chamola, B.P. (Eds) 2001. *Biocontrol Potential and its Exploitation in Sustainable Agriculture*. Kluwer Academic/Plenum Publishers, New York, 421.

Wilson, George Fox, 2010. *Horticultural Pests: Detection and Control*. Biotech Books, New Delhi, xx+240p.

Index

www.ingramcontent.com/pod-product-compliance
Lightning Source LLC
Chambersburg PA
CBHW050505190326
41458CB00005B/1444